Methods in Microbiology
Volume 41

Recent titles in the series

Methods in Microbiology
Volume 41

New Approaches to Prokaryotic Systematics

Edited by

Michael Goodfellow
*School of Biology, Newcastle University,
Newcastle upon Tyne, NE1 7RU, United Kingdom*

Iain Sutcliffe
*Department of Applied Sciences, Northumbria University,
Newcastle upon Tyne, NE1 8ST, United Kingdom*

Jongsik Chun
*School of Biological Sciences, and ChunLab Inc.,
Seoul National University, Seoul, Republic of Korea*

AMSTERDAM · BOSTON · HEIDELBERG · LONDON
NEW YORK · OXFORD · PARIS · SAN DIEGO
SAN FRANCISCO · SINGAPORE · SYDNEY · TOKYO
Academic Press is an imprint of Elsevier

Academic Press is an imprint of Elsevier
32 Jamestown Road, London NW1 7BY, UK
525 B Street, Suite 1800, San Diego, CA 92101-4495, USA
225 Wyman Street, Waltham, MA 02451, USA
The Boulevard, Langford Lane, Kidlington, Oxford OX5 1GB, UK

First edition 2014

ISBN: 978-0-12-800176-9
ISSN: 0580-9517 (Series)

For information on all Academic Press publications
visit our website at www.store.elsevier.com

Working together
to grow libraries in
developing countries

www.elsevier.com • www.bookaid.org

Cover image: Phylogenomics of *Corynebacterium diphtheriae*.
Photo kindly provided by Dr. Vartul Sangal, Northumbria University.
Printed in the United States of America

The editors dedicate this volume to Bob Murray and Larry Wayne as well as to the memory of Peter Sneath (1913–2011), one of the cofounders of numerical taxonomy.

Contents

CHAPTER 5   Revolutionizing Prokaryotic Systematics Through Next-Generation Sequencing

Vartul Sangal, Leena Nieminen, Nicholas P. Tucker, Paul A. Hoskisson

CHAPTER 6   Whole-Genome Analyses: Average Nucleotide Identity

David R. Arahal

CHAPTER 13 MALDI-TOF Mass Spectrometry Applied to Classification and Identification of Bacteria 275

Peter Schumann, Thomas Maier

Contributors

David R. Arahal
Colección Española de Cultivos Tipo (CECT) Parque Científico Universidad de Valencia, Paterna, and Departamento de Microbiología y Ecología, Universidad de Valencia, Burjassot, Valencia, Spain

Julia S. Bennett
Department of Zoology, University of Oxford, Oxford, United Kingdom

Jongsik Chun
School of Biological Sciences, and ChunLab Inc., Seoul National University, Seoul, Republic of Korea

Alison J. Cody
Department of Zoology, University of Oxford, Oxford, United Kingdom

Radhey S. Gupta
Department of Biochemistry and Biomedical Sciences, McMaster University, Hamilton, Ontario, Canada

Volker Gürtler
School of Applied Sciences, RMIT University, Bundoora Campus, Melbourne, Victoria, Australia

Simon R. Harris
Pathogen Genomics, Wellcome Trust Sanger Institute, Cambridge, United Kingdom

Sarah E. Heaps
Institute for Cell and Molecular Biosciences, The Medical School, and School of Mathematics and Statistics, Newcastle University, Newcastle upon Tyne, United Kingdom

Paul A. Hoskisson
Strathclyde Institute of Pharmacy and Biomedical Sciences, University of Strathclyde, Glasgow, United Kingdom

Ying Huang
State Key Laboratory of Microbial Resources, Institute of Microbiology, Chinese Academy of Sciences, Beijing, P.R. China

Olga K. Kamneva
Department of Biology, Stanford University, Stanford, California, USA

Peter Kämpfer
Institut für Angewandte Mikrobiologie, Justus-Liebig-Universität Giessen, Heinrich-Buff-Ring 26, Giessen, Germany

Indrani Karunasagar
Faculty of Biomedical Science, Nitte University Centre for Science Education and Research, University Enclave, Medical Sciences Complex, Deralakatte, Mangalore, Karnataka, India

Mincheol Kim
School of Biological Sciences, Seoul National University, Seoul, Republic of Korea

Martin C.J. Maiden
Department of Zoology, University of Oxford, Oxford, United Kingdom

Thomas Maier
Bruker Daltonics, Bremen, Germany

Biswajit Maiti
Faculty of Biomedical Science, Nitte University Centre for Science Education and Research, University Enclave, Medical Sciences Complex, Deralakatte, Mangalore, Karnataka, India

Raul Munoz
Marine Microbiology Group, Department of Ecology and Marine Resources, Institut Mediterrani d'Estudis Avançats (CSIC-UIB), Esporles, Illes Balears, Spain

Leena Nieminen
Strathclyde Institute of Pharmacy and Biomedical Sciences, University of Strathclyde, Glasgow, United Kingdom

Chinyere K. Okoro
Pathogen Genomics, Wellcome Trust Sanger Institute, Cambridge, United Kingdom

Xiaoying Rong
State Key Laboratory of Microbial Resources, Institute of Microbiology, Chinese Academy of Sciences, Beijing, P. R. China

Vartul Sangal
Faculty of Health and Life Sciences, Northumbria University, Newcastle upon Tyne, United Kingdom

Peter Schumann
Leibniz Institute DSMZ-German Collection of Microorganisms and Cell Cultures, Braunschweig, Germany

Malathi Shekar
UNESCO-MIRCEN for Marine Biotechnology, College of Fisheries, Karnataka Veterinary, Animal and Fisheries Sciences University, Mangalore, Karnataka, India

Gangavarapu Subrahmanyam
Faculty of Biomedical Science, Nitte University Centre for Science Education and Research, University Enclave, Medical Sciences Complex, Deralakatte, Mangalore, Karnataka, India

Nicholas P. Tucker
Strathclyde Institute of Pharmacy and Biomedical Sciences, University of Strathclyde, Glasgow, United Kingdom

Naomi L. Ward
Department of Molecular Biology, University of Wyoming, Laramie, Wyoming, USA

William B. Whitman
Department of Microbiology, University of Georgia, Athens, Georgia, USA

Tom A. Williams
Institute for Cell and Molecular Biosciences, The Medical School, Newcastle University, Newcastle upon Tyne, United Kingdom

Pablo Yarza
Ribocon GmbH, Bremen, Germany

Preface

Prokaryotic systematics began as a largely intuitive science that became increasingly objective with the use of data derived from advances in other scientific fields. Since the subject is markedly data dependent, it is hardly surprising that most of the advances in recent years have resulted from the way data are acquired and handled, as exemplified by developments in chemosystematics and numerical taxonomy. This book is dedicated to three towering figures who not only brought new concepts and practices to the fore in a period of transition but also spelt out the significance of new developments to the scientific community through their selfless and tireless contributions to bodies such as the then *International Committee on Systematic Bacteriology* (now the *International Committee on Systematics of Prokaryotes*).

Prokaryotic systematics is in both an interesting and critical state as, once again, it is in a period of transition. For more than a century, microbial systematists have, out of necessity, relied primarily on the observable phenotype, a product of the genome and cultivation conditions. However, rapid advances in whole-genome sequencing over the last decade provide the platform for a paradigm shift for the systematics community. Consequently, this community needs to respond quickly by establishing how, in the future, the relative contributions of genomics and phenotypes are to be used to classify new taxa and to reanalyse existing ones. Moreover, systematists also need to establish protocols for data storage (similar to those used to store genomic, proteomic and transcriptomic data) that will facilitate data mining and large-scale data analyses.

This volume is intended to provide microbiologists and the broader scientific community with a comprehensive, up-to-date account of methods and data handling techniques that will shape developments in prokaryotic systematics for years to come. We hope that these exciting developments will encourage more young scientists to become engaged in a fascinating and intellectually demanding subject of both theoretical and practical value. The editors, who are all practicing systematists, are indebted to the contributors, all of whom managed to write state-of-the-art chapters despite busy working schedules.

The editors are also grateful to colleagues for their help at various stages of this project, notably Martin Embley, Colin Harwood and Ramon Rosselló-Móra. We are also very much indebted to Jan Fife for her tireless work in helping to "tidy up" manuscripts. One final word of thanks goes to colleagues at Elsevier, not least to Helene Kabes, Surya Narayanan Jayachandran and Mary Ann Zimmerman for seeing the book through from inception to press.

Michael Goodfellow
Iain Sutcliffe
Jongsik Chun
September 2014

The Need for Change: Embracing the Genome

1

William B. Whitman[1]

Department of Microbiology, University of Georgia, Athens, Georgia, USA
[1]Corresponding author: e-mail address: whitman@uga.edu

1 A BRIEF HISTORY OF GENOMIC SEQUENCING OF PROKARYOTES

Because of the small sizes of their genomes and their importance in medical and biological research, prokaryotes were among the first organisms whose genomes were sequenced. Following the sequencing of the first genomes of representatives of the *Bacteria* and *Archaea* in 1995 and 1996, respectively, the first 15 years of microbial genome sequencing yielded more than a thousand complete genome sequences (Liolios et al., 2010). In addition, thousands of draft genome sequences have been prepared. In draft sequencing projects, large numbers of randomly collected sequencing reactions are performed, but the second, more costly step of closing the sequence assembly is not done. These drafts typically contain the sequences of most of the genes in an organism, but their order is not established. Moreover, because gaps still exist in the sequence, it is not possible to know for certain which genes are absent. The end result was that by 2012 more than four thousand genome sequences were deposited in GenBank (Figure 1). Most of these early projects were initiated based on practical applications for selected organisms, often in the fields of medicine (e.g. biopharmaceuticals, drug targets, pathogens and probiotics) or biotechnology (e.g. agriculture, bioenergy, environmental remediation and industrial production of microbial products).

With the development of the Next-Generation Sequencing (NGS) technologies, the costs of genome sequencing became low enough to be performed routinely in many research and clinic laboratories (Didelot, Bowden, Wilson, Peto, & Crook, 2012; Koser et al., 2012; Bertelli & Greub, 2013). Projects were also initiated to sequence prokaryotic genomes more systematically. Prominent efforts include the Genome Encyclopedia of Bacteria and Archaea or GEBA, Human Microbiome Project (HMP) and the 10,000 Genomes Project. A pilot GEBA project was launched in 2007 to systematically explore the genomes of all bacterial and archaeal species with validly published names (Wu et al., 2009). A major goal of GEBA was to capture much of the microbial diversity that was missed in previous work (Hugenholtz, 2002; Krypides, 2009; Pace, 2009). The ultimate goal is to have at least one representative genome sequence of the type strain of every bacterial and archaeal species

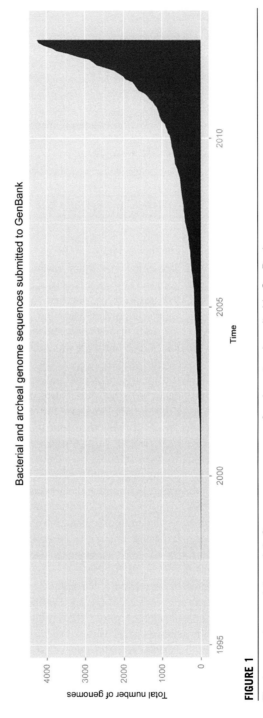

FIGURE 1

The increase in complete and draft genome sequences of prokaryotes deposited in GenBank.

that had been formally named (Lapage et al., 1992). As of 2013, the genomes of 1141 type strains of *Archaea* and *Bacteria* had been sequenced from all sources, including GEBA. An additional two thousand or so genomes have been selected by GEBA for sequencing in the near future. Current progress in this effort can be monitored at the Microbial Earth Project website: http://www.microbial-earth.org/cgi-bin/MEP/in dex.cgi.

The HMP was launched in 2008 to explore the prokaryotes sharing the human body. In addition to sequencing comprehensive rRNA gene and metagenome libraries of the prokaryotes from the human microbiome, this project includes a major effort to sequence genomes from strains isolated from the human body. As of the end of 2013, >1350 genomes of prokaryotes isolated from the gastrointestinal tract, urogenital tract, oral cavity, skin and other human tissues have been sequenced. The 10,000 Genomes Project was led by Prof. Lixin Zhang at the Institute of Microbiology at the Chinese Academy of Sciences in Beijing. Its major goals are to isolate bioactive compounds from marine microorganisms. To this end, marine *Actinobacteria* were isolated from deep sea sediments and other environments. In addition to direct high-throughput screening for novel antibiotics (Zhang et al., 2005), the genomes of the isolates were sequenced to look for genes of biotechnological interest.

2 WHY SEQUENCE THE GENOMES OF PROKARYOTES?

There are a number of very different but equally valid reasons to sequence the genomes of prokaryotes, and genomic sequencing now plays a central role in investigations of a wide variety of questions in prokaryotic biology (Figure 2). One, the genome sequence provides enormous insight into the physiology and ecology of the organism. By identifying genes encoding key steps of important pathways, it is possible to attribute specific properties to the organisms. More generally, on-line tools such as KEGG, SEED and MetaCyc infer the metabolic pathways in an organism based upon the genome sequence (Caspi et al., 2012; Kanehisa et al., 2014; Overbeek et al., 2005). They often provide the first evidence for the pathways of sugar metabolism or the inability to synthesize particular amino acids or vitamins. Specific examples of insights into the metabolic and ecological properties of organisms derived from genomics abound. The importance of H_2 metabolism during *Helicobacter pylori* infections was first realized following recognition of the genes encoding hydrogenases in the genome (Olson & Maier, 2002). Likewise, the abundance of genes for resistance to O_2 toxicity in the rice methanogen *Methanocella conradii* led to the hypothesis that this methanogen is unusually O_2 tolerant (Lu & Lu, 2012a, 2012b). Methanogens are strict anaerobes, and this feature may explain this species' abundance in rice paddies. Among marine bacteria, oligotrophs, which generally only grow slowly in media with extremely low levels of nutrients, can be readily distinguished at the genome level from opportunitrophs, which rapidly grow using a large number of different types of substrates. Oligotrophs typically possess very compact genomes, encoding only a few thousand genes with small

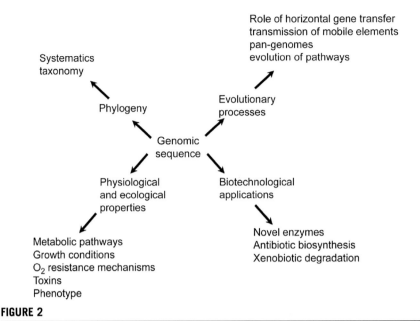

Role of horizontal gene transfer
transmission of mobile elements
pan-genomes
evolution of pathways

Systematics
taxonomy

Phylogeny

Evolutionary
processes

Genomic
sequence

Physiological
and ecological
properties

Biotechnological
applications

Metabolic pathways
Growth conditions
O_2 resistance mechanisms
Toxins
Phenotype

Novel enzymes
Antibiotic biosynthesis
Xenobiotic degradation

FIGURE 2

Central role of genomic sequencing in exploring microbial processes.

intergenic regions. Opportunitrophs possess much larger genomes and encode numerous transport systems for different classes of substrates (Moran et al., 2004). Thus, following the isolation of a new strain of prokaryote with interesting properties, sequencing its genome has now become routine.

Two, the genome sequences of groups of related organisms inform us about the evolutionary processes within a group. For instance, the pan-genome was first recognized by comparing the genomic sequences of many strains of a single prokaryotic species (Medini, Donati, Tettelin, Masignani, & Rappuoli, 2005). The pan-genome is the sum of all the genes found in all strains of a species. It comprises the core genome or the genes that are found in all genomes and the dispensable genome or the pool of genes found in some but not all genomes of the species. The pan-genome results from horizontal gene transfer (HGT) between the strains of a species and with members of other species. For instance, the genomes of each of 17 strains of *Escherichia coli* contain about 5000 genes, but only ~2300 genes are shared among all strains and represent the core genome (Rasko et al., 2008). The pan-genome or the entire set of genes found in any of the *E. coli* strains is ~18,000. The dispensable genome is then ~15,700 genes (the pan-genome minus the core genome). In principle, this concept can be applied to any taxonomic rank, where the pan-genome is composed of the 'extended core', 'character genes' and the 'accessory genes' (Lapierre & Gogarten, 2009). For instance, among 293 genomes of the domain *Bacteria*, the extended core, character and accessory genes comprise 250, 7900 and 139,000 gene families, respectively. The extended core includes highly conserved gene families encoding essential features of replication, transcription and translation. The character gene families define the properties of each physiological or taxonomic group and often encode proteins with diverse substrate specificities. Lastly, the accessory gene families are only present in a few genomes, appear to be associated with plasmids or phages and may have been acquired by recent HGT. Moreover, the 'average' bacterial genome contains all 250 gene families of the extended core, but only 1950 and 855 of the character and accessory gene families, respectively. Lastly, the extended core for *Archaea* and *Bacteria*, two extremes of prokaryotic evolution, only includes about 34 gene families, which emphasizes the enormous difference between these organisms (Makarova, Sorokin, Novichkov, Wolf, & Koonin, 2007). The balance between HGT and vertical evolution and gene invention and loss in formation of modern organisms is only visible because of the availability of numerous genomes for comparative analyses of gene content.

Three, the genome sequence identifies enzymes and biosynthetic pathways of value in biotechnology. Popularly called genome mining or prospecting, methods have been developed to search genome sequences for biosynthetic pathways for novel natural products, such as antibiotics of medical potential (Challis, 2008). Of special interest are enzymes, such as cellulases and other enzymes capable of transforming plant structural polymers to simple sugars, with potential applications in biomass conversion to biofuels. These strategies are based upon the premise that the enormous diversity of prokaryotes has created many more enzyme catalysts than could ever be designed in the laboratory. If they can be discovered by bioinformatic analyses of genomic sequences, reverse genetic engineering can be used to bring these enzymes to commercialization.

Four, genome sequencing provides valuable insights into the phylogeny of prokaryotes and vastly improves our understanding of their systematics. Because the number of genes available for phylogenetic analyses is large, it is possible to calculate robust phylogenetic trees and obtain a wealth of information about the genealogy of an organism. As important, an understanding of the evolutionary process, such as the relative importance of HGT and vertical evolution, can now be included in the descriptions of phylogenetic groups and their classification. As more genomes become available for specific groups, the applications of genome-based systematics will revolutionize the classification of prokaryotes (Coenye, Gevers, Van de Peer, Vandamme, & Swings, 2005; Klenk & Goker, 2010). It will make it possible to use Average Nucleotide Identity (ANI) or Genome-to-Genome-Distance values in defining species boundaries (Goris et al., 2007; Deloger, El Karoui, & Petit, 2009) and replace the imprecise and error prone wet laboratory determinations of DNA–DNA hybridizations. By providing more reliable, complete and portable data (amenable to iterative analyses), it will also allow us to form more accurate groupings of higher taxa. Of specific interest, identification of prokaryotes is still a major challenge that hinders many practical applications. Genomic sequencing of closely related strains provides tools that greatly facilitate identification.

3 THE STATE-OF-THE-ART

From this perspective, this volume is especially timely. Genomics is upon us, but uncertainty remains as to how researchers can effectively apply these new approaches to the questions asked in systematics and evolution. The following chapters describe the state-of-the-art in many areas of microbial genomics and its applications to systematics.

The contributions by Sangal et al. (Revolutionising Prokaryotic Systematics Through Next-Generation Sequencing) and Harris and Okoro (Whole-Genome Sequencing for Rapid and Accurate Identification of Bacterial Transmission Pathways) provide an overview of many of the NGS methodologies. These are important to fully understand because each method has its own strengths and limitations. Since most NGS methods yield small sequences of 20–500 bp, assembling them into a genome sequence of many Mb is often a challenge. Typical results may yield hundreds of contigs, which are continuous regions of the genome covered by overlapping sequences. Some contigs may also be connected into scaffolds, which include neighbouring contigs known to be connected even though there are gaps in sequence between them. Of equal importance is the software for bioinformatic analyses for assembly, annotation and comparative analysis of genomes. In the current environment, the bioinformatics is much more expensive and time-consuming that the actual sequencing and determines the types of questions that can be answered. The strategies for bioinformatic analyses are also illustrated with examples from classification, pathogen identification and determining the genetic bases of phenotypic properties. Arahal (Whole-Genome Analyses: Average Nucleotide Identity) provides additional practical advice for sequencing, including preparation of DNA and sequencing strategies.

Because of its tremendous capacity to reveal the physiological and metabolic properties of an organism, genomic sequencing is a valuable tool for the description of novel species. Arahal (Whole-Genome Analyses: Average Nucleotide Identity) shows how it can also be used to calculate the ANI, which can replace DNA: DNA hybridization for establishing the novelty of a new species. While this approach is currently limited by the availability of genomic sequences of related strains for comparison, as projects like GEBA near completion it is likely to become the preferred method for establishing differences between the genotypes of type strains. This advance will be especially important for describing new species in large and complex genera. Currently, DNA:DNA hybridization techniques require DNA samples from all the members of the genus to establish the novelty of the new species. Because of its difficulty and expense, this is seldom done, and many strains that might represent novel species are not fully described. Once the genome sequences for all the type strains are archived in databases, direct determination of DNA: DNA hybridizations will no longer be necessary.

A major goal of systematics is to discover the natural relationships between various organisms. Genomes provide enormous inventories of sequence data that can be analysed to determine the phylogenies of modern organisms. This can be viewed from the context of the phylogeny of individual genes or the genealogy of the organisms, which are not always the same. The contribution of Williams and Heaps (An Introduction to Phylogenetics and the Tree of Life) examines the steps in calculating the phylogeny of individual genes and demonstrates the process, from selecting the biological question and the sequences to be analysed, to producing an alignment and then considering more complex issues such as the statistical approaches employed during interpretation of phylogenetic trees. Importantly, it shows how different choices at each step can change the answer to the question.

In nature, genes occur in organisms. Due to HGTs, deletions and other genetic events, organisms may have an evolutionary history that is different from those of many of their individual genes. Kamneva and Ward (Reconciliation Approaches to Determing HGT, Duplications and Losses in Gene Trees) describe the concepts and procedures used to reconcile these differences and more fully understand the evolution of the organism. The aims of this approach include (1) prediction of the functions and properties of newly characterized genes and genomes, (2) characterization of the evolutionary history of individual genes, (3) characterization of genome evolution in terms of gene family content and (4) prediction of ancestral gene family composition. Upon analyses of the genomes of a group of organisms, these approaches have the potential of documenting the evolutionary processes that occurred during the formation of the lineage.

Perhaps surprisingly, even with complete genomic sequences, many phylogenetic relationships remain ambiguous because of limitations in tree-building algorithms, the complexity of the evolutionary histories of the organisms and the loss of informative sequences with time. An alternative approach to test phylogenetic relationships and resolve ambiguities utilizes Conserved Signature Indels (CSIs) as described by Gupta (Identification of Conserved Indels that are Useful for Classification and Evolutionary Studies). This valuable approach looks for uniquely shared sequence features, such as insertions or deletions that may lead to clear demarcation of specific genealogies. Because it frequently undergoes insertions and deletions even among closely related strains, the *rrn* operon is particularly useful for analyses of CSIs. The RiboTyping database described by Gürtler et al. (Bacterial Typing and Identification by Genome Analysis of 16S-23S rRNA Intergenic Spacer (ITS) Sequences) shows how to implement this strategy when thousands of sequences need to be compared. In addition to resolving the genealogy of closely related organisms, this database can also provide tools for identification and ribotyping.

The importance of reliable sequences and curated databases is further discussed by Yarza and Munoz (The All-Species Living Tree Project) with regard to the All-Species Living Tree Project. This well-curated database of 16S and 23S rRNA sequences is a source of high quality sequences of the type strains of prokaryotic species. Sequences are also aligned based upon secondary structural constraints to yield robust and informative phylogenetic analyses. Of particular interest is the ability to use the rRNA sequences of type strains to identify similar sequences in the large collection of environmental sequences. For many species, this approach provides one of the best indicators of their distribution.

A second tool for maintaining a reliable database of 16S rRNA sequences is the EzTaxon server described by Kim and Chun (16S rRNA Analysis/EzTaxon). This tool deals with the problem of accurate taxonomic assignment of new organisms based upon their 16S rRNA sequences. It solves this problem in a stepwise manner. First, near neighbours are identified with very rapid database searches using BLASTn. Then, new sequences are aligned with those of their closest relatives for calculation of sequence similarity and for constructing phylogenetic trees. Finally, when possible the sequence is assigned to a pre-existing taxonomic group, such as a species. Because some prokaryotic species possess nearly identical 16S rRNA sequences, the new sequence is not automatically assigned to the species with the highest sequence similarity. Because the process is automated, it is also suitable for determining the taxonomic assignments of libraries of 16S rRNA genes cloned from environment samples.

The numbers of individuals in prokaryotic species are enormous. For instance, *E. coli* probably represents about 10^{20} individual cells (Milkman & Stoltzfus, 1988). The species of the genus *Prochlorococcus* likely represent about 3×10^{27} individuals (Flombaum et al., 2013). Understanding the population dynamics of these huge collections of organisms is an exciting area of investigation made possible by DNA sequencing. Multi-locus sequence analysis (MLSA) and multi-locus sequence typing (MLST) have become major methods to explore this rich diversity. As described by Cody et al. (Multilocus Sequence Typing and the Gene-by-Gene Approach to Bacterial Classification and Analysis of Population Variation), MLST is particularly useful for epidemiological studies. In addition, MLSA has become an alternative to DNA:DNA hybridization in the characterization of new species (Rong and Huang, Multilocus Sequence Analysis: Taking Prokaryotic Systematics to the Next Level). Because of the high conservation of the 16S rRNA gene, MLSA has also proven to be a valuable tool for resolving the phylogeny of closely related species.

The contribution by Rong and Huang also provides an introduction to the practice of MLSA, especially as it is applied to the large genus *Streptomyces*. They show how the diversity revealed by MLSA predicts the functional diversity within a species and the underlying evolutionary processes. Generally, based upon 450–500 bp partial sequences of 5–7 housekeeping genes, even single base changes are important. Thus, MLSA is especially useful for inferring relationships among closely related strains. Rong and Huang provide guidance on the selection of genes for sequencing; strategies of obtaining the sequences, either by designing specific PCR primers or by whole-genome sequencing; strategies and software available for data analyses and the databases used to archive the sequences.

Harris and Okoro (Whole-Genome Sequencing for Rapid and Accurate Identification of Bacterial Transmission Pathways) and Cody et al. (Multilocus Sequence Typing and the Gene-by-Gene Approach to Bacterial Classification and Analysis of Population Variation) extend many of these principles to analyses of draft genomes or mixtures of different types of data including short gene fragments, draft genomes and complete genomes, where the questions to be addressed dictates the extent and type of data analysed. Harris and Okoro also review the hardware and software in NGS. Particular attention is given to the methods of sequence alignments and assembly for the enormous data sets generated by sequencing large numbers of closely related genomes. These methodological concerns are essential for resolving the epidemiology of closely related or slowly evolving pathogens. Cody et al. also describe the methodological developments from the MLST and MLSA approaches to whole-genome sequencing, with draft genome sequences providing an opportunity to greatly increase the number of genes available for the analysis.

In addition to the genetic methods for classifying and identifying prokaryotes, matrix-assisted laser desorption/ionization time-of-flight or MALDI-TOF mass spectroscopy has proven to be valuable tool for rapid and inexpensive identification. As described in the contribution by Schumann and Maier (MALDI-TOF Mass Spectrometry Applied to Classification and Identification of Bacteria), this method has had a major impact in clinical diagnosis, quality control of food production, the pharmaceutical and biotechnological industries, ecological and environmental research and prokaryotic systematics. For these applications, MALDI-TOF MS has proven superior to classical phenotypic methods and nucleic acid sequence technologies due to its low costs in time and expenses. Schumann and Maier also provide a practical guide to cultivation and sample preparation, selection of the matrix for ionization, guidelines for sample preparation from specific types of bacteria or complex substrata and recommendations for optimization.

In conclusion, Kämpfer makes the point that it is the phenotype not the genome sequence that determines the interactions of organisms with their environment and their evolution (Continuing Importance of the "Phenotype" in the Genomic Era). After reviewing the major methodological and conceptual developments in the modern systematics of prokaryotes, he argues that genomics can and should be incorporated into the polyphasic approach, which also includes consideration of growth properties, chemotaxonomic markers and other phenotypic data. Moreover, our ability to interpret the genome sequence is imperfect. Thus, while in principle the entire phenotype is encoded in the genome, the current state of knowledge does not allow a complete understanding of the phenotype based solely on genome sequence. Until that is possible, studies of the phenotype will remain critical to understanding the relationships between organisms.

4 WHERE WE ARE GOING

Since we started with the history of genomic sequencing, it is appropriate to conclude with a few comments about where genomic sequencing is taking microbial systematics. What can we expect in the next decade? We should expect that genomic sequences will become common in descriptions of new species. The sequences are just too valuable not to determine. If the strain is worth isolating, it is worth sequencing. We should also expect better software for analysis of genomes. Currently, a large number of competing programs are available, but a consensus will develop for most routine analyses. We should then expect 'point-and-click' software that is simple enough for nonscientists and undergraduate students to use and reliable enough for professional scientists. An example of this new generation of software is MEGA (http://www.megasoftware.net), which makes many of the tools used for phylogenetic analyses ready available (Tamura, Stecher, Peterson, Filipski, & Kumar, 2013). This software development is necessary to make the data truly public and accessible to scientists with little training in bioinformatics. For instance, experts in an organism's isolation, cultivation and ecology must be able to easily 'read' the genome for its content to be thoroughly explored and its potential fully realized. Moreover, for many fastidious prokaryotes that are difficult to cultivate, we will soon know more about them based on their genome sequences than we will ever be able to directly observe. This will be true for features that might be easily discovered if we could cultivate them, such as amino acid auxotrophies, but also for properties that can only be studied with great difficulty even in model organisms, such as quorum sensing, stress responses and cellular development. This developing wealth of knowledge will provide us an immensely richer understanding of prokaryotic life than we have today.

There will also be opportunities directly impacting on the practice of microbial systematics. Currently, the type strain system is necessary to provide unequivocal identification of prokaryotic species. Even with the molecular methods available prior to NGS, it has been necessary to perform phenotypic and genotypic tests, as well as DNA:DNA hybridization to insure the uniqueness of novel isolates. Thus, a living culture of the type strain of each species had to be preserved to serve as the standard for comparison of all newly described isolates. Preservation comes at enormous expense and requires maintaining large culture collections dedicated to this purpose. However, when the whole-genome sequences are available for all the type strains, there is an opportunity for their whole-genome sequences to serve as a comparison to all future isolates. Importantly, there is no conflict with this approach and the principles of the Bacteriological Code, which only require that species be unique and completely identified prior to naming (Whitman, 2011).

Not only will using the sequences of type strains greatly improve the efficiency of identification of new isolates, but it will also provide an opportunity to reallocate the resources in culture collections to the study of the most important and valuable cultures. Moreover, it will avert a crisis that is otherwise inevitable. The current system of type strains was developed when the number of known species was in the thousands. The number of described species is now about 12,000, and the number of species on earth is probably in the millions (Yarza et al., 2014). The current system where type strains must be preserved would require expansion of the culture collections at least a 100-fold to fully describe the richness of prokaryotic life on earth. The alternative will be to use genome sequences, which can be easily stored on electronic media, as the nomenclatural types for new taxa. Culture collections can then focus of preserving the most biologically interesting cultures that are likely to be of the greatest scientific value.

ACKNOWLEDGEMENT

This work was supported in part by NSF Dimensions in Biodiversity grant OCE-1342694.

REFERENCES

Bertelli, C., & Greub, G. (2013). Rapid bacterial genome sequencing: Methods and applications in clinical microbiology. *Clinical Microbiology and Infection, 19*, 803–813.

Caspi, R., Altman, T., Dreher, K., Fulcher, C. A., Subhraveti, P., Keseler, I. M., et al. (2012). The MetaCyc database of metabolic pathways and enzymes and the BioCyc collection of pathway/genome databases. *Nucleic Acids Research, 40*, D742–D753.

Challis, G. L. (2008). Mining microbial genomes for new natural products and biosynthetic pathways. *Microbiology, 154*, 1555–1569.

Coenye, T., Gevers, D., Van de Peer, Y., Vandamme, P., & Swings, J. (2005). Towards a prokaryotic genomic taxonomy. *FEMS Microbiology Reviews, 29*, 147–167.

Deloger, M., El Karoui, M., & Petit, M. A. (2009). A genomic distance based on MUM indicates discontinuity between most bacterial species and genera. *Journal of Bacteriology, 191*, 91–99.

Didelot, X., Bowden, R., Wilson, D. J., Peto, T. E., & Crook, D. W. (2012). Transforming clinical microbiology with bacterial genome sequencing. *Nature Reviews. Genetics, 13*, 601–612.

Flombaum, P., Gallegos, J. L., Gordillo, R. A., Rincon, J., Zabala, L. L., Jiao, N., et al. (2013). Present and future global distributions of the marine Cyanobacteria *Prochlorococcus* and *Synechococcus*. *Proceedings of the National Academy of Sciences U.S.A., 110*, 9824–9829.

Goris, J., Konstantinidis, K. T., Klappenbach, J. A., Coenye, T., Vandamme, P., & Tiedje, J. M. (2007). DNA-DNA hybridization values and their relationship to whole-genome sequence similarities. *International Journal of Systematic and Evolutionary Microbiology, 57*, 81–91.

Hugenholtz, P. (2002). Exploring prokaryotic diversity in the genomic era. *Genome Biology, 3*, REVIEWS0003.

Kanehisa, M., Goto, S., Sato, Y., Kawashima, M., Furumichi, M., & Tanabe, M. (2014). Data, information, knowledge and principle: Back to metabolism in KEGG. *Nucleic Acids Research, 42*, D199–D205.

Klenk, H. P., & Goker, M. (2010). En route to a genome-based classification of *Archaea* and *Bacteria*? *Systematic and Applied Microbiology, 33*, 175–182.

Koser, C. U., Ellington, M. J., Cartwright, E. J., Gillespie, S. H., Brown, N. M., Farrington, M., et al. (2012). Routine use of microbial whole genome sequencing in diagnostic and public health microbiology. *PLoS Pathogens, 8*, e1002824.

Krypides, N. C. (2009). Fifteen years of microbial genomics: Meeting the challenges and fulfilling the dream. *Nature Biotechnology, 27*, 627.

Lapage, S. P., Sneath, P. H. A., Lessel, E. F., Skerman, V. B. D., Seeliger, H. P. R., & Clark, W. A. (1992). *International code of nomenclature of bacteria*. Washington, DC: American Society for Microbiology.

Lapierre, P., & Gogarten, J. P. (2009). Estimating the size of the bacterial pan-genome. *Trends in Genetics, 25*, 107–110.

Liolios, K., Chen, I. M., Mavromatis, K., Tavernarakis, N., Hugenholtz, P., Markowitz, V. M., et al. (2010). The genomes on line database (GOLD) in 2009: Status of genomic and metagenomic projects and their associated metadata. *Nucleic Acids Research, 38,* D346–D354.

Lu, Z., & Lu, Y. (2012a). Complete genome sequence of a thermophilic methanogen, *Methanocella conradii* HZ254, isolated from Chinese rice field soil. *Journal of Bacteriology, 194,* 2398–2399.

Lu, Z., & Lu, Y. (2012b). *Methanocella conradii* sp. nov., a thermophilic, obligate hydrogenotrophic methanogen, isolated from Chinese rice field soil. *PLoS One, 7,* e35279.

Makarova, K. S., Sorokin, A. V., Novichkov, P. S., Wolf, Y. I., & Koonin, E. V. (2007). Clusters of orthologous genes for 41 archaeal genomes and implications for evolutionary genomics of archaea. *Biology Direct, 2,* 33.

Medini, D., Donati, C., Tettelin, H., Masignani, V., & Rappuoli, R. (2005). The microbial pangenome. *Current Opinion in Genetics and Development, 15,* 589–594.

Milkman, R., & Stoltzfus, A. (1988). Molecular evolution of the *Escherichia coli* chromosome. II. Clonal segments. *Genetics, 120,* 359–366.

Moran, M. A., Buchan, A., Gonzalez, J. M., Heidelberg, J. F., Whitman, W. B., Kiene, R. P., et al. (2004). Genome sequence of *Silicibacter pomeroyi* reveals adaptations to the marine environment. *Nature, 432,* 910–913.

Olson, J. W., & Maier, R. J. (2002). Molecular hydrogen as an energy source for *Helicobacter pylori. Science, 298,* 1788–1790.

Overbeek, R., Begley, T., Butler, R. M., Choudhuri, J. V., Chuang, H. Y., Cohoon, M., et al. (2005). The subsystems approach to genome annotation and its use in the project to annotate 1000 genomes. *Nucleic Acids Research, 33,* 5691–5702.

Pace, N. R. (2009). Mapping the tree of life: Progress and prospects. *Microbiology and Molecular Biological Reviews, 73,* 565–576.

Rasko, D. A., Rosovitz, M. J., Myers, G. S., Mongodin, E. F., Fricke, W. F., Gajer, P., et al. (2008). The pangenome structure of *Escherichia coli*: Comparative genomic analysis of *E. coli* commensal and pathogenic isolates. *Journal of Bacteriology, 190,* 6881–6893.

Tamura, K., Stecher, G., Peterson, D., Filipski, A., & Kumar, S. (2013). MEGA6: Molecular evolutionary genetics analysis version 6.0. *Molecular Biology and Evolution, 30,* 2725–2729.

Whitman, W. B. (2011). Intent of the nomenclatural code and recommendations about naming new species based on genomic sequences. *The Bulletin of BISMiS, 2,* 135–139.

Wu, D., Hugenholtz, P., Mavromatis, K., Pukall, R., Dalin, E., Ivanova, N. N., et al. (2009). A phylogeny-driven genomic encyclopaedia of *Bacteria* and *Archaea. Nature, 462,* 1056–1060.

Yarza, P., Yilmaz, P., Prüße, E., Glöckner, F. O., Ludwig, W., Schleifer, K.-H., et al. (2014). Uniting the classification of cultured and uncultured *Bacteria* and *Archaea* by means of SSU rRNA gene sequences. *Nature Reviews. Microbiology,* in press.

Zhang, L., An, R., Wang, J., Sun, N., Zhang, S., Hu, J., et al. (2005). Exploring novel bioactive compounds from marine microbes. *Current Opinion in Microbiology, 8,* 276–281.

An Introduction to Phylogenetics and the Tree of Life

2

Tom A. Williams[*,1], **Sarah E. Heaps**[*,†]

Institute for Cell and Molecular Biosciences, The Medical School, Newcastle University, Newcastle upon Tyne, United Kingdom
†*School of Mathematics and Statistics, Newcastle University, Newcastle upon Tyne, United Kingdom*
[1]*Corresponding author: e-mail address: tom.williams2@ncl.ac.uk*

1 INTRODUCTION

Phylogenetic trees are fundamental to organising, understanding and testing hypotheses about the evolution of biological diversity. Early phylogenies were based on morphology: useful for multicellular eukaryotes, but much less so when inferring relationships among prokaryotes or among the different branches of the tree of life, most of which is microbial. Although comparisons of biochemical properties provided some insight into bacterial relationships, they proved unreliable at deeper taxonomic levels, and by 1960, it seemed that a universal phylogeny was out of reach, with the only unambiguous division in the microbial world separating the eukaryotes from the structurally simpler prokaryotes (Stanier & van Niel, 1962). This situation changed completely with the advent of molecular sequencing, which provided biologists with a rich new source of information about evolutionary history (Zuckerkandl & Pauling, 1965) that was just as relevant for prokaryotes and microbial eukaryotes as for animals, plants and fungi. The greatest early success of the sequencing era came when Carl Woese and colleagues showed that the ribosomal RNA (rRNA) sequences of prokaryotes clustered into two groups that were at least as divergent from each other as they were from the rRNA genes of eukaryotes, demonstrating that the prokaryotes comprised two distantly related lineages, the Bacteria and Archaea (Woese & Fox, 1977; Woese, Kandler, & Wheelis, 1990).

The discovery of the Archaea demonstrated the power of sequence data for investigating relationships among prokaryotes, and in the intervening years, analyses of rRNA and, more recently, whole genome sequences have become standard approaches in molecular evolution and systematics. The advantages of sequences over other types of data—such as morphology, physiology and biochemistry—for inferring phylogenies are clear, particularly in the case of prokaryotes, microbial eukaryotes and viruses. Sequence data are highly informative, and today, millions of characters can be analysed simultaneously in single-gene or concatenated multiple sequence alignments. With contemporary (i.e. "next-generation") sequencing technologies, new sequences are cheap and relatively easy to obtain, and the number of sequences in public databases is so large that the data needed to address many unanswered or new evolutionary questions is already available. From the biological point of view, one of the greatest strengths of sequence-based phylogenies is the capability they provide for inferring relationships among organisms for which other meaningful points of comparison do not really exist. For example, all cellular organisms synthesise proteins on a ribosome, so a tree based on rRNA can include the bacterium *Escherichia coli*, the archaeon *Sulfolobus solfataricus* and the eukaryotes *Saccharomyces cerevisiae* and *Homo sapiens*, organisms which would otherwise be difficult or impossible to fit into a single, meaningful classification. Much of the early excitement around sequence-based phylogenies was due to their potential use in constructing a universal tree of life that would include all cellular organisms (Woese et al., 1990). In fact, much progress has been made on this issue in the sequencing era (Embley & Martin, 2006), although the relationships among the major lineages of cellular life remain actively debated (Ciccarelli et al., 2006; Cox, Foster, Hirt, Harris, & Embley, 2008; Foster, Cox, & Embley, 2009; Gribaldo, Poole, Daubin, Forterre, & Brochier-Armanet, 2010; Williams, Foster, Cox, & Embley, 2013; Williams, Foster, Nye, Cox, & Embley, 2012), as will be discussed in more detail below.

Another major advantage of sequence data is that it is unambiguously categorical: there are 4 possible states (A, C, G and T) for each nucleotide position, and 20 for each amino acid. As a result, sequences are considerably more amenable to rigorous statistical analysis than phenotypic characters, whose states must often be encoded in a somewhat arbitrary way (Stevens, 1991). This categorical character of sequence data is important—sequences may represent the richest source of information about prokaryotic evolution currently available, but as with other kinds of data, they can be positively misleading (Felsenstein, 1978) if analysed using inappropriate methods. Thus, while obtaining sequences is easier than ever before, careful phylogenetic analysis using the best available methods remains a time-consuming and potentially challenging task. With the right tools in hand, the process of building phylogenies can be relatively straightforward, but it is not automatic—each step (Figure 1), from collecting and aligning sequences to choosing the most appropriate phylogenetic model and building the trees, involves making decisions that may change the outcome. The aim of this chapter is to provide a practical guide to each of these steps and to introduce some of the best and most frequently used software for phylogenetic analysis. In order to make our discussion more concrete, we will work through an attempt to resolve one of the most interesting and controversial questions in phylogenetics—the relationship between Bacteria, Archaea and Eukarya, the three major lineages of cellular life.

Following Woese's discovery of the Archaea, the question naturally arose as to which of the prokaryotic groups (Bacteria or Archaea), if either, was more closely related to the eukaryotes. This question is complex because of the symbiogenic origins of eukaryotic cells (Sagan, 1967): all eukaryotes have a mitochondrion or

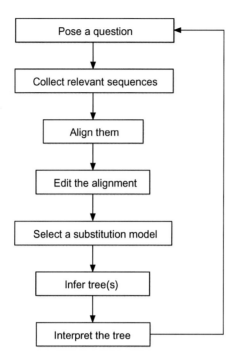

FIGURE 1

A workflow for phylogenetic analysis. The outline of a generic approach that can be used to address many questions in phylogenetics. In this chapter, we decided to investigate the relationship between Archaea and eukaryotes. This decision motivated our selection of SSU ribosomal RNA sequences for analysis, and the properties of that dataset suggested a particular approach to alignment and phylogenetic modelling. The resulting trees were then interpreted in the light of the original question, helping to focus discussion on their most relevant features.

mitochondria-related organelle that descends from a free-living alphaproteobacterium (Andersson et al., 1998; Esser et al., 2004), and many also possess a plastid descended from cyanobacteria (Martin et al., 2002). Thus, different compartments of eukaryotic cells have different phylogenetic origins. However, the genetic and ultrastructural similarities between mitochondria and plastids and their bacterial relatives are sufficiently strong that a broad consensus now exists on the origins of these organelles. Instead, contemporary debate focuses on the phylogenetic affinity of the eukaryotic nucleocytoplasmic lineage, which is often taken to represent the original host cell for these bacterial partners (Embley & Martin, 2006). Early analyses of rRNA led by Woese and coworkers (Woese, 1987; Woese & Fox, 1977) suggested that each of the three "domains" of life—Bacteria, Archaea and Eukarya—were monophyletic; in other words, that all Archaea, for example, are more closely related to each other than any of them are to Bacteria or eukaryotes. Combined with evidence from analyses of ancient gene duplications which suggested that the root of this "universal tree" lay on the branch leading to Bacteria (Gogarten et al., 1989; Iwabe, Kuma, Hasegawa, Osawa, & Miyata, 1989), these results led to the now-famous rooted three-domains tree (Woese et al., 1990), in which the Eukarya and Archaea form monophyletic sister groups to the exclusion of Bacteria. This tree represents the dominant hypothesis for the deepest branches of the tree of life and as such plays an important role in modern evolutionary biology. In this chapter, our goal will be to investigate whether it remains the most strongly supported hypothesis given currently available sequence data and statistical models.

2 STEP 1: POSING A QUESTION

The first step in any phylogenetic analysis is to frame the question you are attempting to answer, or the hypothesis you wish to test. This provides a rationale for choosing the sequences to analyse and a framework for interpreting the results. In the present case, our aim is to test whether the three-domains tree is supported from contemporary sequence data. Consulting the literature, we can see that a number of alternatives to the three-domains tree have been proposed. Several of these involve the placement of the eukaryotes (or at least, the set of conserved eukaryotic genes encoding the ribosome and related cellular components) within the Archaea, as the sister group to the Crenarchaeota (Lake, Henderson, Oakes, & Clark, 1984), the Thaumarchaeota (Kelly, Wickstead, & Gull, 2011), or the Thermoplasmatales (Pisani, Cotton, & McInerney, 2007). It may be worth keeping some of these alternative hypotheses in mind as we analyse and interpret our results.

3 STEP 2: CHOOSING RELEVANT SEQUENCES

Since our question addresses the relationships between domains, we will need to include sequences from all three domains of life—Bacteria, Archaea and Eukarya. A pervasive problem that affects all attempts to resolve inter-domain trees, as well as many smaller-scale phylogenetic analyses, is that individual genes often do not contain sufficient phylogenetic information (or signal) to produce a well-resolved species tree. As a result, many modern analyses attempt to combine signal from multiple genes using either "supermatrices" or "supertrees". In the supermatrix approach (de Queiroz & Gatesy, 2007), alignments from individual genes are simply concatenated and analysed as if they represented one large gene, although some aspects of the evolutionary model may be allowed to vary among the constituent genes. In the supertree approach (Bininda-Emonds, 2004), individual trees are inferred separately for each gene, and the information in these trees is then combined to produce a consensus estimate of the species tree. These methods have a number of advantages when the goal is to infer a species tree: for example, trees inferred from supermatrices are usually very well resolved, with high support values (see Section 5.3 and 5.4.) for most or all branches. However, these methods also add an additional layer of complexity to phylogenetic analyses, and they introduce a number of difficulties and caveats. In particular, the supermatrix approach necessarily assumes that all the genes in the matrix are evolving on the same underlying species tree, and violation of this assumption (e.g. due to horizontal gene transfer in some genes) can lead to the recovery of trees that are strongly supported but incorrect (see, e.g., Moreira & Lopez-Garcia, 2005). For more information, see chapter 'Reconciliation Approaches to Determining HGT, Duplications, and Losses in Gene Trees' by Kamneva and Ward, in this volume provides a discussion dealing with these cases. Here, we will sidestep these issues by focusing on the phylogenetic analysis of just one gene—that encoding the RNA component of the small subunit of the ribosome (16S rRNA). This is the most frequently used gene in prokaryotic phylogeny (see, for instance, chapter 'The All-Species Living Tree Project' by Yarza and Munoz, in this volume) and is also well suited for analysis of inter-domain relationships because of its ubiquity and very slow evolutionary rate. The phylogenetic methods we will apply can be easily extended to model the evolution of protein-coding sequences; for those interested in building supermatrices or supertrees, we recommend first consulting the extensive literature on these methods, to which Rannala and Yang (2008) provide an excellent entry point. Finally, it is important to bear in mind that the analysis of any single gene, no matter how broadly distributed or well conserved, provides only one perspective on the evolution of the organisms that encode it. Our aim here is to thoroughly analyse a single gene in order to introduce some of the most important concepts in phylogenetic analysis; state-of-the-art work typically involves a much larger sample of genes to provide a much more robust estimate of species phylogenies—the interested reader should consult chapter 'Reconciliation Approaches to Determining HGT, Duplications, and Losses in Gene Trees' by Kamneva and Ward in this volume.

3.1 OBTAINING 16S rRNA SEQUENCES FOR BACTERIA, ARCHAEA AND EUKARYA

Due to their historical importance as phylogenetic markers in the era before complete genome sequencing, 16S rRNA sequences are available for a very wide range of Bacteria, Archaea and Eukarya. Here, we will make use of a publicly available dataset of 36 sequences that was analysed by one of the present authors in Williams et al. (2012). The sequences can be obtained from the public repository Dryad (see Table 1, which provides links to the data, software and Web resources referenced in this chapter). This dataset is useful for our purposes here because it is relatively small, and so each step of the analysis can be performed quickly. It is also an interesting dataset for illustrating the impact that different decisions made during the analysis can have on the inferred phylogeny, as we will see in section 7 below. From the relevant Dryad page, download and extract the archive "rrna.tar". Our re-analyses will require the "ssu.fa" and "ssu_all.fa" files inside this archive. These files are partially redundant: "ssu.fa" (hereafter the SSU dataset) contains 32 small subunit rRNA sequences from Bacteria, Archaea and Eukarya; "ssu_all.fa" contains the same 32 sequences and 4 new sequences from recently sequenced Archaea (Thaumarchaeota, Aigarchaeota and Korarchaeota; hereafter

Table 1 The Freely Available Resources Referenced in This Chapter

Resource	Description	Link
16S ribosomal RNA sequence dataset	36 rRNA sequences from Bacteria, Archaea and eukaryotes analysed in Williams et al. (2012)	http://datadryad.org/resource/doi:10.5061/dryad.0hd1s
Muscle	Popular alignment tool	http://www.drive5.com/; http://www.ebi.ac.uk/Tools/msa/muscle/
Jalview	Alignment viewer	http://www.jalview.org/
TrimAl	Alignment masking (editing) program	http://trimal.cgenomics.org/; http://phylemon2.bioinfo.cipf.es/
jModelTest2	Model comparison	https://code.google.com/p/jmodeltest2/
RAxML	Maximum likelihood inference of phylogeny	http://www.exelixis-lab.org/; http://www.phylo.org/sub_sections/portal/ (Web server)
PhyloBayes, PhyloBayes-MPI	Bayesian inference of phylogeny; implements the CAT and CAT+GTR models	http://www.phylobayes.org; http://www.phylo.org/sub_sections/portal/
Dendroscope	Tree viewer	http://ab.inf.uni-tuebingen.de/software/dendroscope/
FigTree	Tree viewer	http://tree.bio.ed.ac.uk/software/figtree/
AWTY (Nylander, Wilgenbusch, Warren, & Swofford, 2008)	Graphical exploration of MCMC convergence in Bayesian phylogenetic inference	http://king2.scs.fsu.edu/CEBProjects/awty/awty_start.php

the SSU+TAK dataset). We will analyse these two datasets using exactly the same protocol, in order to investigate whether slight changes in taxon sampling can influence the inferred tree.

3.2 A NOTE ON THE AVAILABILITY AND USE OF DATA AND METHODS

Openness and reproducibility are fundamental to scientific progress. In principle, ensuring reproducibility in phylogenetic analyses should be straightforward because sequences and alignments can be easily shared over the internet. Further, the analyses are all computational, which should help to limit the role of human bias or error. Unfortunately, reproducibility in phylogenetics is generally low because researchers often fail to make their datasets publicly available (Drew et al., 2013). One of the many benefits of publishing the raw materials of your phylogenetic analyses in a public repository is that it allows others to build on or refine your work; thus, beyond the immediate context of this practical, we encourage you to explore and experiment with the datasets we use here. In the same spirit, all of the methods used in this tutorial are freely available and open source; this ensures that the scientific community can investigate and verify the algorithms used, and represents another important component of ensuring reproducibility in phylogenetic research.

Most phylogenetic software is designed to run on Unix-based systems, of which the most popular today are OS X and the various distributions of Linux. If you have access to one of these systems, you will find installing and running the tools used in our tutorial much more straightforward than if you are limited to a computer running Windows. As the purpose of this chapter is to introduce the basic principles of phylogenetics, and not to provide a Unix handbook, we have designed our tutorial so that all the analyses can be run using a Web browser and one or more of the Web servers that various laboratories generously provide for free online (see Table 1). However, we strongly encourage the interested reader to run the analyses locally, as familiarity with Unix and the ability to compile and run academic software is a prerequisite for any serious phylogenetic work. Throughout the rest of this chapter, we generally do not provide the exact commands you will need to run the analyses—for information on the basic operation of the different tools, refer to the available documentation on their Web sites (Table 1).

If you want to run the analyses locally but only have access to a Windows machine, one option is to set up a virtual machine running a variant of Linux, such as Ubuntu (http://www.ubuntu.com/). VirtualBox (https://www.virtualbox.org/) is a free and relatively easy-to-use virtualisation program available for Windows and other operating systems that will help you to set up a virtual machine. If you are planning to do a lot of bioinformatic analysis on your computer, installing a Linux distribution directly to your hard drive (i.e. dual-booting with Windows) may be the best option, as this will allow your analyses to make full use of the underlying hardware.

4　STEP 3: ALIGNING SEQUENCES AND EDITING THE ALIGNMENT

With the SSU rRNA datasets in hand, the next step in our analysis is to align the sequences. Sequence alignment is critical because although we assume that the rRNA genes of contemporary organisms are all descended from an ancestral gene that was present in their common ancestor, these genes have experienced not only point mutations (i.e. one nucleotide being substituted for another) but also insertions and deletions (gains or losses of one or more nucleotides) during their evolutionary history. As a result, the sequences from different organisms may have different lengths, and so (for example) the 100th nucleotide in the *H. sapiens* sequence is not necessarily equivalent to the 100th nucleotide of the *E. coli* sequence, or to the 100th nucleotide of the ancestral gene. Our inference of the phylogenetic tree will be based on comparisons of *homologous* nucleotides—that is, the nucleotides in contemporary genes that are descended from an ancestral nucleotide in their common ancestor (see Figure 2 for a visual clarification of this important idea). The purpose of sequence alignment is to arrange these homologous nucleotides into columns, or sites, in the final alignment file, ready for subsequent phylogenetic analysis.

Since the evolutionary history of a gene family is almost never known with certainty, the alignment that we infer will represent a hypothesis (or rather, a set of hypotheses) about that history that is likely to be wrong in some respects. Clearly, the quality of the sequence alignment is absolutely critical to the reliability of any subsequent phylogenetic analysis. Consideration of even the very small-scale example in Figure 2 indicates that the number of possible alignments and substitution histories

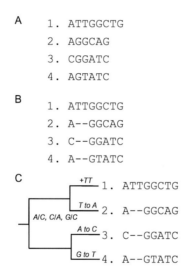

A
1. ATTGGCTG
2. AGGCAG
3. CGGATC
4. AGTATC

B
1. ATTGGCTG
2. A--GGCAG
3. C--GGATC
4. A--GTATC

C
+TT — 1. ATTGGCTG
T to A — 2. A--GGCAG
A/C, C/A, G/C
A to C — 3. C--GGATC
G to T — 4. A--GTATC

FIGURE 2

Sequence alignments represent inferences of homology, and as such contain information about the evolutionary history of a gene family. (A) Four related gene sequences sampled from contemporary organisms. (B) One possible alignment of the three sequences, identifying eight homologous sites (alignment columns). If this alignment is correct, then it is likely that sequence 1 contains a "TT" insertion after the first position; thus, its fourth nucleotide (G) is equivalent to the second position G in the three other sequences. (C) One possible evolutionary history for these sequences, based on the alignment in (B). The homologous nucleotides in columns 6 and 8 of the alignment are C, G in sequences 1 and 2, but A, C in sequences 3 and 4. This provides some evidence that sequences 1 and 2 are more closely related to each other than either is to 3 or 4 and vice versa. Inferred ancestral sequences and an associated substitutional history are mapped onto the interior branches of the phylogenetic tree. Inferences of homology can be made with reference to the reconstructed ancestral sequences: for example, the nucleotides T, A, T and T in column 7 of the alignment in (B) are homologous because they all descend from the same nucleotide in position 5 of the ancestral sequence.

for any real sequence dataset is likely to be astronomical. Given this context, the aim of sequence alignment is generally not to find the best global alignment, but rather to produce the best alignment possible within a reasonable computational time, and also to identify (and somehow deal with) the regions of this alignment that are likely to be problematic or unreliable. A large number of sequence alignment programs have been developed for this purpose; their relative strengths and weaknesses are reviewed in Thompson, Linard, Lecompte, and Poch (2011). In this tutorial, we will use the Muscle aligner (Edgar, 2004), a popular choice that performs reasonably well in alignment benchmarks. Muscle can either be downloaded or run locally, or via online Web servers such as Phylemon—see Table 1 for details. An excellent alternative to using a single alignment tool such as Muscle is to align your sequences with a number of different methods and then produce a consensus alignment; this is a slightly more elaborate and time-consuming approach, and so we will not use it here. Nonetheless, it does tend to produce better alignments and may be something to consider for your own analyses in the future (see Wallace, O'Sullivan, Higgins, & Notredame, 2006 for a discussion of this approach). An alternative strategy, which is generally more complex than sequential sequence alignment and tree building, is to simultaneously infer the alignment and phylogenetic tree from unaligned sequence data (see, e.g., Redelings & Suchard, 2005 for further details).

For the time being, align both the SSU and SSU+TAK datasets using Muscle with the default parameters, either on your own computer or online. If you open the outfile file in a text editor, you will see that the program has introduced gaps ("-") in the sequences in order to align the homologous positions against each other. This will be most obvious if you use an alignment viewing program such as Jalview (Waterhouse, Procter, Martin, Clamp, & Barton, 2009; Table 1), which also allows you to colour the nucleotides by type or sequence identity in order to visualise the parts of the alignment that are more or less similar to each other (more or less *conserved*). Viewing the alignment in this way, it will soon become obvious that some regions are much more conserved than others. Conserved regions are relatively easy to align because few changes have taken place and homologous positions are usually straightforward to identify. However, you may notice other parts of the alignment that contain many more gaps, and where the case for positional homology looks much weaker. Since SSU rRNA is one of the most conserved genes known, the alignment will look quite reasonable overall, although you may notice more ambiguity near the 3'-end of the molecule (e.g. beyond position 1100 in the alignment); often alignments will contain many more of these disordered regions. This observation raises the question of how these less reliable parts of the alignment should be treated. Opinions differ on this issue: on the one hand, including these regions in your analysis might introduce noise or other non-historical signal, potentially reducing the reliability of the inferred tree (Talavera & Castresana, 2007). On the other hand, over-zealous removal of the less conserved regions of the alignment is likely to remove genuine signal that could help to resolve the tree. In this tutorial, we will remove these regions of the alignment, but we encourage you to investigate other perspectives on this issue (see, e.g., Lee, 2001).

Once the decision to remove poorly aligning regions has been made, the next question is how these regions should be selected. This is sometimes done manually, by visualising the alignment in a program such as Jalview and simply deleting the columns or regions that look unreliable, and to which no obvious manual improvements can be made. The alternative to this manual approach is to use an automated "masking" program that uses some set of rules to decide which parts of the alignment to remove. In principle, expert opinion may be as good as or better than the heuristics employed by these programs. However, use of masking programs has the great benefit of reproducibility: if you include the program and settings used in the Methods section of your paper, then others will be able to see exactly what you did. For that reason, we will use an automated masking program here, although—as with the issue of whether the alignment should be edited at all—opinions differ on this subject. We will use trimAl (Capella-Gutierrez, Silla-Martinez, & Gabaldon, 2009), a masking program which uses the level of conservation of the sequence alignment to determine which of several modes (of varying strictness) should be used to select the regions to be deleted. As with the other steps in this tutorial, it is well worth exploring the alternatives to this program in your own research. Some of the most widely used examples include Gblocks (Castresana, 2000) and BMGE (Criscuolo & Gribaldo, 2010). TrimAl can be run locally or via the Phylemon Web server (Table 1); you should select the "automated1" mode when trimming the SSU and SSU+TAK datasets and then inspect the edited alignments using Jalview.

5 STEP 4: THE THEORY OF FITTING AND SELECTING A PHYLOGENETIC MODEL

Steps 4 and 5 in our analysis involve fitting phylogenetic models in one of two inferential frameworks (frequentist or Bayesian) and then comparing these models in a principled fashion. We first provide an introduction to the statistical background of the models and the basis for statistical inference and model comparison. We then illustrate these ideas through analyses of our SSU and SSU+TAK datasets. The impatient reader may wish to skip forward to the practical guidelines in section 6, dipping back into the statistical theory to gain a better understanding of the underpinning ideas.

5.1 MARKOV NUCLEOTIDE SUBSTITUTION MODELS

Statistical models for molecular evolution provide a structured framework for describing the complex and uncertain relationships between the molecular sequences for a collection of species. They are generally based on a set of simplifying assumptions about the evolutionary process that allows these relationships to be explained using a reasonably small number of parameters. When the values of these parameters, including the phylogeny, are fixed at one possible set of values they might take, the model attaches probabilities to the different alignments that could be observed, and these probabilities will be different for different sets of parameters. When we fit models to data, we then learn which parameter values are more or less consistent with the observed data according to the model we have assumed.

Most phylogenetic models assume that substitutions in molecular sequences can be modelled using continuous time Markov processes (CTMPs). A crucial property of these models, called the Markov property, is that the future state of the process (i.e. the remaining time before the next substitution and the character state resulting from the next substitution) depends only on the current state of the process and not on its past given this current state. Consider a single site of our sequence of nucleotides evolving over time on one branch of the tree. The CTMP describing the substitutions along that branch can be characterised by an instantaneous rate matrix Q. Given the length of the branch in question, Q determines the transition probabilities of the process. A transition probability is the probability of being in state j at the end of the branch given that the process was in state i at the start of the branch. These probabilities, for all possible combinations of states i and j, form a transition matrix. Standard models assume that the CTMP on any particular branch of the tree is time reversible and stationary. The assumption of reversibility means that the transition matrix would be the same if the process ran forwards or backwards in time. The assumption of stationarity means that (i) the probability of the character state being A, C, T or G does not change over time, that is, over the duration of the branch (these probabilities are called the stationary distribution, or probabilities, of the process) and (ii) the probability of transitioning from one state to another in a window of time depends only on the size of the window and not on its position in time. A consequence of making these assumptions is that the likelihood function of a tree is not affected by the position of the root. This makes it impossible to learn about the root position. Nevertheless, the assumptions are made in order to simplify the mathematics underpinning the model. In particular, they allow the instantaneous rate matrix Q to be decomposed in the following form:

$$Q = \begin{pmatrix} - & \rho_{AG}\pi_G & \rho_{AC}\pi_C & \rho_{AT}\pi_T \\ \rho_{AG}\pi_A & - & \rho_{GC}\pi_C & \rho_{GT}\pi_T \\ \rho_{AC}\pi_A & \rho_{GC}\pi_G & - & \rho_{CT}\pi_T \\ \rho_{AT}\pi_A & \rho_{GT}\pi_G & \rho_{CT}\pi_C & - \end{pmatrix}$$

where the dashes on the diagonal signify that these elements are defined to ensure the rows sum to zero. The terms ρ_{AG}, ρ_{AC}, ρ_{AT}, ρ_{GC}, ρ_{GT} and ρ_{CT} are called exchangeability parameters and they can be interpreted as the instantaneous rates of change between the different character states. The terms π_A, π_G, π_C and π_T are the probabilities from the stationary distribution of the process. Note that there is a different instantaneous rate of change between all possible pairs of nucleotides. This is called the general time-reversible (GTR) model (Tavaré, 1986). Other commonly used models for DNA are special cases. For example, the HKY85 model (Hasegawa, Kishino, & Yano, 1985) is a special case with only two distinct exchangeability parameters: the transition rate α and the transversion rate β. In other words, under the HKY85 model, $\rho_{AG}=\rho_{CT}=\alpha$ and $\rho_{AC}=\rho_{AT}=\rho_{GC}=\rho_{GT}=\beta$. The K80 model (Kimura, 1980), like the HKY85 model, assumes different rates for transitions and transversions. However, it imposes the additional constraint that all the stationary probabilities are equal, that is, $\pi_A=\pi_G=\pi_C=\pi_T=1/4$. This constraint on the stationary probabilities is also imposed by the JC69 model (Jukes & Cantor, 1969) which additionally assumes that the rates of change between all nucleotides are the same. A more detailed description of these and other Markov models of nucleotide substitution can be found in Yang (2006, Chapter 1).

So far we have only discussed the process on one branch of the tree. Under standard phylogenetic models, a transition matrix governed by the same instantaneous rate matrix Q applies to every branch of the tree. In order to extend the model to allow for all of the sites in our molecular sequence, standard models then assume that the same Markov process applies to every site and that sites are independent of each other. This assumption of independence can be interpreted as follows. Consider a small fixed tree on four taxa and assume that we know the branch lengths and the values of all the substitution model parameters in our model. Suppose we want to use our model to find the probability of having the nucleotides A, A, T, A at site number i (for any i). Independence simply means that this probability would not be affected by what happens at any other site(s). As you may already have guessed, this assumption is made for mathematical convenience, rather than to represent any real biological understanding.

In most molecular sequences, there will be heterogeneity in the extent to which different sites are conserved due to varying selective (or functional) constraints. This can be accommodated in our model by modifying it to allow each site to evolve at its own rate. In order to share information on rates between sites, these rates are assumed to come from a probability distribution, typically a gamma distribution with mean equal to one, which is often denoted as $\Gamma(\alpha, \alpha)$. The shape of this distribution influences the degree of rate heterogeneity between sites and is determined by the single parameter α. For computational convenience, the gamma distribution is often replaced by a discrete approximation with M categories. Although the continuous distribution is only recovered in the limit as M approaches infinity, a small number of categories generally provides a good model for the heterogeneity in the data, for example, $M=4$ is a popular choice (Yang & Rannala, 2012). Models that allow gamma rate heterogeneity are typically denoted by, for example, HKY85$+\Gamma$ or GTR$+\Gamma$. Note that these extended models still assume that the processes at different sites are independent of each other.

The standard phylogenetic model described so far makes a number of simplifying assumptions about the evolutionary process. For example, we assume that there is a single composition vector, $\pi = (\pi_A, \pi_G, \pi_C, \pi_T)$, which applies to all branches of the tree and at all sites. In other words, we are assuming that sequence composition (the proportion of A, G, C and T bases) tends to remain constant across sites and over evolutionary time. However, these assumptions are violated in many real datasets. For example, the GC content of 16S rRNA varies from 45% to 74% across the diversity of sampled Bacteria, Archaea and Eukarya (Cox et al., 2008). Although simplifying assumptions make statistical models simpler and inference more computationally tractable, they can also impact on inferences about the underlying phylogeny, as we will see in our analysis of the SSU and SSU$+$TAK datasets. Motivated by such inferential concerns, models have been developed which allow sequence composition to vary across sites (Lartillot & Philippe, 2004) and across branches of the tree, that is, over time (Foster, 2004). Models with the latter property can have the additional benefit of allowing data to be informative about the root position. In this tutorial, we will use the CAT model (Lartillot & Philippe, 2004) which allows sequence composition to vary across sites. The CAT model assigns each site to one of K groups, where K is less than or equal to the total number of sites in the alignment. Each group of sites has its own composition vector, and there is a probability vector of length K which determines the probability of a site belonging to each of the K groups. The number of groups K is unknown and treated as another parameter of the model. In statistical terms, this is a mixture model with an unknown number of mixture components.

Although we have focused here on the analysis of nucleotide data, protein evolution is also generally modelled using CTMPs and so the same principles apply. However, because there are many more amino acids than nucleotides, the most general reversible, stationary model (the GTR model) contains many more parameters than the equivalent model for nucleotides. As a result, the most commonly used models for amino acid substitutions are empirical models in which fixed values, inferred from large protein datasets, are assumed for the exchangeability parameters or stationary probabilities or both (see, e.g., Le & Gascuel, 2008). However, the very large concatenated protein datasets used in modern phylogenomic analyses are often analysed using the same models discussed here.

5.2 INFERRING PHYLOGENIES UNDER MARKOV SUBSTITUTION MODELS

There are two main schools of thought regarding the appropriate framework for statistical inference of phylogenies using Markov models: frequentist and Bayesian. The two schools differ fundamentally in their interpretations of probability and parameters. Frequentist inference is based on the idea that probability represents a long run relative frequency. For example, the probability of obtaining a head on a coin toss would be interpreted as the limit of the proportion of heads obtained if the coin toss was repeated infinitely many times under identical conditions. In contrast, in Bayesian inference, probability is regarded as "degree of belief" and it is subjective. In the coin toss example, *your* probability of obtaining a head is interpreted as a measurement of how strongly *you* believe a head will occur. Consequently, probability is still a meaningful concept in the case of one-off uncertain events such as whether Species A diverged from Species B. Applying the frequentist definition of probability in the context of non-repeatable events presents conceptual difficulties.

The frequentist interpretation of parameters is that they are fixed but unknown constants and so cannot have probabilities associated with them. The only probabilities considered come from repeated realisations of the model. In contrast, Bayesian inference treats parameters and all other unknowns as random variables to which we assign probabilities. This facilitates straightforward quantification of uncertainty through probability statements. A more detailed comparison between the two schools of thought can be found in our Online Supplement (http://dx.doi.org/10.1016/bs.

mim.2014.05.001). For now, we note that one great practical advantage of the Bayesian approach is that it provides a very natural framework for formulating complex hierarchical (multi-level) models. In a phylogenetic context, this is why the CAT and other highly structured models are generally formulated in the Bayesian framework. Such models often perform very well on large datasets.

5.3 FREQUENTIST INFERENCE

Phylogenetic software for inference based in the frequentist paradigm, such as RAxML (Stamatakis, 2006) and PhyML (Guindon et al., 2010), generally fits models using the method of maximum likelihood. Let $y = (y_{ij})$ denote an alignment of molecular sequences in which y_{ij} is the character state (e.g. nucleotide) at the jth site for species i. Given a particular Markov substitution model, we can write down a likelihood function $p(y|\theta)$ which depends on the data y and unknowns θ. These unknowns include the parameters of the substitution model, the branch lengths and the tree topology. The likelihood function provides a measure of how likely the data y are given the unknowns θ. For the purposes of maximum likelihood estimation, however, it is regarded as a function of the unknowns θ given the data y and is often written as $L(\theta|y)$ rather than $p(y|\theta)$. The maximum likelihood estimate $\hat{\theta}$ of θ is the value of θ which maximises the likelihood function. Finding the maximum is an optimisation problem which must be solved numerically if no closed form solution exists. For standard phylogenetic models, the likelihood must be maximised numerically. For a fixed tree, standard numerical optimisation routines are used to find the model parameters and branch lengths which maximise the likelihood. In principle, this procedure should be repeated for all possible trees to find that which leads to the greatest (maximised) likelihood. However, the number of trees on n taxa grows very rapidly with n. For example, there are 15 unrooted trees on 5 taxa, over 2 million unrooted trees on 10 taxa and over 2×10^{20} trees on 20 taxa. This means that an exhaustive search is only feasible for alignments based on a small number of taxa. For larger alignments, the search over tree space is generally heuristic and search algorithms may become stuck in a local maximum. This can occur if the maximised likelihood for a tree is not the global maximum, but it is greater than the maximised likelihoods for the neighbouring trees which are explored during the tree search. As a result, there is no guarantee that the search will find the maximum likelihood tree, that is, the global maximum.

In a frequentist analysis, the quantification of uncertainty about the maximum likelihood tree is usually based on nonparametric bootstrapping (Felsenstein, 1985). Suppose there are N sites in the alignment. Boostrapping involves resampling the sites (alignment columns) with replacement N times in order to generate a bootstrap sample of the same size as the original alignment. The maximum likelihood estimation procedure is then repeated to compute the bootstrap tree as the maximum likelihood tree for the bootstrap alignment. Typically, 100–1000 bootstrap samples are generated and analysed in this way. For every clade in the original maximum likelihood tree, the percentage of bootstrap trees that contain that clade are then computed. These percentages are called the bootstrap support values, and they are generally marked on phylograms visualising the maximum likelihood tree, or a so-called consensus tree which only includes the most commonly occurring clades in the bootstrap samples. Bootstrap support values are difficult to interpret beyond an intuitive sense that a high bootstrap support value indicates consistent support for a clade across sites. More rigorous interpretations are not easy to defend, although see Yang (2006, Chapter 6).

5.4 BAYESIAN INFERENCE

As remarked above, in the Bayesian approach to inference, probability distributions are assigned to parameters (and, more generally, unknowns). Consider data y which are modelled as having arisen from a statistical model which depends on unknowns θ. Given the chosen model, we can write down a likelihood function $p(y|\theta)$ which provides a measure of how likely the data y are conditional on the unknowns θ. We quantify our initial uncertainty about θ through a probability distribution $\pi(\theta)$ called the prior distribution which indicates how likely we believe each possible value of θ is before seeing the data. The likelihood is then used to update the prior distribution in light of the data y. This yields a posterior distribution $\pi(\theta|y)$ which summarises all of our uncertainty about θ after seeing the data. The rule for updating the prior distribution is called Bayes' Theorem which can be written as

$$\pi(\theta|y) = \frac{\pi(\theta)p(y|\theta)}{p(y)}$$

where

$$p(y) = \int \pi(\theta)p(y|\theta)\,\mathrm{d}\theta$$

is a normalising constant, called the *marginal likelihood*, which will be discussed further in Section 5.7. In general, this integral cannot be evaluated in closed form. It can also be very difficult to approximate numerically, especially for complex models where θ involves a large number of unknown quantities. In fact, difficulty in computing this integral had been one of the main objections to the Bayesian approach until the early 1990s when computational algorithms called Markov chain Monte Carlo (MCMC) methods effectively solved the computational problem.

MCMC methods work by generating samples $\theta^{[1]}, \theta^{[2]}, \ldots$ from the posterior distribution $\pi(\theta|y)$ in such a way that knowledge of the normalising constant $p(y)$ is not required. A large number of samples then yields an approximation to the posterior distribution $\pi(\theta|y)$. There is, however, an ambiguity in this statement: what constitutes "a large number" of samples? The problem is that the samples drawn from the posterior distribution are not independent of each other. Instead, they form a Markov chain (hence the name Markov chain Monte Carlo) whose stationary distribution is equal to the posterior distribution. It is therefore necessary to run the chain for long enough that it can "forget" the value for θ at which it was initialised and thereby reach its stationary distribution. Sometimes, movement towards and around the stationary distribution can be very slow. In such cases, we say the chain suffers from slow convergence and poor mixing properties. Unfortunately, these are common problems in phylogenetic analyses because it can be very difficult for the chain to move around tree space. It is therefore good practice to run two or more MCMC chains, initialised at different starting points, and to check that they converge to the same distribution. This will be explained further in the application to the SSU datasets.

Most Bayesian phylogenetic software uses a Metropolis-within-Gibbs sampler in which the unknowns (e.g. model parameters, branch lengths, the tree) in θ are updated one at a time from their conditional posteriors given the current values of all other unknowns. Most, if not all, of these updates cannot be performed as direct simulations from the required distribution. Samples are therefore generated using Metropolis Hastings steps. In a Metropolis Hastings step, a new value for the unknown is generated from a distribution called a proposal distribution, which we choose. This proposed value is then accepted or rejected with a certain probability called the acceptance probability. The acceptance probability is carefully formulated so that the resulting Markov chain has an equilibrium distribution equal to the posterior with density $\pi(\theta|y)$. If the proposal is accepted, then the chain moves to the proposed new value. If it is rejected, the chain remains at its current value.

In a Bayesian analysis, the post-data uncertainty about the parameters and other unknowns is represented completely by the posterior distribution. From this, we can compute any posterior quantities of interest, for example, Bayesian credible intervals for parameters or the posterior probability in support of a particular hypothesis. Given an MCMC sample from the posterior, histograms are commonly used to visualise the shapes of marginal posterior distributions for model parameters. The latter information can also be presented more concisely by using the MCMC samples to compute posterior summaries, such as posterior means and Bayesian credible intervals. Similarly, in a phylogenetic context, the posterior distribution for the tree can be computed (approximately) by the proportion of times each tree was sampled during MCMC. The posterior probability of a clade can be approximated likewise by the proportion of sampled trees which contained that clade. As noted above, these posterior probabilities have unambiguous definitions. A posterior probability of 0.93 for a clade is the posterior probability that the "true" tree contains that clade given the chosen model and prior. A word of warning here: the statement "given the chosen model" represents the very strong assumption that the model we have chosen is the actual mechanism by which the data were generated.

5.5 MODEL COMPARISON AND ASSESSMENT

GEP Box famously wrote "All models are wrong but some are useful" (Box, 1979). It is important to realise that we can never hope to find the "true" model by which our data were generated. At best we hope to find a model which is sufficiently flexible to allow us to capture the most important aspects of the evolutionary process.

Implicitly, we hope that models that provide a good fit to the data also provide biologically reasonable inferences about the underlying phylogeny. In the statistical literature, model choice is guided by the principle that we should seek out models which provide a good fit to the data without incorporating redundant or meaningless parameters. In the following sections, we consider model choice in the frequentist and Bayesian inferential frameworks. We then apply some of the principles discussed in this section to a comparison between the GTR model and the CAT model for the SSU and SSU+TAK alignments.

5.6 FREQUENTIST METHODS

In a frequentist framework, a variety of likelihood-based methods are available to help in deciding which model provides the best fit to the data. These include likelihood ratio tests and penalised likelihood criteria.

The likelihood ratio test can be applied in comparisons of nested models. Models M_0 and M_1 are nested if the null model M_0 can be derived from the alternative model M_1 by placing constraints on the values of the parameters of M_1. For example, the JC69 model is nested within the K80 model because setting $\alpha = \beta$ in the K80 model produces the JC69 model. The likelihood ratio test statistic is defined as

$$LR = -2\ln\left\{ L\left(\hat{\theta}_0|y,M_0\right)/L\left(\hat{\theta}_1|y,M_1\right)\right\} = -2\left\{ \ln L\left(\hat{\theta}_0|y,M_0\right) - \ln L\left(\hat{\theta}_1|y,M_1\right)\right\}$$

where $L\left(\hat{\theta}_0|y,M_0\right)$ is the likelihood of the null model evaluated at the maximum likelihood estimate $\hat{\theta}_0$ of its parameters and similarly for $L\left(\hat{\theta}_1|y,M_1\right)$. Under certain regularity conditions, if the null model is true, the likelihood ratio statistic asymptotically (for large samples) follows a χ^2 distribution on r degrees of freedom where r is the number of constraints imposed on the alternative model M_1 to recover the null model M_0. In the JC69/K80 example, $r=1$. In phylogenetics, an immediate problem with the likelihood ratio test is that the nesting condition demands that the same tree is used for both models, even though the tree with the highest maximised likelihood may differ between models. In practice, however, Posada and Crandall (2001) found that the model chosen by likelihood ratio tests was generally robust against different choices of the common tree.

An alternative approach that does not require models to be nested is to use penalised likelihood criteria which attempt to provide an indication of model fit (as measured by the maximised log-likelihood), adjusted by a term which penalises model complexity. Examples of such model selection tools are the Akaike Information Criterion (AIC) (Akaike, 1974) and the Bayesian Information Criterion (BIC) (Schwarz, 1978) defined as

$$\text{AIC} = -2\ln L\left(\hat{\theta}|y,M\right) + 2n \quad \text{and} \quad \text{BIC} = -2\ln L\left(\hat{\theta}|y,M\right) + n\ln N$$

Here, $L\left(\hat{\theta}|y,M\right)$ is the likelihood of the model evaluated at the maximum likelihood estimate $\hat{\theta}$ of its parameters, n is the number of free parameters in the model and N is the size of the observed sample, usually taken to be the number of sites in the alignment. For a comparison between the various methods of model selection in phylogenetics, see Posada and Buckley (2004).

For standard models of nucleotide substitution, likelihood ratio tests and comparisons via AIC and BIC can be carried out using jModelTest 2 (Darriba, Taboada, Doallo, & Posada, 2012; see Table 1).

5.7 BAYESIAN MODEL CHOICE

The Bayesian framework offers a principled approach to dealing with model uncertainty. A central role is played by the marginal likelihood $p(y)$, defined earlier as the normalising constant in Bayes' Theorem, which is involved in the computation of Bayes factors and posterior model probabilities. Note that it depends on the prior distribution as well as the model. Bayes factors are often used to compare pairs of models and have an interpretation in terms of the posterior odds of one model in comparison with the other. Posterior model probabilities can be interpreted as the posterior probability that each model is correct if it is assumed that the collection of models being compared contains the "true" model, that is, the true data generating mechanism. For further details on formal Bayesian model choice and the related concept of Bayesian model averaging, see our Online Supplement (http://dx.doi.org/10.1016/bs.mim.2014.05.001).

Despite its attractive framework, formal Bayesian model choice is often hampered by the difficulty of computing the marginal likelihood, which is generally a very difficult numerical integration problem. Accordingly, most Bayesian phylogenetic programs do not provide functions to perform this calculation. An exception is MrBayes (Ronquist et al., 2012), which allows users to approximate marginal likelihoods using either the harmonic mean or the (greatly superior) stepping-stone method (Xie, Lewis, Fan, Kuo, & Chen, 2011). For practical guidance on how to perform the calculation, see Section 4.4 of the MrBayes Manual (http://mrbayes.sourceforge.net/).

In situations where it is not possible to compute the marginal likelihood for all models under consideration, there are a variety of informal methods that allow comparisons to be made. These include cross-validation and posterior predictive model checking. Both can be carried out using the Bayesian program PhyloBayes (Lartillot, Lepage, & Blanquart, 2009) through the cvrep/pb/readcv/sumcv commands and the ppred command, respectively. In PhyloBayes, cross-validation involves dividing the data into a training set and a validation set and then approximating the probability of the validation data given the training data under each possible model. Ideally, this procedure would be repeated for every possible partition of the data into training and validation sets, averaging the probability across partitions. However, this is computationally infeasible and so in practice only a small random sample from the set of possible partitions is used. The model with the highest cross-validation score is judged to be best-fitting. Unfortunately, cross-validation can often be computationally prohibitive because the posterior needs to be recomputed for every partition of the data. Consequently, we will not use it in this chapter. Posterior predictive model checking provides a less computationally demanding approach to model choice, and a practical example can be found in the next section. It is based on the posterior predictive distribution of a hypothetical replicate y_{rep} of the data y which could have been observed under the model M

$$p(y_{rep}|y,M) = \int p(y_{rep}|\theta,M)p(\theta|y,M)d\theta$$

These model checks can be based on test quantities $T(y)$ which are simple summaries designed to capture aspects of the data that we want the model to capture adequately. For example, in PhyloBayes, one of the predefined test quantities is the biochemical specificity, measured by the mean number of distinct character states in a column. In general, it will not be possible to compute the posterior predictive distribution of a test quantity $T(y)$ analytically; however, given MCMC draws $\theta^{[1]},\theta^{[2]},\ldots,\theta^{[B]}$ from the posterior distribution of a model M, it is easy to build up a numerical approximation by simulating a replicated dataset $y_{rep}^{[i]}$ for each draw from the posterior and computing $T(y_{rep}^{[i]})$. The idea is that if the model fits well then the observed data should look plausible under the posterior predictive distribution. A simple way of checking this graphically is to plot the posterior predictive distribution for the test quantity (for example in a histogram) and then to examine the position of the observed test quantity $T(y)$ in that distribution. Poor model fit would be indicated by the observed test quantity lying far into the tails of the posterior predictive distribution. When comparing two or more models, the model for which the observed test quantity lies most centrally provides the best fit to the data in terms of capturing the feature of the data summarised by $T(y)$. In PhyloBayes, the position of the observed test quantity relative to the posterior predictive distribution is summarised using a P-value; very small or very large P-values indicate that the observed value lies in one of the tails of the predictive distribution, suggesting poor model fit. We note that even when only one model is being considered, it is good practice to assess its fit to the data using posterior predictive checks like those discussed here.

6 STEP 5: INFERRING TREES—PRACTICAL GUIDELINES FOR FITTING AND COMPARING MARKOV SUBSTITUTION MODELS

6.1 ALIGNMENT FORMATS FOR PHYLOGENY PROGRAMS

While alignments are often generated and viewed in formats such as FASTA (where each sequence is prefaced with a one-line header, starting with a ">") or ClustalW (in which the alignment is formatted for easy visual inspection or printing), most phylogeny packages, including the two we will introduce here, require that the input alignment be in PHYLIP format, perhaps for historical reasons. Converting data between formats is one of the most irritating and mundane aspects of phylogenetics (and, more generally, bioinformatics), but a number of conversion tools are available to streamline the process. One option is readAl, which is part of the trimAl package that we used in Step 3 to edit the alignment. It can be run locally or on the Phylemon Web server (Table 1). Use readAl to convert your alignments to the standard PHYLIP format using the "-phylip" option and then open the resulting files in a text editor. You will see that all these different alignment formats—FASTA, Clustal, PHYLIP—contain exactly the same information, but presented differently. With PHYLIP-formatted versions of your SSU and SSU + TAK alignments, you are ready to run the phylogeny packages. We will analyse these datasets using two of the models introduced in section 5.1 GTR+Γ and CAT+Γ (i.e. the GTR or CAT models with a gamma distribution for modelling across-site rate variation), and compare the results. In general, when comparing two or more models, they should be fitted in the same inferential framework. Inference for the CAT model is only implemented in the Bayesian paradigm and so we will fit and compare the models in the Bayesian framework. The program we will use does not perform marginal likelihood calculations and so we will compare models using posterior predictive checks. Although unnecessary in this analysis, for the purpose of illustration, we will also fit the GTR model using maximum likelihood.

6.2 INFERRING MAXIMUM LIKELIHOOD PHYLOGENIES USING RAxML

In this tutorial, we will use RAxML (Stamatakis, 2006), one of the most popular maximum likelihood packages. One of the great strengths of RAxML is that it is optimised for modern multi-core processors; if you compile and run it locally, make sure to choose a version appropriate for your system so that you can take advantage of all of the cores in your desktop or laptop processor to run the analyses more quickly. RAxML can be run either locally or online through a Web server such as CIPRES (see Table 1). Whichever option you choose, the most important parameters to set are the model specification and the number of bootstrap replicates. In this case, these will be the GTR + Γ model, and 100 rapid bootstraps. As discussed in section 5.3, maximum likelihood methods rely on heuristic tree search algorithms to look for the maximum likelihood tree and so it is possible that the optimisation algorithm will get stuck in a local maximum. Therefore, it is good practice to rerun RAxML analyses several times using different starting values for the random seed (-x and -p options).

6.3 BAYESIAN ANALYSES WITH PHYLOBAYES

PhyloBayes (Lartillot et al., 2009) is a Bayesian phylogenetic package. Its speciality is the implementation of CAT and other complex mixture models that often fit alignments of highly divergent sequences (such as the rRNAs from across the tree of life) much better than GTR and other single Q-matrix models. PhyloBayes can be installed locally or run on the CIPRES Web server (Table 1); more so than for the other packages we discuss here, you will have much more flexibility and control over your PhyloBayes analyses if you run them locally. Two versions of PhyloBayes are currently available—a serial (single-core) and parallelised (multi-core) version, the latter called PhyloBayes-MPI (Lartillot, Rodrigue, Stubbs, & Richer, 2013). We will use the single-core version here, but bear in mind that the MPI version can be very useful for speeding up analyses of larger datasets or inference under more complex models. As discussed in section 5.4, the strategy for a Bayesian analysis is to run multiple MCMC chains and then check for convergence. To start a chain running with PhyloBayes, you invoke the "pb" command with the appropriate options for model and prior specification:

```
pb -s -d myAlignment.phy -cat myChainName &
```

Here, -cat indicates that the model to be fitted is the CAT model. Replace this with -gtr for the GTR model. By default, PhyloBayes uses the discrete gamma model with four categories to describe across-site rate heterogeneity. We accept this default here, but it can be changed by the inclusion of appropriate options. The posterior distribution is formed by combining information from the data (the likelihood) with information from the prior and so the choice of prior distribution can and will influence our posterior inferences about the phylogenetic tree. In theory, the prior should be chosen to reflect prior beliefs about the model parameters, branch lengths and tree. However, eliciting prior information is a challenging topic in its own right, and provision of guidelines is well beyond the scope of this tutorial. Here, we simply accept the default prior, but encourage the interested reader to explore the effect on the posterior distribution of changing the prior. Again, this can be achieved by the inclusion of appropriate options with the "pb" command.

Assessing whether MCMC chains have converged is a difficult problem and you are likely to need to generate *at least* tens of thousands of iterations. As discussed previously, we suggest running two chains for each model and comparing the MCMC output. PhyloBayes provides two programs that implement numerical convergence diagnostics for pairs of chains—bpcomp and tracecomp. See the PhyloBayes manual (http://www.phylobayes.org) for some guidelines on how to interpret the results of these diagnostics. The .trace files produced by the "bp" command record a few parameter summaries (e.g. the log-likelihood) for each sample from the posterior distribution, and the numerical diagnostics invoked by the tracecomp program work on this MCMC output. As an alternative, we encourage the reader to examine the convergence of these statistics graphically using the guidelines available in our Online Supplement (http://dx.doi.org/10.1016/bs.mim.2014.05.001), which also provides further comments on assessing mixing in tree space.

The first time you examine the numerical and/or graphical diagnostics, you may judge that the MCMC chains have not converged. If this is the case, you should continue to run the chains until there is no evidence of any lack of convergence. Note that convergence and model checking are completely different things—once your MCMC chains appear to have converged, you can proceed to comparing the fit of the GTR and CAT models.

6.4 POSTERIOR PREDICTIVE CHECKS

The "ppred" program can be used to do posterior predictive checks once your analyses have converged. We will focus on the site-specific biochemical diversity test, which can be performed by invoking ppred with the -sat flag (other tests, such as for compositional homogeneity, are often useful but will not be considered here; see Foster, 2004). Site-specific biochemical diversity refers to the number of different nucleotides (or amino acids) per alignment column—see Figure 3 for more details. Run this test for one of your GTR and one of your CAT chains. Our results for the SSU dataset (no TAK) are plotted in Figure 3: you can see that the range of site-specific diversity predicted under the CAT model overlaps the observed value, whereas GTR predicted levels of diversity that were much too high. This result is one indication that the GTR model does a poor job of capturing the site-specific evolutionary dynamics observed in this alignment. Are the results similar for the SSU +TAK dataset? Before moving on to the next step, we note that PhyloBayes also implements a very popular and flexible model that combines the across-site mixture model of CAT with a GTR matrix for exchangeabilities between nucleotides (or amino acids): CAT+GTR. This model is somewhat more realistic than the CAT model on its own because it also allows the rates of change between different

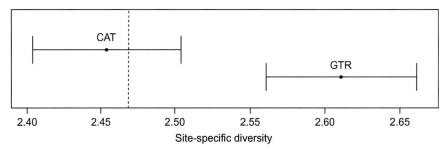

FIGURE 3

Posterior predictive checks for site-specific biochemical diversity on the SSU alignment under the GTR and CAT models. Site-specific diversity is simply the mean number of different nucleotides observed per alignment column. In real datasets, this value tends to be rather low: when you look at an alignment, you will notice that most columns largely consist of only one or two of the four possibilities (the same pattern is usually observed with amino acid data). The observed (i.e. "real") value is plotted as a dashed line; posterior predictive means and an interval two standard deviations either side are also plotted for the GTR and CAT models. (This interval represents a symmetric 95% Bayesian credible interval if it can be assumed that the posterior predictive distribution is approximately normal.) The values predicted under GTR are too high, suggesting that GTR does not adequately model the site-specific nature of the evolutionary process—this may well be because GTR averages nucleotide frequencies and exchangeabilities over the whole alignment. The observed value is, however, consistent with the posterior predictive distribution for the CAT model, indicating that CAT is a more realistic model for this alignment, at least for this aspect of the evolutionary process.

nucleotides to differ. You may like to try fitting the CAT+GTR model to one or both of our alignments and comparing your results with those we discuss below—if you can, use PhyloBayes-MPI for significant speed improvements.

7 STEP 6: INTERPRETING THE PHYLOGENETIC TREE

The phylogeny packages discussed here write out the trees as plain text files in Newick format, which is rather difficult to parse by eye. Instead, we recommend the use of tree viewing software such as Dendroscope or FigTree (Table 1) to visualise the phylogenies. Open and compare the trees inferred for the two datasets under the GTR and CAT models (four trees in total). You will notice some differences between them, both in terms of topology (branching order) and the support values (maximum likelihood bootstraps or Bayesian posterior probabilities) for different groups. Remember that the question we wanted to address with these analyses was the position of the Eukarya relative to the Archaea: do they form monophyletic sister groups, as under the three-domains tree, or do the Eukarya emerge from within the Archaea, as proposed by some alternative hypotheses, such as the eocyte hypothesis of Lake et al. (1984)? To answer this question, we will need to root the trees, in order to establish the direction of ancestor–descendant relationships. Like almost all phylogenetic models in current use, GTR and CAT are stationary and time reversible, meaning that the position of the root has no effect on the probability of the tree, and so cannot be inferred as part of the analysis (see section 5.1 for further details). This means that the position of the root must be established based on independent data—a far from ideal but very commonly encountered situation in phylogenetics. The usual strategy is to root the tree on a known outgroup—a sequence (or group) whose divergence from all others is known to represent the earliest split in the tree; the root can then be placed on the branch that joins the outgroup to the rest of the tree. For example, a phylogenetic analysis of birds might include mammal sequences as an outgroup because the split between birds and mammals is known to predate the radiation of birds. As mentioned in Section 1, analyses of this type place the root of the tree of life on the branch leading to the Bacteria. We will assume that bacterial root here, in order to polarise our trees and investigate the relationship between Eukarya and Archaea. To do this using your tree viewing software, you will first need to confirm that all the Bacteria cluster together (form a clan; Wilkinson, McInerney, Hirt, Foster, & Embley, 2007) in the tree. If this is the case, select the branch joining this cluster to the rest of the tree and reroot; this can be done from the Edit menu in Dendroscope. The tree needs to be "rerooted" rather than simply rooted because phylogeny packages often output trees with an arbitrary root; this can be safely ignored if the analysis was performed using a stationary, time-reversible model, as was the case here. You can now investigate the position of the Eukarya relative to the Archaea in your rooted trees. You may find Figures 4 and 5, which summarise the results that we obtained from our Bayesian analyses, a helpful point of comparison. The analysis of the SSU dataset with the GTR model gave a strongly supported three-domains tree, with maximal posterior support (posterior probability (PP)=1) for the

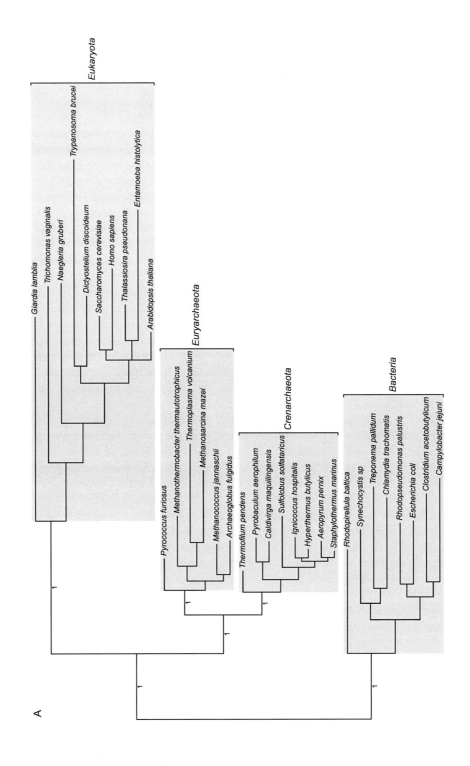

A

Eukaryota

Giardia lamblia
Trichomonas vaginalis
Naegleria gruberi
Trypanosoma brucei
Dictyostelium discoideum
Saccharomyces cerevisiae
Homo sapiens
Thalassiosira pseudonana
Entamoeba histolytica
Arabidopsis thaliana

Euryarchaeota

Pyrococcus furiosus
Methanothermobacter thermautotrophicus
Thermoplasma volcanium
Methanosarcina mazei
Methanococcus jannaschii
Archaeoglobus fulgidus

Crenarchaeota

Thermofilum pendens
Pyrobaculum aerophilum
Caldivirga maquilingensis
Sulfolobus solfataricus
Ignicoccus hospitalis
Hyperthermus butylicus
Aeropyrum pernix
Staphylothermus marinus

Bacteria

Rhodopirellula baltica
Synechocystis sp
Treponema pallidum
Chlamydia trachomatis
Rhodopseudomonas palustris
Escherichia coli
Clostridium acetobutylicum
Campylobacter jejuni

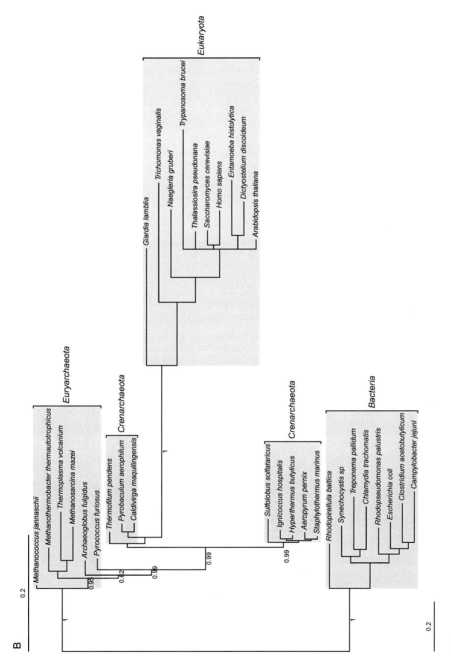

FIGURE 4

Phylogenies inferred from the SSU dataset under (A) the GTR and (B) the CAT models. Branch lengths measure the expected number of substitutions per site. Support values on branches are Bayesian posterior probabilities. In the GTR analysis of this dataset (A), the eukaryotes emerge as the sister group to a monophyletic Archaea with maximal posterior support (PP = 1)—the classic "three-domains" (Woese et al., 1990) tree. When the same alignment was analysed using the CAT model (B), the eukaryotes are placed within the Crenarchaeota, also with very high support (PP = 0.99); this topology corresponds to the eocyte hypothesis (Lake et al., 1984). Both trees were rooted on the branch leading to Bacteria, consistent with analyses of ancient paralogous genes (Gogarten et al., 1989; Iwabe et al., 1989).

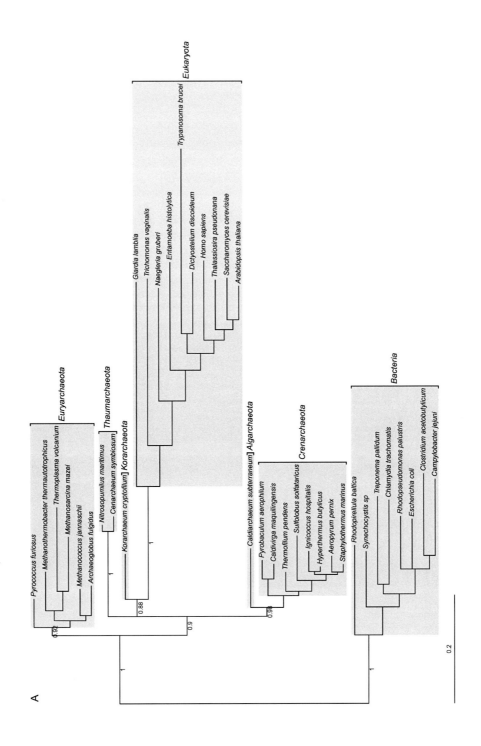

A

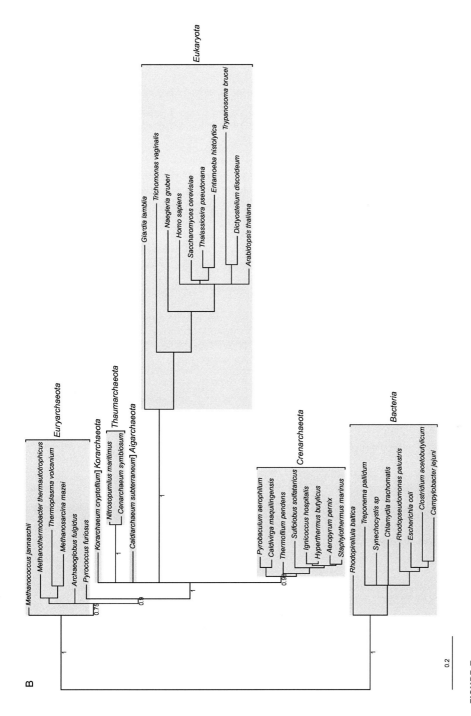

FIGURE 5

Phylogenies inferred from the SSU + TAK dataset under (A) the GTR and (B) the CAT models. Support values on branches are Bayesian posterior probabilities. With the addition of the new sequences from the TAK Archaea (Thaumarchaeota, Aigarchaeota and Korarchaeota), both the GTR (A) and CAT (B) models recover a topology in which the eukaryotes emerge from within the Archaea; the support for a clade of eukaryotes and TACK Archaea is highest (PP=1) under the CAT model. As in Figure 4, both trees were rooted on the branch leading to Bacteria, consistent with analyses of ancient paralogous genes (Gogarten et al., 1989; Iwabe et al., 1989).

monophyly of the Archaea. *Monophyly* indicates that all Archaea share a common ancestor to the exclusion of eukaryotes and vice versa. If you are unsure which sequences belong to Archaea and which to Eukarya, try a Web search for their names. In our analyses, all other combinations of models and datasets gave an eocyte tree, in which the Eukarya emerge from within the Archaea; that is, some Archaea are more closely related to eukaryotes than they are to other Archaea. The support for this relationship was reasonable in the GTR analysis of the SSU+TAK dataset (PP=0.9 for a clade of eukaryotes plus TACK), rising to PP=0.99 and PP=1 in the analyses of both datasets using the CAT model (PP of 1). Given that our posterior predictive tests suggested that the CAT model fit the data better than the GTR model, and that adding the additional archaeal sequences leads to the recovery of the same topology under both models, these analyses support the eocyte tree over the three-domains tree and suggest that the small subunit rRNA genes of eukaryotes, at least, may descend from within the Archaea—in particular, the "TACK" group containing the Thaumarchaeota, Aigarchaeota, Crenarchaeota and Korarchaeota.

Having performed these analyses yourself, we hope that you are now more aware of the caveats and limitations that necessarily accompany results of this type. In this tutorial, we saw that both taxon sampling (the choice of species to be included in the analysis) and the phylogenetic model used can influence the final result. More generally, decisions made at any of the steps of the analysis have the potential to influence the inferred tree, not least of which is the choice of gene to analyse in the first place. Here, we focused on SSU rRNA as one of the most frequently used phylogenetic markers, but any comprehensive attempt to address the three domains/eocyte debate would involve analyses of a much broader range of genes—at the very least, the set of 30–40 other genes, mostly encoding protein components of the ribosome, that are conserved on the genomes of most cellular life forms. As mentioned above, simultaneous analysis of multiple genes raises a number of complications that we did not deal with here, including how to handle the all-too-frequent cases in which the genes being analysed disagree as to the underlying species tree. If you are interested in exploring the three domains/eocyte question further, a good starting point may be some of the recent literature on this topic, which attempts to engage with these issues (Guy & Ettema, 2011; Lasek-Nesselquist & Gogarten, 2013; Williams et al., 2013, 2012), as well as chapter 'Reconciliation Approaches to Determining HGT, Duplications, and Losses in Gene Trees' by Kamneva and Ward in this volume.

CONCLUSIONS

In this chapter, we have attempted to introduce some of the most important aspects of phylogenetic analysis, focusing in particular on sequence alignment, taxon sampling and the selection of an appropriate phylogenetic model. Here, we have concentrated on a hands-on approach; for an excellent, up-to-date theoretical treatment, see Yang and Rannala (2012). All of these decisions have the potential to greatly influence the results of your analysis; as a result, phylogenies must be interpreted tentatively and with their caveats and limitations in mind. Ultimately, phylogenies are hypotheses of evolutionary relationships that help us to organise biological diversity and to understand the evolution of traits. This is not to say that the process of phylogenetics is arbitrary: there are benchmarks and best practices available for most of the steps, and following these as carefully as possible will increase the robustness of the resulting phylogeny. The ideal situation is when a variety of methods agree on the result; when this is not the case, tests of model fit such as those discussed in step 4 may indicate which phylogeny should be tentatively preferred. We hope that these analyses have provided some sense of the fascination of phylogenetics. The idea that it is possible to make inferences, however cautious, about events that took place billions of years ago through the statistical analysis of sequences from contemporary organisms can be intoxicating, and the advent of next-generation sequencing, combined with continuing advances in statistical methodology, makes this an exciting time to enter the field!

ACKNOWLEDGEMENTS

We thank Cymon J. Cox, Malcolm Farrow, Peter G. Foster and T. Martin Embley for helpful feedback on an earlier draft of this chapter. This work was supported by a Marie Curie postdoctoral fellowship to T. A. W. S. E. H. is supported under a European Research Council Advanced Investigator grant awarded to T. Martin Embley.

REFERENCES

Akaike, H. (1974). New look at statistical-model identification. *IEEE Transactions on Automatic Control*, *19*(6), 716–723.

Andersson, S. G., Zomorodipour, A., Andersson, J. O., Sicheritz-Ponten, T., Alsmark, U. C., Podowski, R. M., et al. (1998). The genome sequence of Rickettsia prowazekii and the origin of mitochondria. *Nature*, *396*(6707), 133–140.

Bininda-Emonds, O. R. (2004). The evolution of supertrees. *Trends in Ecology & Evolution*, *19*(6), 315–322.

Box, G. E. P. (1979). Robustness in the strategy of scientific model building. In R. L. Launer, & G. N. Wilkinson (Eds.), *Robustness in statistics*. Waltham, Massachusetts: Academic Press.

Capella-Gutierrez, S., Silla-Martinez, J. M., & Gabaldon, T. (2009). trimAl: A tool for automated alignment trimming in large-scale phylogenetic analyses. *Bioinformatics*, *25*(15), 1972–1973.

Castresana, J. (2000). Selection of conserved blocks from multiple alignments for their use in phylogenetic analysis. *Molecular Biology and Evolution*, *17*(4), 540–552.

Ciccarelli, F. D., Doerks, T., von Mering, C., Creevey, C. J., Snel, B., & Bork, P. (2006). Toward automatic reconstruction of a highly resolved tree of life. *Science*, *311*(5765), 1283–1287.

Cox, C. J., Foster, P. G., Hirt, R. P., Harris, S. R., & Embley, T. M. (2008). The archaebacterial origin of eukaryotes. *Proceedings of the National Academy of Sciences of the United States of America*, *105*(51), 20356–20361.

Criscuolo, A., & Gribaldo, S. (2010). BMGE (Block Mapping and Gathering with Entropy): A new software for selection of phylogenetic informative regions from multiple sequence alignments. *BMC Evolutionary Biology*, *10*, 210.

Darriba, D., Taboada, G. L., Doallo, R., & Posada, D. (2012). jModelTest 2: More models, new heuristics and parallel computing. *Nature Methods*, *9*(8), 772.

de Queiroz, A., & Gatesy, J. (2007). The supermatrix approach to systematics. *Trends in Ecology & Evolution*, *22*(1), 34–41.

Drew, B. T., Gazis, R., Cabezas, P., Swithers, K. S., Deng, J., Rodriguez, R., et al. (2013). Lost branches on the tree of life. *PLoS Biology*, *11*(9), e1001636.

Edgar, R. C. (2004). MUSCLE: A multiple sequence alignment method with reduced time and space complexity. *BMC Bioinformatics*, *5*, 113.

Embley, T. M., & Martin, W. (2006). Eukaryotic evolution, changes and challenges. *Nature*, *440*(7084), 623–630.

Esser, C., Ahmadinejad, N., Wiegand, C., Rotte, C., Sebastiani, F., Gelius-Dietrich, G., et al. (2004). A genome phylogeny for mitochondria among alpha-proteobacteria and a predominantly eubacterial ancestry of yeast nuclear genes. *Molecular Biology and Evolution*, *21*(9), 1643–1660.

Felsenstein, J. (1978). Cases in which parsimony or compatibility methods will be positively misleading. *Systematic Zoology*, *27*, 401–410.

Felsenstein, J. (1985). Confidence-limits on phylogenies—An approach using the bootstrap. *Evolution*, *39*(4), 783–791.

Foster, P. G. (2004). Modeling compositional heterogeneity. *Systematic Biology*, *53*(3), 485–495.

Foster, P. G., Cox, C. J., & Embley, T. M. (2009). The primary divisions of life: A phylogenomic approach employing composition-heterogeneous methods. *Philosophical Transactions of the Royal Society of London Series B, Biological sciences*, *364*(1527), 2197–2207.

Gogarten, J. P., Kibak, H., Dittrich, P., Taiz, L., Bowman, E. J., Bowman, B. J., et al. (1989). Evolution of the vacuolar H+-ATPase: Implications for the origin of eukaryotes. *Proceedings of the National Academy of Sciences of the United States of America*, *86*(17), 6661–6665.

Gribaldo, S., Poole, A. M., Daubin, V., Forterre, P., & Brochier-Armanet, C. (2010). The origin of eukaryotes and their relationship with the Archaea: Are we at a phylogenomic impasse? *Nature Reviews. Microbiology*, *8*(10), 743–752.

Guindon, S., Dufayard, J. F., Lefort, V., Anisimova, M., Hordijk, W., & Gascuel, O. (2010). New algorithms and methods to estimate maximum-likelihood phylogenies: Assessing the performance of PhyML 3.0. *Systematic Biology*, *59*(3), 307–321.

Guy, L., & Ettema, T. J. (2011). The archaeal 'TACK' superphylum and the origin of eukaryotes. *Trends in Microbiology*, *19*(12), 580–587.

Hasegawa, M., Kishino, H., & Yano, T. (1985). Dating of the human-ape splitting by a molecular clock of mitochondrial DNA. *Journal of Molecular Evolution*, *22*(2), 160–174.

Iwabe, N., Kuma, K., Hasegawa, M., Osawa, S., & Miyata, T. (1989). Evolutionary relationship of archaebacteria, eubacteria, and eukaryotes inferred from phylogenetic trees of duplicated genes. *Proceedings of the National Academy of Sciences of the United States of America*, *86*(23), 9355–9359.

Jukes, T. H., & Cantor, C. R. (1969). *Evolution of protein molecules*. New York: Academic Press.

Kelly, S., Wickstead, B., & Gull, K. (2011). Archaeal phylogenomics provides evidence in support of a methanogenic origin of the Archaea and a thaumarchaeal origin for the eukaryotes. *Proceedings. Biological Sciences/The Royal Society*, *278*(1708), 1009–1018.

Kimura, M. (1980). A simple method for estimating evolutionary rates of base substitutions through comparative studies of nucleotide sequences. *Journal of Molecular Evolution*, *16*(2), 111–120.

Lake, J. A., Henderson, E., Oakes, M., & Clark, M. W. (1984). Eocytes: A new ribosome structure indicates a kingdom with a close relationship to eukaryotes. *Proceedings of the National Academy of Sciences of the United States of America*, *81*(12), 3786–3790.

Lartillot, N., Lepage, T., & Blanquart, S. (2009). PhyloBayes 3: A Bayesian software package for phylogenetic reconstruction and molecular dating. *Bioinformatics*, *25*(17), 2286–2288.

Lartillot, N., & Philippe, H. (2004). A Bayesian mixture model for across-site heterogeneities in the amino-acid replacement process. *Molecular Biology and Evolution*, *21*(6), 1095–1109.

Lartillot, N., Rodrigue, N., Stubbs, D., & Richer, J. (2013). PhyloBayes MPI: Phylogenetic reconstruction with infinite mixtures of profiles in a parallel environment. *Systematic Biology*, *62*(4), 611–615.

Lasek-Nesselquist, E., & Gogarten, J. P. (2013). The effects of model choice and mitigating bias on the ribosomal tree of life. *Molecular Phylogenetics and Evolution*, *69*(1), 17–38.

Le, S. Q., & Gascuel, O. (2008). An improved general amino acid replacement matrix. *Molecular Biology and Evolution*, *25*(7), 1307–1320.

Lee, M. S. Y. (2001). Unalignable sequences and molecular evolution. *Trends in Ecology & Evolution*, *16*(12), 681–685.

Martin, W., Rujan, T., Richly, E., Hansen, A., Cornelsen, S., Lins, T., et al. (2002). Evolutionary analysis of Arabidopsis, cyanobacterial, and chloroplast genomes reveals plastid phylogeny and thousands of cyanobacterial genes in the nucleus. *Proceedings of the National Academy of Sciences of the United States of America*, *99*(19), 12246–12251.

Moreira, D., & Lopez-Garcia, P. (2005). Comment on "The 1.2-megabase genome sequence of Mimivirus" *Science*, *308*(5725), 1114, author reply 1114.

Nylander, J. A., Wilgenbusch, J. C., Warren, D. L., & Swofford, D. L. (2008). AWTY (are we there yet?): A system for graphical exploration of MCMC convergence in Bayesian phylogenetics. *Bioinformatics*, *24*(4), 581–583.

Pisani, D., Cotton, J. A., & McInerney, J. O. (2007). Supertrees disentangle the chimerical origin of eukaryotic genomes. *Molecular Biology and Evolution*, *24*(8), 1752–1760.

Posada, D., & Buckley, T. R. (2004). Model selection and model averaging in phylogenetics: Advantages of akaike information criterion and Bayesian approaches over likelihood ratio tests. *Systematic Biology*, *53*(5), 793–808.

Posada, D., & Crandall, K. A. (2001). Selecting the best-fit model of nucleotide substitution. *Systematic Biology*, *50*(4), 580–601.

Rannala, B., & Yang, Z. (2008). Phylogenetic inference using whole genomes. *Annual Review of Genomics and Human Genetics*, *9*, 217–231.

Redelings, B. D., & Suchard, M. A. (2005). Joint Bayesian estimation of alignment and phylogeny. *Systematic Biology*, *54*(3), 401–418.

Ronquist, F., Teslenko, M., van der Mark, P., Ayres, D. L., Darling, A., Hohna, S., et al. (2012). MrBayes 3.2: Efficient Bayesian phylogenetic inference and model choice across a large model space. *Systematic Biology*, *61*(3), 539–542.

Sagan, L. (1967). On the origin of mitosing cells. *Journal of Theoretical Biology*, *14*(3), 255–274.

Schwarz, G. E. (1978). Estimating the dimension of a model. *Annals of Statistics*, *6*(2), 461–464. http://dx.doi.org/10.1214/aos/1176344136.

Stamatakis, A. (2006). RAxML-VI-HPC: Maximum likelihood-based phylogenetic analyses with thousands of taxa and mixed models. *Bioinformatics*, *22*(21), 2688–2690.

Stanier, R. Y., & van Niel, C. B. (1962). The concept of a bacterium. *Archives of Microbiology*, *42*, 17–35.

Stevens, P. F. (1991). Character states, morphological variation, and phylogenetic analysis—A review. *Systematic Botany*, *16*(3), 553–583.

Talavera, G., & Castresana, J. (2007). Improvement of phylogenies after removing divergent and ambiguously aligned blocks from protein sequence alignments. *Systematic Biology*, *56*(4), 564–577.

Tavaré, S. (1986). Some probabilistic and statistical problems in the analysis of DNA sequences. In *Lectures on Mathematics in the Life Sciences: 17*, (pp. 57–86).

Thompson, J. D., Linard, B., Lecompte, O., & Poch, O. (2011). A comprehensive benchmark study of multiple sequence alignment methods: Current challenges and future perspectives. *PLoS One*, *6*(3), e18093.

Wallace, I. M., O'Sullivan, O., Higgins, D. G., & Notredame, C. (2006). M-Coffee: Combining multiple sequence alignment methods with T-Coffee. *Nucleic Acids Research*, *34*(6), 1692–1699.

Waterhouse, A. M., Procter, J. B., Martin, D. M., Clamp, M., & Barton, G. J. (2009). Jalview Version 2—A multiple sequence alignment editor and analysis workbench. *Bioinformatics*, *25*(9), 1189–1191.

Wilkinson, M., McInerney, J. O., Hirt, R. P., Foster, P. G., & Embley, T. M. (2007). Of clades and clans: Terms for phylogenetic relationships in unrooted trees. *Trends in Ecology & Evolution*, *22*(3), 114–115.

Williams, T. A., Foster, P. G., Cox, C. J., & Embley, T. M. (2013). An archaeal origin of eukaryotes supports only two primary domains of life. *Nature*, *504*(7479), 231–236.

Williams, T. A., Foster, P. G., Nye, T. M., Cox, C. J., & Embley, T. M. (2012). A congruent phylogenomic signal places eukaryotes within the Archaea. *Proceedings. Biological Sciences/The Royal Society*, *279*(1749), 4870–4879.

Woese, C. R. (1987). Bacterial evolution. *Microbiological Reviews*, *51*(2), 221–271.

Woese, C. R., & Fox, G. E. (1977). Phylogenetic structure of the prokaryotic domain: The primary kingdoms. *Proceedings of the National Academy of Sciences of the United States of America*, *74*(11), 5088–5090.

Woese, C. R., Kandler, O., & Wheelis, M. L. (1990). Towards a natural system of organisms: Proposal for the domains Archaea, Bacteria, and Eucarya. *Proceedings of the National Academy of Sciences of the United States of America*, *87*(12), 4576–4579.

Xie, W., Lewis, P. O., Fan, Y., Kuo, L., & Chen, M. H. (2011). Improving marginal likelihood estimation for Bayesian phylogenetic model selection. *Systematic Biology*, *60*(2), 150–160.

Yang, Z. (2006). *Computational molecular evolution*. Oxford: Oxford University Press.

Yang, Z., & Rannala, B. (2012). Molecular phylogenetics: Principles and practice. *Nature Reviews. Genetics*, *13*(5), 303–314.

Zuckerkandl, E., & Pauling, L. (1965). Molecules as documents of evolutionary history. *Journal of Theoretical Biology*, *8*(2), 357–366.

The All-Species Living Tree Project

3

Pablo Yarza[*,1,2], **Raul Munoz**[†,2]

Ribocon GmbH, Bremen, Germany

†*Marine Microbiology Group, Department of Ecology and Marine Resources, Institut Mediterrani d'Estudis Avançats (CSIC-UIB), Esporles, Illes Balears, Spain*

¹*Corresponding author: e-mail address: pyarza@ribocon.com*

1 INTRODUCTION

Data acquisition, data processing and scientific developments are leading to a rapid accumulation of digital information for microbiological purposes. The fact that the *16S rRNA* gene sequence repository has grown exponentially since the early 1990s, currently exceeding ~3,500,000 entries, is a good example of this tendency (Figure 1). In recent times, the study of microbiomes has enhanced activities related to the classification and identification of microorganisms, boosting the number of sequence submissions to public databases of the International Nucleotide Sequence Database Collaboration (INSDC). This ever-increasing accumulation of information has promoted the active networking of microbiologists and computer scientists leading to the development of tools and infrastructure to store, access and analyse data.

The classification of *Archaea* and *Bacteria* provides a sound foundation for microbiology. It has a major application in all information databases with respect to the need to handle valid taxon identifiers. Users and providers of taxonomic information (i.e. classification and nomenclature) constitute a large community ranging from individual researchers (e.g. taxonomists) to computerized systems (e.g. DNA sequence repositories). For practical reasons related to informative content and availability, the current classification of *Archaea* and *Bacteria* is based on genealogical patterns inferred from comparative analyses of *16S rRNA* genes (Ludwig, Glöckner, & Yilmaz, 2011). Nevertheless, a shift in the general standards for prokaryotic classification is expected when the catalogue of complete genomes becomes sufficiently comprehensive (Klenk & Göker, 2010; Richter & Rosselló-Móra, 2009), though this is not yet the case (Figure 1). The often incomplete or absent taxonomic information attached to the gene sequence (e.g. *organism name*) has propagated from primary nucleotide repositories (i.e. INSDC), highlighting the need for curation.

---

[2]For the LTP consortium.

Methods in Microbiology, Volume 41, ISSN 0580-9517, http://dx.doi.org/10.1016/bs.mim.2014.07.006

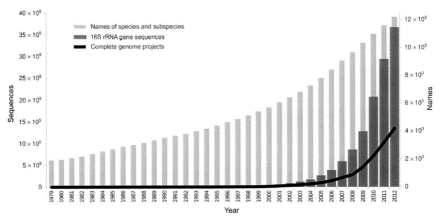

FIGURE 1

Annual cumulative growth of databases. 16S ribosomal RNA (light grey, to the left axis) was obtained from SILVA release 114 (http://www.arb-silva.de). Genome projects (black, to the right axis) refer to complete genome projects on *Archaea* and *Bacteria* according to the GOLD database (http://www.genomesonline.org). Names of prokaryotic species and subspecies with standing in nomenclature since 1980 (dark grey, to the right axis) were obtained from LPSN (http://www.bacterio.net). The total number of names is around 2000 greater than the total number of distinct type strains due to homotypic synonyms, new combinations, nomina nova and later heterotypic synonyms, or illegitimate names.

The 'All-Species' Living Tree Project (LTP) started in 2007, given the demand for a reliable taxonomic resource based on rRNA gene sequence data (Yarza et al., 2008). The LTP is an initiative coordinated by the executive editors of the journal *Systematic and Applied Microbiology* for the development of a taxonomic tool encompassing updated sequence databases, alignments and phylogenetic trees of the type strains of hitherto described species of *Archaea* and *Bacteria*. The LTP project counted on the help of four additional partners to set up an informatics infrastructure and a training environment for curators. These partners are (1) LPSN (Euzéby, 1997; www.bacterio.net) providing support on taxonomic nomenclature; (2) ARB (Ludwig et al., 2004; www.arb-home.de) for support on phylogenetics and classification; (3) SILVA (Quast et al., 2013; www.arb-silva.de) supplying sequence databases, computational resources and Web hosting; and (4) Ribocon (www.ribocon.com) for training and database management.

2 SOURCES OF INFORMATION

2.1 CLASSIFICATION OF MICROBIAL DATABASES

Overall, the multiple kinds of data together with the distinct requirements of particular users have defined a network through which information flows, gaining specificity and integration (Figure 2). The process happens among collaborating microbial

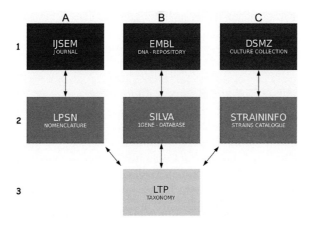

FIGURE 2

Data flow between microbial databases. Three main activities in microbiology are the description of taxa (A); provision of associated sequence data (B); and the storage of strains and information into Biological Resource Centres (C). These tasks constitute the primary-level information (1). Through data selection and curation, specific information becomes available as secondary resources (2). Further integration of higher resources leads to tertiary-level information, with even more specific and optimized databases (3). Arrows indicate main information flows.

databases which can be classified into three categories according to the quantity and refinement of their data.

At the first level, big infrastructures exist to preserve the data generated by three main activities in microbiology: (a) description of microorganisms and microbial communities, (b) sequencing of strains and (c) their preservation in Biological Resource Centres (BRCs) (Figure 2). The INSDC-member databases, for example, can be regarded as primary repositories of sequence data.

Secondary infrastructures have arisen to provide high quality and specific software platforms and databases, for example, rRNA gene databases like SILVA (Figure 2). One of the main aims of these rRNA resources is to continuously upgrade data according to changes in primary repositories (e.g. new submissions to INSDC) while maintaining quality and reliability. These kinds of secondary databases have already become so large that most of their resources are invested in system's maintenance and the development of tools to increase usability and analysis capability; hence, manually supervised tasks are increasingly devolved to independent dedicated teams.

Tertiary resources have narrowed their focus and reduced the size and complexity of their databases. Here, the largest investment is given to manual supervision tasks performed by expert curators in order to complete or correct information gathered from primary and secondary databases. The curation of rRNA gene sequence data (e.g. LTP project), for instance, comprises a search of interesting sequence entries for a specific purpose (e.g. taxonomy → type strains) and the incorporation of added value like curated metadata and sequence alignment. The ultimate goal of tertiary resources is twofold: to provide a reference tool for a given user community and to have an impact on secondary- and primary-level databases which acquire the curated data and thereby improve their service quality (Figure 2).

2.1.1 Taxonomy (LPSN and Bergey's Manual)

The existence of an official nomenclature for *Archaea* and *Bacteria* is one of the most important achievements in microbiology in recent times. It represents a global agreement for the naming of prokaryotes, with strong implications for scientific communication. According to the International Committee on Systematics of Prokaryotes, the *International Journal of Systematic and Evolutionary Microbiology* (IJSEM) is the official journal for the publication of validly published archaeal and bacterial names, thereby providing a primary resource for microbial systematics. Up to December 2012, nearly 15,000 names of prokaryotic taxa (of any rank) had been published; this number has grown at a constant rate of about 750 names per year since 2006 (Figure 1). In 1997, the List of Prokaryotic Names with Standing in Nomenclature (LPSN; http://www.bacterio.net/; Euzéby, 1997) was created to cover the past and present nomenclature of each published prokaryotic taxon into one single Web resource. LPSN became a highly respectable and specialized secondary resource which substantially improves access to taxonomic information. The LPSN provides information on the latest valid nomenclature for each taxon, the nomenclatural type and its taxonomic classification, related publications and taxonomic opinions. Dr. Aidan Parte is the current curator responsible for the LPSN (Parte, 2014).

Although classification of *Archaea* and *Bacteria* is not officially regulated, the inference of genealogical relationships based on the 'molecular clock' concept, particularly the gene of the small subunit of ribosomal RNA (*SSU* or *16S rRNA*), provided the key for the classification of prokaryotes based on natural relationships (Amann, Ludwig, & Schleifer, 1995; Fox, Pechman, & Woese, 1977; Ludwig & Schleifer, 1994). In 2001, the second edition of the *Bergey's Manual of Systematic Bacteriology* gave the phylogenetic backbone of the prokaryotes, by providing an updated and emended framework for prokaryotic classification based on rRNA (Garrity & Holt, 2001). Bergey's Taxonomic Outlines have been subsequently updated to include new published species and additional sequence data (Ludwig et al., 2012; Ludwig, Euzéby, & Whitman, 2010; Ludwig, Schleifer, & Whitman, 2009).

2.1.2 Type-strain information (StrainInfo database)

BRCs act as long-term reservoirs for cultivable microorganisms (e.g. DSMZ – Deutsche Sammlung von Mikroorganismen und Zellkulturen). They deal with the authentication, safe preservation and supply of deposited cell material and operate according to international laws regarding health and safety requirements, quarantine regulations, intellectual property rights and classification of microorganisms into hazard groups (see www.wfcc.info, for more information). The name of each archaeal and bacterial species has to be validly published to confirm to the Bacteriological Code of Nomenclature (Lapage et al., 1992) and has to be represented by a nomenclatural type, that is, a viable and culturable strain to which the name is permanently linked. This is the reason why the circumscription of a new species needs to be based on a careful comparison between the new isolated strains and the type strains of genealogically related species. In order to allow exploration and unlimited access to the taxon's phenotypic and genotypic characteristics, it is mandatory that type strains should be deposited into two internationally recognized service culture collections, in two different countries (Tindall, Kämpfer, Euzéby, & Oren, 2006).

Finding information (e.g. sequence data) for a particular type strain can be hampered for two main reasons: (I) different stocks of the same type strain held in different collections are cited differently (e.g. ATCC 9001, CECT 515, DSM 30083 for the *Escherichia coli* type strain), leading to syntactical variation, synonymy and homonymy between culture identifiers and (II) type strains coexist within a catalogue of nearly 1 million *Archaea* and *Bacteria* available in ~600 BRCs worldwide (June 2013; http://www.wfcc.info/ccinfo/statistics/). In 2005, the StrainInfo database was created with the aim of integrating information of all strains held in BRCs into a single online catalogue (Dawyndt, Vancanneyt, De Meyer, & Swings, 2005) with the public sequence entries available for those cultures. As a secondary information resource, the main features of StrainInfo are the capability to automatically resolve equivalent strain identifiers and to link them with external taxonomic resources of sequence data and publications, thereby allowing efficient integrated studies of type strains.

2.1.3 Sequences and alignments (ARB and SILVA)

Three independent research groups in Europe, the United States and Australia (SILVA, Pruesse et al., 2007; RDP-II, Cole et al., 2007; greengenes, DeSantis et al., 2006) emerged with the aim to (1) provide the scientific community with updated universal alignments for optimal and comparable phylogenetic reconstructions; (2) produce and maintain curated datasets of nearly full-length rRNA sequences to be used for in-depth phylogenetic analyses; and (3) develop a set of bioinformatic tools for online sequence data management and analyses.

SILVA databases were generated to meet the need for reference, comprehensive, quality checked and regularly updated datasets of aligned 16S/18S (*SSU*) and 23S/28S (*LSU*) rRNA gene sequences of *Archaea*, *Bacteria* and *Eukarya* (Quast et al., 2013). SILVA maintains the universal rRNA alignments for the ARB software package (Ludwig et al., 2004), one of the most comprehensive tools for phylogenetic analysis. These alignments have been manually revised taking into account the functional and evolutionary constraints given by the molecule's secondary structure (helix- and stem-loop structures) (Ludwig & Schleifer, 1994), while maximizing the positional orthology (i.e. needed to obtain reliable and comparable phylogenies) (Peplies, Kottmann, Ludwig, & Glöckner, 2008; Schloss, 2010).

Two SILVA databases (PARC and REF) are available with distinct quality standards. Critical quality parameters include sequence length, ambiguities, homopolymers, chimaera probability and alignment quality criteria. In the SSU PARC database (~3.8 million entries in SILVA release 115), all sequences with lengths above 300 bp are retained, whereas in SSU REF databases ($\sim$1.5 million entries) the cutoff is 900 bp for archaeal and 1200 bp for bacterial sequences. In addition, sequence entries in SILVA are enriched with additional metadata obtained from other resources, for instance, strain information (EMBL, LTP, StrainInfo), taxonomic classification (EMBL, greengenes, RDP-II, LTP) and curated habitat descriptors (megx.net) (see Quast et al., 2013, for more details). The complete *SSU* rRNA gene sequence dataset has grown at a nearly exponential rate since the early 1990s, compared to the arithmetic growth of species descriptions (Figure 1).

3 DATABASE CREATION AND UPDATING

The LTP started as a manually supervised process designed to merge existing sources of information into a new curated type-strain database (Yarza et al., 2008):

1. From LPSN, a non-redundant list of names was created to represent all the hitherto described species and subspecies of *Archaea* and *Bacteria*. In the process, later heterotypic synonyms (e.g. *Streptomyces parvisporogenes*), *Cyanobacteria* not validly published according to the Bacteriological Code (e.g. all species within the genus *Anabaena*) and the Candidatus category (e.g. *Candidatus Baumannia cicadellinicola*) were not included. From the SILVA database, a dataset of candidate type-strain sequences was selected. SILVA entries had already been mapped to the StrainInfo database, which distinguishes type from non-type strains. All *16S rRNA* gene sequence entries from the SILVA REF database tagged as type (T) or cultured (C) were selected as candidates in the preliminary dataset. Then, these sequences were manually assigned to validly published names of species or subspecies (LPSN) by means of the manual verification of type-strain identifiers. This task was hampered by the abundance of outdated sequence-associated information like the species name and/or the strain identifier. These sequence-associated metadata fields were scarcely supervised at the respective INSDC database, thus justifying the manual supervision of the complete procedure. StrainInfo was often consulted to learn updated equivalences between culture collections.

2. Several hundred type strains were missing in the initial dataset and their sequences had to be manually sought in other resources (Bergey's Manual, EMBL, IJSEM). Indeed, even after such searches, the complete set of type strains was not fully represented. The type strains of some of these species had never been sequenced; others were represented by low-quality sequences. The species without good-quality sequences for their type strain were dubbed 'orphan' species.

3. At the pre-final stage, the sequence dataset under consideration was redundant because some type strains had either been repeatedly sequenced or their sequenced genomes included multiple copies of the ribosomal operon. This redundancy is not required for classification and identification purposes. Subsequently, the LTP team decided to retain only one sequence per type strain in the final dataset, namely, the one considered to be the best in terms of length,

ambiguities, homopolymers, chimaera probability and alignment quality. However, in cases of doubt, the earliest submission to an INSDC partner was chosen. The 'one sequence per type strain' standard was only overlooked for a small minority of genomes with highly divergent (<98% sequence identity) copies of the rRNA operon (e.g. *Haloarcula marismortui* DSM 3752).

The LTP database is periodically updated (Table 1) to account for new taxa and other taxonomic changes that are published monthly in the IJSEM (Munoz et al., 2011). All new sequences belong to the type strains of the species they are representing. This information is initially taken from LPSN and complemented with IJSEM and StrainInfo resources. While sequencing and submission of *23S rRNA* sequence data still occasionally occur, the descriptions of almost all new species of *Archaea* and *Bacteria* include the accession number of a public *16S rRNA* gene entry for the type strain, which can be downloaded from an INSDC database. All new entries included in the database are manually improved in terms of alignment and metadata as described above. In summary, the LTP database is updated according to the following criteria: (1) new names are given to an existing type strain when new combinations or earlier homotypic synonyms are published; (2) new sequence entries are used to represent existing type strains when they have been proven to be of better quality; (3) new type strains or neotypes and their respective sequence entries are added to the database in the case of new species and subspecies descriptions; and (4) entries are deleted from the database when their names should not be used to reference any type strain, such as later heterotypic synonyms or rejected names.

4 FEATURES OF THE DATABASE
4.1 OPTIMIZED SSU AND LSU ALIGNMENTS

Although all sequences in the LTP originally come from the SILVA database, they are submitted to an extra manual supervision of the alignment, further resolving positions that have been previously misaligned. The *SSU* alignment is of better quality than the *LSU* alignment, given the ample experience with a much larger dataset. In this regard, the LTP release 102 involved a major improvement of the SILVA's *LSU*-seed alignment (i.e. a reference core of trusted sequences for database creation), including a manual inspection of 5000 new sequences which were appended to the 2800 existing in the seed. Moreover, the alignment was extended to a final size of 100,000 positions to better accommodate insertions contributed by different taxa (Yarza et al., 2010).

4.2 CURATED HIERARCHICAL CLASSIFICATION

All sequences in the LTP are complemented with the hierarchic classification (from genus to phylum) available at LPSN, which includes merged information from the NCBI taxonomy, Taxonomic Outlines of Bergey's Manual, TOBA and suggestions made by the LTP. The full classification is available in the database fields *hi_tax_ltp*

Table 1 Summary of LTP Releases

Release	Type	Date	Total Entries	New Entries	Deleted Entries	Net Increase	% Incorrect	IJSEM-Sync	SILVA-Sync	EMBL-Sync
LTPs93	SSU	Aug. 2008	6728	6728	0	6728	22	Dec. 2007	Feb. 2008	Dec. 2007
LTPs95	SSU	Oct. 2008	7006	299	21	278	45	Jun. 2008	Jul. 2008	Jun. 2008
LTPs100	SSU	Sep. 2009	7710	750	46	704	50	Aug. 2009	Aug. 2009	Jun. 2009
LTPs102	SSU	Sep. 2010	8029	363	44	319	58	Feb. 2010	Feb. 2010	Nov. 2009
LTPs102	LSU	Sep. 2010	792	792	0	792	6	Feb. 2010	Feb. 2010	Nov. 2009
LTPs104	SSU	Mar. 2011	8545	545	29	516	74	Dec. 2010	Oct. 2010	May. 2010
LTPs106	SSU	Aug. 2011	8815	279	9	270	77	May. 2011	Apr. 2011	Dec. 2010
LTPs108	SSU	Jul. 2012	9279	490	26	464	60	Dec. 2011	Sep. 2011	Jun. 2011
LTPs111	SSU	Feb. 2013	9701	422	7	415	73	Aug. 2012	Jul. 2012	Mar. 2012

'Sync' fields correspond to EMBL, IJSEM, SILVA and release dates. 'Net Increase' of a release is the number of new entries minus the number of deleted entries. '% Incorrect' refers to the percentage of new entries whose INSDC records had incorrect information in the organism name field.

Table 2 Description of LTP Specific Fields

Field Name	Description
fullname_ltp	Corrected species name according to LPSN (http://www.bacterio.net)
rel_ltp	Name of the LTP release where a sequence entry appears for the first time
hi_tax_ltp	Name of the family where the taxon is classified or, for unclassified genera, the name of the next available high taxon above genus (e.g. 'Unclassified Clostridiales' for *Blautia stercoris*; Park, Kim, Roh, & Bae, 2012)
type_ltp	Type species receive the label 'type sp.' in this field
riskgroup_ltp	Risk-group classification of microorganisms according to the Federal Institute for Occupational Safety and Health (BAuA) in Germany
tax_ltp	Taxonomic classification into high-taxonomic ranks according to LPSN
url_lpsn_ltp	URL information to access LPSN's species file

and *tax_ltp* (Table 2). As a complement, information for nomenclatural types (i.e. type species of genera and higher ranks) has been retrieved from LPSN. All sequences belonging to nomenclatural types are labelled as such in another specific field *type_ltp* (Table 2).

In addition, the correct species name (*fullname_ltp*, Table 2) according to LPSN is given to each sequence to replace the 'organism name' information which has appeared mistakenly in more than 50% of newly deposited type-strain *16S rRNA* gene sequences (Table 1). The reason for these inconsistencies in primary data is a reflection of a delay between sequence submission and species publication (Yarza et al., 2008). In practise, a new isolate can be named differently until it is published according to the rules of the Bacteriological Code (Lapage et al., 1992) in one of the journals publishing new taxa. But the final name stands after its publication in IJSEM, either in an article or a validation list. For example, the nucleotide entry HE613447 submitted in 2011 reads '*Achromobacter* sp. R-46660', when it really refers to the species name *Achromobacter spiritinus* effectively published in 2013. Wrong data stored in primary resources unavoidably spread to the whole network of databases (Figure 2). To improve this situation, authors are encouraged to update information even after the original submissions. In addition, scientific journals might play a role by requiring an update of taxonomic metadata of, at least, the sequences of type strains.

4.3 RISK-GROUP CLASSIFICATION

Microbial species are classified by the Federal Institute for Occupational Safety and Health (BAuA) in Germany, according to the risk they present to humans, animals and plants. This information is regularly updated and made public in the Technische Regeln für Biologische Arbeitsstoffe (TRBA) document 'Einstufung von Bakterien in Risikogruppen' (TRBA 466). These data are initially implemented by the DSMZ, which serves as a source for mapping with the field *riskgroup_ltp* into the LTP database (Table 2).

4.4 TAXONOMIC THRESHOLDS

The LTP dataset is an analytic tool which allows us to understand the meaning of numerical taxa boundaries based on sequence identity. The lowest identity found within each taxon of every rank is based on a distance matrix calculated with the LTP alignment. Considering all taxa at the levels of genus, family and phylum, a general lower-cutoff value for each rank was obtained. In general, *16S rRNA* gene sequence identities lower than $94.9\% \pm 0.4$, $87.5\% \pm 1.3$, $78.4\% \pm 2.0$ lead to the circumscription of new genera, families and phyla, respectively (Yarza et al., 2008). For *23S rRNA* gene sequences, these cutoffs differ slightly: $93.2\% \pm 1.3$ (genus), $87.7\% \pm 2.5$ (family) and 75.3% (phylum) (Yarza et al., 2010). The low errors observed above (i.e. 95% confidence interval of the mean) indicate that taxonomists have historically circumscribed in a coherent way independently of the taxonomic group under study. Therefore, the application of numerical thresholds can be a complementary aid when describing new taxa or revising existing ones. Likewise, the use of these taxa boundaries may be useful for prospective studies (i.e. OTU abundances) on *SSU* and *LSU* environmental datasets.

5 PHYLOGENETIC TREES

The *SSU*-based phylogenetic tree offered by the LTP is calculated using the complete dataset of type strains plus an additional selection of $\sim$3000 supporting sequences to stabilize certain under-represented groups (e.g. *Cyanobacteria*). The algorithm of choice is maximum likelihood implemented in the RAxML program (Stamatakis, 2006), as it is a robust method which, in addition, produces similar topologies to the neighbour-joining and maximum parsimony reconstructions (Yarza et al., 2008). To reduce noise caused by poorly resolved areas of alignments (i.e. due to hypervariability and sequencing errors), a 40% maximum frequency filter was applied and 1390 total alignment positions were considered. Calculation is performed with the GTRGAMMA model (Stamatakis, 2006) and 100 bootstrap replicates. In some releases (e.g. LTP release 111), newly added sequences are incorporated into the existing tree using the ARB parsimony tool with the option for keeping the initial topology while inserting additional sequences. Every 1–2 years, a new full reconstruction is calculated from scratch using the methodology explained above.

The phylogenetic tree calculated for the *LSU* dataset followed different guidelines due to the extreme shortage of taxa in many groups (Yarza et al., 2008). Initially, a highly stringent filtering approach enabled 2000 high quality (and non-redundant) *LSU* sequences (type and non-type strain) from SILVA to stand in the initial core dataset. This set of sequences was subject to a maximum likelihood reconstruction in combination with a 50% maximum frequency filter, allowing 2463 positions of the entire alignment. The missing partial or lower quality type-strain sequences were added to the existing tree using the ARB parsimony tool.

The resulting tree topologies are carefully examined to evaluate the monophyly assumption of every taxon. Clades are identified and named when the nomenclatural type is present (e.g. type species of genera, type genera of families). A comparison against many other taxon-specific and broad-range trees, using different sequence datasets and methods, has supported the main genealogical patterns inferred by the LTP (Yarza et al., 2010).

LTP's *SSU* and *LSU* phylogenies provide a way to readily examine classifications all along the current taxonomy. The *SSU* tree highlights a low significance with respect to the relative branching order of phyla, including some considerable 'jumps' (*Acidobacteria*, *Cyanobacteria*, *Fusobacteria*), as indicated by relatively short branches. In contrast, some taxa, such as the phylum *Bacteroidetes–Chlorobi* and the phylum *Chlamydia–Verrucomicrobia–Lentisphaerae* confirm well-supported higher order structures. In addition, it is interesting to see that the *Deltaproteobacteria* and *Epsilonproteobacteria* are separated from the rest of the *Proteobacteria*, resembling the weak union of the two sister classes (*Bacilli* and *Clostridia*) from the *Firmicutes* (Yarza et al., 2010). Although there is a tendency to show higher phylogenetic coherence at lower ranks, there are several well-known paraphyletic groups, like the *Bacillaceae*, *Clostridiaceae* and *Pseudomonadaceae*, which still require further taxonomic revision. Additionally, the LTP trees contribute new evidence for completing the classification of certain misclassified species, for example, *Spongiispira norvegica* (Kaesler et al., 2008) which was originally described as a member of the order *Oceanospirillales* but not associated with a family. However, current data show a *SSU* sequence similarity of 97% against *Oceaniserpentilla haliotis* and a clear affiliation with the family *Oceanospirillaceae* (LTP release 102; Yarza et al., 2010) (Figure 3).

6 LTP AS A TAXONOMIC TOOL

The All-Species LTP is not an attempt to reconstruct the species phylogeny with total confidence, but is designed to provide the scientific community with a curated taxonomic tool. As such, the LTP is a collection of reference material (Table 3) that is publicly available and regularly updated. The package includes (i) the sequence database of *SSU* and *LSU* sequences from archaeal and bacterial type strains complemented with curated metadata in ARB and CSV formats, both including the LTP-specific fields (Table 2); (ii) the complete dataset of aligned type-strain sequences in FASTA format; (iii) a single phylogenetic tree containing all archaeal and bacterial species; and (iv) a set of descriptive tables and list of changes; see Table 3 for details. The regular corrections performed by the LTP on rRNA alignments and metadata contribute to the SILVA platform, thereby improving its quality over time (Quast et al., 2013). This transference of information exemplifies the network's success in storing and providing reliable microbial information (Figure 2). Some examples of LTP usage in research projects include facilitating the collection of reference sequences for the reconstruction of taxa genealogies (e.g. Cousin et al.,

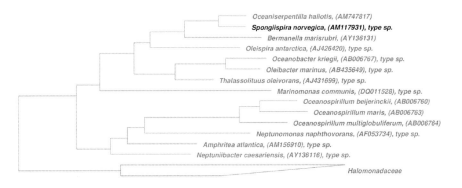

Oceaniserpentilla haliotis, (AM747817)
Spongiispira norvegica, (AM117931), type sp.
Bermanella marisrubri, (AY136131)
Oleispira antarctica, (AJ426420), type sp.
Oceanobacter kriegii, (AB006767), type sp.
Oleibacter marinus, (AB435649), type sp.
Thalassolituus oleivorans, (AJ431699), type sp.
Marinomonas communis, (DQ011528), type sp.
Oceanospirillum beijerinckii, (AB006760)
Oceanospirillum maris, (AB006763)
Oceanospirillum multiglobuliferum, (AB006764)
Neptunomonas naphthovorans, (AF053734), type sp.
Amphritea atlantica, (AM156910), type sp.
Neptuniibacter caesariensis, (AY136116), type sp.
Halomonadaceae

0.01

FIGURE 3

Phylogenetic position of *Spongiispira norvegica* Gp_4_7.1[T] based on *16S rRNA* gene sequences. Tree topology extracted from the LTP release 111. Scale bar indicates estimated sequence divergence. *Spongiispira norvegica* was described as a member of the order *Oceanospirillales* without a family affiliation being given. However, it shows a *SSU* sequence similarity of 97% against *Oceaniserpentilla haliotis* and a clear affiliation with the family *Oceanospirillaceae*.

Table 3 Suite of Downloadable Materials

Release information	– Description of new release
	– Description of LTP fields
	– Table with INSDC entries with incorrect organism name information
	– Table with list of changes from last release (updated, deleted, new)
Database exports	– ARB format including all aligned sequences, metadata and trees
	– CSV export including metadata only
Tree exports	– Full expanded tree in PDF
	– Full expanded tree in NEWICK
	– Collapsed overview in PDF
Sequence exports	– All aligned type-strain sequences in FASTA
	– All unaligned type-strain sequences in FASTA
Scripts	– ARB filter used to export LTP sequences in FASTA format

LTP release 111 (http://www.arb-silva.de/projects/living-tree).

2012), performing fast and reliable taxonomic affiliations in rRNA surveys (e.g. Santamaria et al., 2012) and serving as reference datasets for testing bioinformatic procedures (e.g. Mizrahi-Man, Davenport, & Gilad, 2013). In addition, a contact address (living-tree@arb-silva.de) enables the exchange of suggestions or particular requests by the user community. LTP should continue to improve the speed of release production by improving data retrieval from secondary resources (LPSN, StrainInfo and SILVA) and by optimizing post-production analysis. Clear tendencies towards standardization, automation and quality management have been initiated, resulting in a better digital communication with the other microbial information resources.

ACKNOWLEDGEMENTS

We acknowledge contributions from the LTP project leader, Ramon Rosselló-Móra, and the scientific support of Rudolf Amann, Jean Euzéby, Wolfgang Ludwig, Karl-Heinz Schleifer, Frank Oliver Glöckner, Michael Richter and Jörg Peplies. We are also grateful for the feedback received from LTP users. Funding was provided by the Max Planck Society, Elsevier and the EU project SYMBIOMICS (grant EU-264774).

REFERENCES

Amann, R. I., Ludwig, W., & Schleifer, K. H. (1995). Phylogenetic identification and *in situ* detection of individual microbial cells without cultivation. *Microbiology Reviews, 59,* 143–169.

Cole, J. R., Chai, B., Farris, R. J., Wang, Q., Syed-Mohideen, A. S. K., McGarrell, D. M., et al. (2007). The ribosomal database project (RDP-II): Introducing myRDP space and quality controlled public data. *Nucleic Acids Research, 35*(Database issue), D169–D172.

Cousin, S., Gulat-Okalla, M.-L., Motreff, L., Gouyette, C., Bouchier, C., Clermont, D., et al. (2012). *Lactobacillus gigeriorum* sp. nov., isolated from chicken crop. *International Journal of Systematic and Evolutionary Microbiology, 62,* 330–334.

Dawyndt, P., Vancanneyt, M., De Meyer, H., & Swings, J. (2005). Knowledge accumulation and resolution of data inconsistencies during the integration of microbial information sources. *IEEE Transactions on Knowledge and Data Engineering, 17,* 1111–1126.

DeSantis, T. Z., Hugenholtz, P., Larsen, N., Rojas, M., Brodie, E. L., Keller, K., et al. (2006). Greengenes, a chimera-checked 16S rRNA gene database and workbench compatible with ARB. *Applied and Environmental Microbiology, 72,* 5069–5072.

Euzéby, J. P. (1997). List of bacterial names with standing in nomenclature: A folder available on the Internet. *International Journal of Systematic and Evolutionary Microbiology, 47,* 590–592.

Fox, G. E., Pechman, K. R., & Woese, C. R. (1977). Comparative cataloguing of 16S ribosomal ribonucleic acid: Molecular approach to prokaryotic systematics. *International Journal of Bacteriology, 27,* 44–57.

Garrity, G. M., & Holt, J. G. (2001). The road map to the manual. In D. R. Boone, R. W. Castenholz, G. M. Garrity, et al. (Eds.), *The Archaea and the deeply branching and phototrophic bacteria: Vol. 1. Bergey's manual of systematic bacteriology.* New York: Springer.

Kaesler, I., Graeber, I., Borchert, M. S., Pape, T., Dieckmann, R., von Döhren, H., et al. (2008). *Spongiispira norvegica* gen. nov., sp. nov., a marine bacterium isolated from the boreal sponge Isops phlegraei. *International Journal of Systematic and Evolutionary Microbiology, 58,* 1815–1820.

Klenk, H.-P., & Göker, M. (2010). En route to a genome-based classification of *Archaea* and *Bacteria*? *Systematic and Applied Microbiology, 33,* 175–182.

Lapage, S. P., Sneath, P. H. A., Lessel, E. F., Skerman, V. B. D., Seeliger, H. P. R., & Clark, W. A. (1992). *International code of nomenclature of bacteria (1990 revision).* Washington: Bacteriological Code, American Society for Microbiology.

Ludwig, W., Euzéby, J., Schumann, P., Busse, H.-J., Trujillo, M. E., Kämpfer, P., et al. (2012). Road map of the phylum *Actinobacteria.* In M. Goodfellow, P. Kämpfer, H.-J. Busse, M. E. Trujillo, K.-i Suzuki, W. Ludwig, W. B. Whitman, et al. (Eds.), *The Actinobacteria. part A: Vol. 5. Bergey's manual of systematic bacteriology* (pp. 1–28). New York: Springer.

Ludwig, W., Euzéby, J., & Whitman, W. B. (2010). Road map of the phyla *Bacteroidetes, Spirochaetes, Tenericutes (Mollicutes), Acidobacteria, Fibrobacteres, Fusobacteria, Dictyoglomi, Gemmatimonadetes, Lentisphaerae, Verrucomicrobia, Chlamydiae*, and Planctomycetes. In N. R. Krieg, J. T. Staley, D. Brown, B. P. Hedlund, B. J. Paster, N. Ward, W. Ludwig, & W. B. Whitman (Eds.), *Bergey's manual of systematic bacteriology: Vol. 4* (2nd ed., pp. 1–19). New York: Springer.

Ludwig, W., Glöckner, F. O., & Yilmaz, P. (2011). The use of rRNA gene sequence data in the classification and identification of prokaryotes. In F. Rainey A. Oren, *Taxonomy of prokaryotes: Vol. 38. Methods in microbiology* (pp. 349–384). Amsterdam: Academic Press.

Ludwig, W., & Schleifer, K. H. (1994). Bacterial phylogeny based on 16S and 23S rRNA sequence analysis. *FEMS Microbiology Reviews*, *15*, 155–173.

Ludwig, W., Schleifer, K.-H., & Whitman, W. B. (2009). Revised road map to phylum *The Firmicutes*. In P. De Vos, G. M. Garrity, D. Jones, N. R. Krieg, W. Ludwig, F. A. Rainey, K.-H. Schleifer, W. B. Whitman, et al. (Eds.), *Vol. 3. Bergey's manual of systematic bacteriology* (pp. 1–13). *New York: Springer.*

Ludwig, W., Strunk, O., Westram, R., Richter, L., Meier, H., Kumar, Y., et al. (2004). ARB: A software environment for sequence data. *Nucleic Acids Research*, *32*, 1363–1371.

Mizrahi-Man, O., Davenport, E. R., & Gilad, Y. (2013). Taxonomic classification of bacterial 16S rRNA genes using short sequencing reads: Evaluation of effective study designs. *PLoS One*, *8*, e53608.

Munoz, R., Yarza, P., Ludwig, W., Euzéby, J., Amann, R., Schleifer, K.-H., et al. (2011). Release LTPs104 of the All-Species Living Tree. *Systematic and Applied Microbiology*, *34*, 169–170.

Park, S. K., Kim, M. S., Roh, S. W., & Bae, J. W. (2012). *Blautia stercoris* sp. nov., isolated from human faeces. *International Journal of Systematic and Evolutionary Microbiology*, *62*, 776–779.

Parte, A. C. (2014). LPSN-list of prokaryotic names with standing in nomenclature. *Nucleic Acids Research*, *42*, D613–D616.

Peplies, J., Kottmann, R., Ludwig, W., & Glöckner, F. O. (2008). A standard operating procedure for phylogenetic inference (SOPPI) using (rRNA) marker genes. *Systematic and Applied Microbiology*, *31*, 251–257.

Pruesse, E., Quast, C., Knittel, K., Fuchs, B. M., Ludwig, W., Peplies, J., et al. (2007). SILVA: A comprehensive online resource for quality checked and aligned ribosomal RNA sequence data compatible with ARB. *Nucleic Acids Research*, *35*, 7188–7196.

Quast, C., Pruesse, E., Yilmaz, P., Gerken, J., Schweer, T., Yarza, P., et al. (2013). The SILVA ribosomal RNA gene database project: Improved data processing and web-based tools. *Nucleic Acids Research*, *41*(Database issue), D590–D596.

Richter, M., & Rosselló-Móra, R. (2009). Shifting the genomic gold standard for the prokaryotic species definition. *Proceedings of the National Academy of Sciences of the United States of America*, *106*, 19126–19131.

Santamaria, M., Fosso, B., Consiglio, A., Caro, G. D., Grillo, G., Licciulli, F., et al. (2012). Reference databases for taxonomic assignment in metagenomics. *Briefings in Bioinformatics*, *13*, 682–695.

Schloss, P. D. (2010). The effects of alignment quality, distance calculation method, sequence filtering, and region on the analysis of 16S rRNA gene-based studies. *PLoS Computational Biology*, *6*, e1000844.

Stamatakis, A. (2006). RAxML-VI-HPC: Maximum likelihood-based phylogenetic analyses with thousands of taxa and mixed models. *Bioinformatics*, *22*, 2688–2690.

Tindall, B. J., Kämpfer, P., Euzéby, J. P., & Oren, A. (2006). Valid publication of names of prokaryotes according to the rules of nomenclature: Past history and current practice. *International Journal of Systematic and Evolutionary Microbiology, 56,* 2715–2720.

Yarza, P., Ludwig, W., Euzéby, J., Amann, R., Schleifer, K.-H., Glöckner, F. O., et al. (2010). Update of the All-Species Living Tree Project based on 16S and 23S rRNA sequence analyses. *Systematic and Applied Microbiology, 33,* 291–299.

Yarza, P., Richter, M., Peplies, J., Euzeby, J., Amann, R., Schleifer, K.-H., et al. (2008). The All-Species Living Tree project: A 16S rRNA-based phylogenetic tree of all sequenced type strains. *Systematic and Applied Microbiology, 31,* 241–250.

16S rRNA Gene-Based Identification of *Bacteria* and *Archaea* using the EzTaxon Server

Mincheol Kim*, Jongsik Chun*,†,1

*School of Biological Sciences, Seoul National University, Seoul, Republic of Korea
†ChunLab Inc., Seoul National University, Seoul, Republic of Korea
¹Corresponding author: e-mail address: jchun@snu.ac.kr

1 INTRODUCTION

Ribosomal RNA genes have been used as standard phylogenetic markers in molecular taxonomic studies since the pioneering studies on the tree of life by Woese and Fox (1977). Their ubiquitous distribution across all archaeal and bacterial lineages, evolutionarily conserved nature, and a wide range of variable regions facilitated the use of rRNA genes for a variety of taxonomic purposes. The small subunit ribosomal RNA (=16S rRNA in prokaryotes) was the phylogenetic marker of choice from an early stage and has been used extensively to date (Woese, 1987). Earlier difficulties in the determination of the whole 16S rRNA primary structure were overcome soon after the invention of PCR and improvements in Sanger DNA sequencing technology. At present, sequencing of 16S rRNA genes costs as little as $5 for partial or $25 for full-length sequences, thereby allowing the routine use of 16S rRNA gene sequencing for the classification and identification of prokaryotes in clinical and environmental surveys.

The use of 16S rRNA gene sequences for the classification and identification of prokaryotes is mainly dependent on comparisons against a database of known sequences. Currently, the sequences of type strains of ~99% of prokaryotic species with validly published names are available in public databases (Chun & Rainey, 2014), indicating the extent of information available for the identification of unknown *Bacteria* and *Archaea*. In general, 16S rRNA gene sequences are used in two ways in microbial systematics, namely for calculating pairwise sequence similarities and for phylogenetic analyses following multiple sequence alignments. The 16S rRNA gene sequence similarity between two strains provides a simple yet very robust criterion for the identification of newly isolated strains, whereas phylogenetic analyses can be used to elucidate overall evolutionary relationships between related

Methods in Microbiology, Volume 41, ISSN 0580-9517, http://dx.doi.org/10.1016/bs.mim.2014.08.001

taxa. The former is a critical checkpoint for species-level identification, while the latter is better for genus or suprageneric classification.

DNA–DNA hybridization (DDH) has served as a standard molecular method for species delineation in the classification of prokaryotes. Over the last 50 years, the 70% DDH value has served as a rigid boundary for circumscribing species despite some exceptions and drawbacks (Wayne et al., 1987). Taxonomists have searched for alternative genotypic standards due to the labour-intensive and error-prone nature of the DDH methodology (Gevers et al., 2005). Stackebrandt and Goebel (1994) suggested that the 16S rRNA gene should be used as a potential substitute for DDH by showing that there is a strong correlation between 16S rRNA gene sequence similarities and DDH values. The 70% DDH value was considered to be equivalent to a 97% 16S rRNA gene sequence similarity; consequently, this threshold has been widely used in prokaryotic systematics. This threshold was subsequently raised to 98.2–99.0% depending on the taxa under study (Stackebrandt & Ebers, 2006); a comparable similarity threshold (98.65%) was recently supported by the results of a large-scale comparative study between 16S rRNA gene sequence similarities and genome-driven average nucleotide identity values (Kim, Oh, Park, & Chun, 2014).

Among currently used methods available for the classification and identification of prokaryotes, the analysis of 16S rRNA gene sequences offers a reproducible and technically easy procedure that is also scalable. The procedure is universally applicable, does not require specialist knowledge and is cost-effective; individual isolates can be identified for around $10. However, easy, simple and scientifically sound bioinformatics are required for the assembly and comparison of 16S rRNA gene sequences. The EzTaxon server (Chun et al., 2007; Kim et al., 2012) was introduced to meet these requirements along with additional functionalities and tools. In this chapter, we provide the scientific background to EzTaxon, a taxonomy-oriented 16S rRNA gene database, and show how it can be used to identify members of the domains *Bacteria* and *Archaea*.

2 USE OF 16S rRNA GENE SEQUENCES IN PROKARYOTIC SYSTEMATICS

2.1 SEQUENCING OF 16S rRNA GENES

Various primers targeting conserved regions within the 16S rRNA gene have been developed as a result of the widespread use of 16S rRNA gene sequences in evolutionary and phylogenetic studies of prokaryotes; the most widely used ones are given in Table 1. Innumerable full-length and partial regions of 16S rRNA genes have been sequenced using the Sanger method and more recently with next-generation sequencing platforms. The deposition of ever-increasing numbers of 16S rRNA gene sequences in public databases without quality control filters inevitably leads to the accumulation of many poor quality sequences (Ashelford, Chuzhanova, Fry, Jones, & Weightman, 2005). Uncertainties during PCR and sequencing processes always create a certain level of inherent errors which are exacerbated when

Table 1 Commonly Used Primers for Amplifying Variable Regions of 16S rRNA Gene Sequences

Primer	Sequence (5′–3′)	Target Group	E. coli Numbering	References
8f (27f)	AGAGTTTGATCMTGGCTCAG	Bacteria	8–27	Lane (1991)
341f	CCTACGGGRSGCAGCAG	Bacteria, Archaea	341–357	Baker, Smith, and Cowan (2003)
519f	CAGCMGCCGCGGTAATWC	Universal	519–536	Wang and Qian (2009)
968f	AACGCGAAGAACCTTAC	Bacteria	968–984	Nübel et al. (1996)
338r	TGCTGCCTCCCGTAGGAGT	Bacteria	337–355	Fierer, Hamady, Lauber, and Knight (2008)
518r	ATTACCGCGGCTGCTGG	Bacteria	518–534	Muyzer, De Waal, and Uitterlinden (1993)
907r (926r)	CCGTCAATTCCTTTRAGTTT	Bacteria	907–926	Liu, Marsh, Cheng, and Forney (1997)
1392r	ACGGGCGGTGTGTRC	Universal	1392–1406	Lane (1991)
1492r	TACGGYTACCTTGTTACGACTT	Bacteria	1492–1513	Lane (1991)

sequencing is carried out only once in a single direction. Furthermore, low sequencing depths of coverage can also result in erroneous sequences. When the Sanger method is used, four to five combinations of sequencing reactions with different primers are generally necessary to generate high-quality full-length 16S rRNA gene sequences. All bases should be sequenced at least twice in both directions. In addition, primer regions should not be included in the final sequence as primer binding regions are not sequenced during the Sanger sequencing process.

2.2 CALCULATION OF NUCLEOTIDE SEQUENCE SIMILARITY VALUES OF 16S rRNA GENE SEQUENCES

A pairwise sequence similarity value is calculated as a fraction of a matched region between two sequences after sequence alignment and can be achieved by adding alignment gaps to line up all homologous nucleotide positions. Since the true evolutionary history of 16S rRNA genes is not known, all pairwise sequence alignment processes are hypothetical in nature. Many algorithms have been developed and used in prokaryotic systematics. Each algorithm is based on different assumptions and optimizations, so the output, i.e., alignment, is often different. Thus, sequence similarity values also vary depending on the algorithms and parameters used. It is therefore important to select theoretically and practically sound algorithms and parameters. The EzTaxon server implemented the robust algorithm developed by Myers and Miller (1988), while the initial search for potential phylogenetic neighbours is carried out using the BLASTn program (Altschul, Gish, Miller, Myers, & Lipman, 1990). This algorithm was also used in a study by Kim et al. (2014) in which a similarity threshold of 98.65% was proposed as the species boundary. The BLAST program alone should not be used as a similarity calculation for taxonomic purposes as it looks for local optimizations (Tindall, Rossello-Mora, Busse, Ludwig, & Kampfer, 2010).

3 IDENTIFICATION OF BACTERIA USING THE EZTAXON DATABASE

3.1 EZTAXON DATABASE

Unlike animals and plants, taxonomic assignment of prokaryotic strains to known species involves comparisons with related type strains and hence the sequences of type strains of all known archaeal and bacterial species should be available for comparison. At present, over 5 million 16S rRNA sequences are deposited in the GenBank database though the only ones that matter for identification of prokaryotes are the type strains of species with validly published names; presently there are about 11,000 validly named species of *Bacteria* and *Archaea*.

The original EzTaxon database was designed to be a complete compilation of 16S rRNA gene sequences of type strains with accessory bioinformatic functions, including homology search and calculation of sequence similarity (Chun et al., 2007).

Since its launch, this Web-based service has been widely used for the classification and identification of *Bacteria* and *Archaea* (Logan et al., 2009; Tindall et al., 2010), including clinically significant bacteria (Park et al., 2012).

While the original EzTaxon database held information on type strains, the second version of the database, named EzTaxon-e, has an extended repertoire to target microbial diversity beyond the formal nomenclatural system by including sequences representing hitherto unclassified or uncultured phylotypes (Kim et al., 2012). The names of such phylotypes were arbitrarily chosen; in many cases, they are named after GenBank accession numbers of representative sequences. For example, phylotype AB177176_s, where the suffix s (_s) denotes "species", was named after the GenBank accession number of its representative sequence, AB177176. This sequence was recovered from methane hydrate-bearing sub-seafloor sediment at the Peru margin (Inagaki et al., 2006) and represents not only a novel species but also a novel phylum in our phylogenetic analysis. Consequently, this phylotype has been designated as AB177176_p (phylum); AB177176_c (class); AB177176_o (order); AB177176_f (family); AB177176_g (genus) and AB177176_s (species) in the hierarchical system of EzTaxon. This informal nomenclatural system allowed the addition of >40,000 unclassified/uncultured phylotypes to the database (as of June 2014). These tentative names will be replaced by valid names when a strain representing phylotype AB177176_s is isolated and formally named. In such a case, the name AB177176_s will be recorded as the old name for the newly proposed species, so these tentative names can still be traced retrospectively. All of the sequences from environmental sources were carefully checked for chimera formation (Kim et al., 2012).

Another important attribute of the database is that it is built on a complete hierarchical system where all sequences have six-level taxonomic ranks (from species to phylum). When a novel species is proposed, it is mandatory to give it a genus name and a species epithet. However, the assignment of a species to known families or higher ranking taxa is optional according to the International Code of Nomenclature of Bacteria (Lapage et al., 1992). Since many novel species have been proposed without their assignment to suprageneric taxa, we generated numerous phylogenetic trees to assign all species/phylotypes to their correct ranks in EzTaxon's hierarchical classification. Many misclassified taxa discovered in the course of this process have been correctly classified in the EzTaxon hierarchical system (see http://www.ezbiocloud.net/eztaxon/hierarchy). These misclassified taxa require attention and should be reclassified in future.

3.2 ALGORITHM FOR "EZTAXON SEARCH"

Identification of bacterial isolates can be carried out by calculating 16S rRNA gene sequence similarities against the type strains of all known prokaryotic species. Calculating sequence similarities against all known species takes a lot of computing time; hence, it is more effective to find their closest phylogenetic neighbours first by using a faster search algorithm and then by calculating pairwise similarities.

Consequently, the identification process in the EzTaxon server involves the following two steps, cumulatively named an "EzTaxon search":

> Step 1: Initial search to find closely related sequences (phylogenetic neighbours) using the BLASTn program (Altschul et al., 1990)
> Step 2: Pairwise sequence alignment to calculate the sequence similarity values between the query sequence and hit sequences identified in Step 1.

To find the phylogenetic neighbouring species, four BLASTn searches are executed with four different query sequences (Figure 1). First, the most similar sequences in the EzTaxon database are selected using the whole (full-length) query sequence as a query in the BLASTn search. The full-length query sequence is then divided into three equal length fragments, and each fragment is then used as a query sequence in each subsequent BLASTn search. The top hits obtained from all four searches are then combined and subjected to a robust pairwise sequence alignment (Myers & Miller, 1988) against the original full-length query sequence. Using this

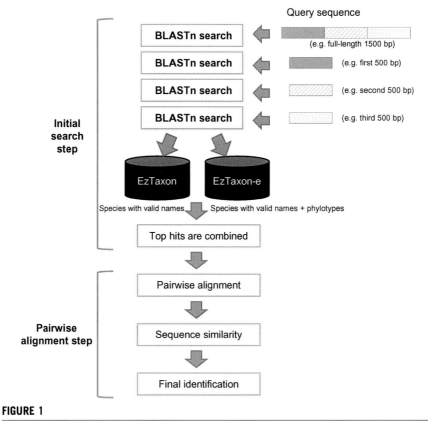

FIGURE 1

Algorithm for the "EzTaxon search". (See the colour plate.)

algorithm, users can compare the query sequence with closely related, albeit short, target sequences in the database, which is otherwise not possible. We recommend that the top 30 hits from each of the four BLASTn searches be saved and used for pairwise sequence similarity calculations. Since the top hits are the compilation of four different BLASTn searches, the final results usually contain more than 30 hits. It should be noted that the final order of hits in the EzTaxon search is based on percentage pairwise sequence similarity, not BLASTn scores.

Another useful statistical measure provided by EzTaxon is the "completeness" values which indicates how complete the query sequence is with respect to its full-length counterpart in the EzTaxon database (Kim et al., 2012). Given the high accessibility and affordability of the Sanger DNA sequencing method, full-length 16S rRNA gene sequences should be used for scientific purposes. Indeed, the cutoff proposed for species demarcation by Kim et al. (2014) is based on full-length sequence comparisons. This means that completeness value given as a percent can be used to evaluate the suitability of sequences for taxonomic use.

3.3 OVERALL WORKFLOW FROM SANGER DNA SEQUENCE DATA

The overall workflow for the bacterial identification process is summarized in Figure 2, together with the required bioinformatics tools. This approach can be used to identify either partial or full-length 16S rRNA gene sequences, which can be obtained from single or multiple Sanger DNA sequencing reactions, respectively. This workflow includes a combination of automated and manual operations to ensure maximum accuracy while also providing an efficient connection between the different software tools.

3.4 ASSEMBLY AND TRIMMING OF SEQUENCES

Usually, results of a Sanger DNA sequencing reaction are stored in a file with a special format, called *ab1* (raw data file defined by Applied Biosystems, CA). This file contains not only nucleotide sequence information but also quality values of each base calls, dubbed PHRED scores (Ewing & Green, 1998). PHRED scores of 10, 20, 30 and 40 indicate 90%, 99%, 99.9% and 99.99% accuracy, respectively. These values are mainly used by computer programs and rarely by humans. In general, both ends of sequences obtained from Sanger sequencers contain low-quality regions. Consequently, the first step in the whole process is to extract sequences from *ab1* files and then trim the low-quality terminal regions. This can be easily achieved at the EzTaxon Web site.

> *For single ab1 file*:
> Step 1: Go to the "Identify" page at http://www.ezbiocloud.net/eztaxon/identify.
> Step 2: Select an *ab1* file and upload. Make sure that you enter the right direction of the sequence, i.e., $5' \rightarrow 3'$ or $3' \rightarrow 5'$.
> Step 3: Your query sequence, which is trimmed at both ends using a PHRED score of 20 as cutoff, should appear on the same web page.

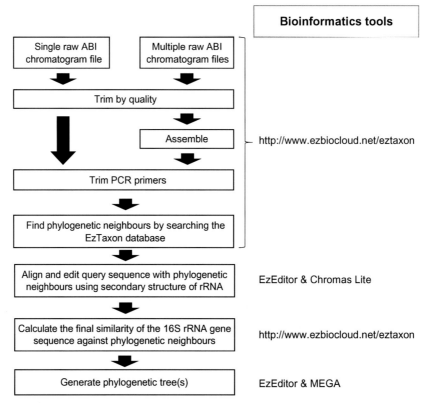

| Bioinformatics tools |

Single raw ABI chromatogram file / Multiple raw ABI chromatogram files → Trim by quality → Assemble → Trim PCR primers → Find phylogenetic neighbours by searching the EzTaxon database — http://www.ezbiocloud.net/eztaxon

Align and edit query sequence with phylogenetic neighbours using secondary structure of rRNA — EzEditor & Chromas Lite

Calculate the final similarity of the 16S rRNA gene sequence against phylogenetic neighbours — http://www.ezbiocloud.net/eztaxon

Generate phylogenetic tree(s) — EzEditor & MEGA

FIGURE 2

Overall workflow from Sanger DNA sequence data to phylogenetic analysis using EzTaxon and other tools.

Step 4: Simply click [Identify] to run the EzTaxon search with the EzTaxon-e option. Please note that the sequence obtained at this stage is likely to contain many errors and should be edited later; hence, identification results at this stage should not be considered final.

Step 5: Go to the "Results" page, move to the detailed identification page of the query sequence and click on [EzEditor file] to obtain the EzEditor data file. This file includes sequences of top hits (EzTaxon's reference sequences) as well as the query sequence.

The EzTaxon server also provides an assembly function for multiple *abl* files that were generated from a single PCR amplicon.

For multiple abl files:

Step 1: Compress all *abl* files into a zip file. The zip format is an archive file format that supports lossless data compression. Software tools for zip-archiving are freely available (e.g. from http://www.7-zip.org/) or included in native computer operating systems.

Step 2: Go to the "Assemble" page of the EzTaxon site (http://www.ezbiocloud.net/eztaxon/assemble).

Step 3: Select a zip file (from Step 1) and upload.

Step 3: EzTaxon will assemble multiple overlapping *abl* files into a single contig. The final contig is trimmed by PHRED quality scores and both PCR primers are also removed.

Step 4: Download the EzEditor data file which includes sequences of top hits as well as the final contig sequence and the original Sanger sequencing reads.

3.5 MANUAL EDITING OF SEQUENCES USING THE SECONDARY STRUCTURE INFORMATION

The sequences obtained from *abl* files may contain errors and should be carefully edited prior to any subsequent analysis. Checking individual bases manually in each chromatogram is a tedious and laborious task. It makes sense, therefore, to focus on areas of a sequence that may: (i) be inconsistent among different sequencing reactions; (ii) show unusual base differences from its phylogenetic neighbours; or (iii) exhibit an inadequate secondary structure, i.e., mismatches in rRNA hairpin stem positions.

There are several software tools that can be used to view and browse chromatograms from *abl* files. Among the freely available ones, the Chromas Lite (http://technelysium.com.au/?page_id=13) and Bioedit (http://www.mbio.ncsu.edu/bioedit/bioedit.html) programs provide sufficient functionality for browsing chromatograms near bases of interest. When there is an ambiguity in determining a base in the final sequence, examining each chromatogram provides the critical data for users to make decisions.

EzEditor is a software tool that allows comparison of multiple alignments while visualizing rRNA secondary structures (Jeon et al., 2014). It is an improved version of jPHYDIT software (Jeon et al., 2005) and is available at http://www.ezbiocloud.net/sw/ezeditor. In our laboratory, EzEditor and Chromas Lite are simultaneously used to correct errors in assembled contigs. The current pipeline (Figure 3) assumes that the 16S rRNA genes were amplified using the 27f and 1492r primers (Table 1), and that five or six sequencing reactions were carried out for each sequence using the primers given in Table 1.

3.5.1 Manual editing of sequences with EzEditor

Step 1: Open the downloaded EzEditor data file using the EzEditor program, and the *abl* chromatogram file(s) using the Chormas Lite program.

Step 2: Open the "Align Window" of EzEditor. The multiple sequence alignment screen will be displayed where sequences of phylogenetic neighbouring species are listed in ascending order of sequence similarity to the query sequence, and the unaligned query sequence is placed in the last row (Figure 3).

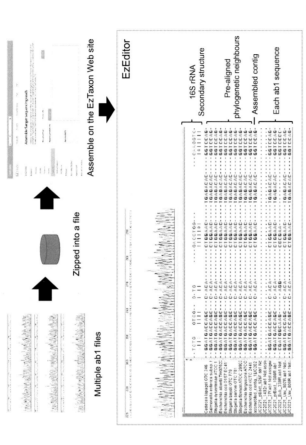

Multiple ab1 files

Zipped into a file

Assemble on the EzTaxon Web site

EzEditor

16S rRNA
Secondary structure

Pre-aligned
phylogenetic neighbours

Assembled contig

Each ab1 sequence

Edit the assembled contig by comparing phylogenetic neighbours, original sequencing reads and the secondary structure

Final 16S rRNA sequence

16S rRNA sequence similarity search with EzTaxon

Phylogenetic analysis using the EzEditor and MEGA programs

FIGURE 3

Editing and correcting contig sequences using EzEditor and Chromas Lite. (See the colour plate.)

Step 3: The query sequence is aligned against the top hit sequence (placed at the right top row in the graphics) using the "Pairwise Alignment" function (Ctrl+P key) in EzEditor. For this special function, gaps are inserted into the query sequence in order to add it to the prealigned format of EzTaxon's 16S rRNA database. Also, align the sequences of the original *abl* files against the assembled contig using the same function.

Step 4: Correct and edit the alignment of the query sequence while checking the integrity of the secondary structure and consider the sequence differences between the query sequence and those of its phylogenetic neighbours. Consult the original chromatograms (from *abl* files), if necessary; manual inspection of chromatograms is often the most reliable way forward.

Step 5: The final sequence after editing and correction is obtained from EzEditor using the "Copy as FASTA" function. This can be fed to the EzTaxon search.

3.6 IDENTIFICATION OF STRAINS USING THE EZTAXON SERVER

Sequence similarity is a simple, but powerful, statistic in prokaryotic identification. In theory, even 100% similarity against known species does not guarantee correct identification (Fox, Wisotzkey, & Jurtshuk, 1992). However, in routine laboratories, high similarities (e.g. >99%) are considered as correct identifications. In a recent large-scale study, 98.65% was recommended as the species boundary cutoff for 16S rRNA gene sequence similarities (Kim et al., 2014); this means that if two sequences show similarity levels lower than this value, they are likely to belong to different species. An EzTaxon search is thus a valuable way to determine whether a sequence is likely to have been derived from a putative novel taxon.

Because of the lack of 16S rRNA gene sequence variation within species and among closely related species, the top best hits from the EzTaxon search need to be carefully interpreted. For example, let us assume that the EzTaxon search of an isolate results in a 99.7% similarity to species A and a 99.6% to species B. The top hit (species A) shows the highest similarity, but in this case, identification cannot be confidently made as species B is also very closely related to the query sequence. In many cases such as this, species A and B share identical or almost identical 16S rRNA gene sequences. To resolve this problem, the concept of the "taxonomic group" has been introduced. This term refers to a group of species that show very similar sequence similarities (>99.7%). Currently defined taxonomic groups can be accessed at http://www.ezbiocloud.net/eztaxon/taxonomic_group. A typical example is the case of *Escherichia coli* and *Shigella* spp. which share almost identical 16S rRNA gene sequences. The "*E. coli* taxonomic group" contains *E. coli*, *Escherichia albertii*, *Shigella boydii*, *Shigella dysenteriae*, *Shigella flexneri* and *Shigella sonnei*. Query sequences that show high similarity to any of these species will be identified as belonging to the *E. coli* taxonomic group, but not as a particular species (e.g. *S. sonnei*). The way in which taxonomic groups are displayed in an EzTaxon search is shown in Figure 4.

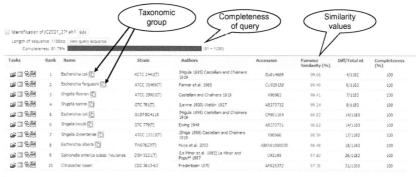

FIGURE 4

"Taxonomic group" shown as the identification result of the EzTaxon search. (See the colour plate.)

3.7 PHYLOGENETIC ANALYSIS

Phylogenetic analyses can be achieved for multiple sequence alignments stored in the EzEditor file. Within the EzEditor program, users can mask ambiguously aligned regions for subsequent analysis and run the MEGA software (Tamura et al., 2011), which includes the neighbour-joining (Saitou & Nei, 1987), maximum-parsimony (Fitch, 1972) and maximum-likelihood (Felsenstein, 1981) methods for generating phylogenetic trees.

CONCLUDING REMARKS

Thanks to the revolutionary next-generation DNA sequencing technology, genome sequencing has become more affordable and within reach of many microbiologists (Padmanabhan, Mishra, Raoult, & Fournier, 2013). However, its use in the identification of *Bacteria* and *Archaea* is greatly hampered by the lack of available reference genome data representing validly named species (Chun & Rainey, 2014). It is expected to take several years or more until a comprehensive database of genome sequences of all known species is built for routine use of genome data in microbial systematics. Until then, 16S rRNA gene sequencing will play a key role as the first choice method for identification in many microbiological disciplines due to its universality and technical easiness. The workflow proposed in this chapter provides a seamless and semi-automated way of performing similarity-based identification from raw Sanger sequencing data and should be useful in many clinical and general microbiology laboratories where prokaryotes are routinely isolated and identified.

ACKNOWLEDGEMENT

We thank Jenny Tan for editing the chapter.

REFERENCES

Altschul, S. F., Gish, W., Miller, W., Myers, E. W., & Lipman, D. J. (1990). Basic local alignment search tool. *Journal of Molecular Biology, 215*, 403–410.

Ashelford, K. E., Chuzhanova, N. A., Fry, J. C., Jones, A. J., & Weightman, A. J. (2005). At least 1 in 20 16S rRNA sequence records currently held in public repositories is estimated to contain substantial anomalies. *Applied and Environmental Microbiology, 71*, 7724–7736.

Baker, G., Smith, J., & Cowan, D. A. (2003). Review and re-analysis of domain-specific 16S primers. *Journal of Microbiological Methods, 55*(3), 541–555.

Chun, J., Lee, J. H., Jung, Y., Kim, M., Kim, S., Kim, B. K., et al. (2007). EzTaxon: A web-based tool for the identification of prokaryotes based on 16S ribosomal RNA gene sequences. *International Journal of Systematic and Evolutionary Microbiology, 57*, 2259–2261.

Chun, J., & Rainey, F. A. (2014). Integrating genomics into the taxonomy and systematics of the *Bacteria* and *Archaea*. *International Journal of Systematic and Evolutionary Microbiology, 64*, 316–324.

Ewing, B., & Green, P. (1998). Base-calling of automated sequencer traces using phred. II. Error probabilities. *Genome Research, 8*, 186–194.

Felsenstein, J. (1981). Evolutionary trees from DNA sequences: A maximum likelihood approach. *Journal of Molecular Evolution, 17*, 368–376.

Fierer, N., Hamady, M., Lauber, C. L., & Knight, R. (2008). The influence of sex, handedness, and washing on the diversity of hand surface bacteria. *Proceedings of the National Academy of Sciences of the United States of America, 105*, 17994–17999.

Fitch, W. M. (1972). Towards defining the course of evolution: Minimum change for a specific tree topology. *Systematic Zoology, 20*, 406–416.

Fox, G. E., Wisotzkey, J. D., & Jurtshuk, P. J. (1992). How close is close: 16S rRNA sequence identity may not be sufficient to guarantee species identity. *International Journal of Systematic Bacteriology, 42*, 166–170.

Gevers, D., Cohan, F. M., Lawrence, J. G., Spratt, B. G., Coenye, T., Feil, E. J., et al. (2005). Re-evaluating prokaryotic species. *Nature Reviews Microbiology, 3*, 733–739.

Inagaki, F., Nunoura, T., Nakagawa, S., Teske, A., Lever, M., Lauer, A., et al. (2006). Biogeographical distribution and diversity of microbes in methane hydrate-bearing deep marine sediments on the Pacific Ocean Margin. *Proceedings of the National Academy of Sciences of the United States of America, 103*, 2815–2820.

Jeon, Y. S., Chung, H., Park, S., Hur, I., Lee, J. H., & Chun, J. (2005). jPHYDIT: A JAVA-based integrated environment for molecular phylogeny of ribosomal RNA sequences. *Bioinformatics, 21*, 3171–3173.

Jeon, Y. S., Lee, K., Park, S. C., Kim, B. S., Cho, Y. J., Ha, S. M., et al. (2014). EzEditor: A versatile sequence alignment editor for both rRNA- and protein-coding genes. *International Journal of Systematic and Evolutionary Microbiology, 64*, 689–691.

Kim, O. S., Cho, Y. J., Lee, K., Yoon, S. H., Kim, M., Na, H., et al. (2012). Introducing EzTaxon-e: A prokaryotic 16S rRNA gene sequence database with phylotypes that represent uncultured species. *International Journal of Systematic and Evolutionary Microbiology, 62*, 716–721.

Kim, M., Oh, H. S., Park, S. C., & Chun, J. (2014). Towards a taxonomic coherence between average nucleotide identity and 16S rRNA gene sequence similarity for species demarcation of prokaryotes. *International Journal of Systematic and Evolutionary Microbiology, 64*, 346–351.

Lane, D. (1991). 16S/23S rRNA sequencing. In E. Stackebrandt & M. Goodfellow (Eds.), *Nucleic acid techniques in bacterial systematics* (pp. 115–175). Chichester: John Wiley & Sons.

Lapage, S. P., Sneath, P. H. A., Lessel, E. F., Skerman, V. B. D., Seeliger, H. P. R., & Clark, A. C. (1992). *International Code of Nomenclature of Bacteria.* Washington, DC: ASM Press.

Liu, W.-T., Marsh, T. L., Cheng, H., & Forney, L. J. (1997). Characterization of microbial diversity by determining terminal restriction fragment length polymorphisms of genes encoding 16S rRNA. *Applied and Environmental Microbiology, 63*, 4516–4522.

Logan, N. A., Berge, O., Bishop, A. H., Busse, H. J., De Vos, P., Fritze, D., et al. (2009). Proposed minimal standards for describing new taxa of aerobic, endospore-forming bacteria. *International Journal of Systematic and Evolutionary Microbiology, 59*, 2114–2121.

Muyzer, G., De Waal, E. C., & Uitterlinden, A. G. (1993). Profiling of complex microbial populations by denaturing gradient gel electrophoresis analysis of polymerase chain reaction-amplified genes coding for 16S rRNA. *Applied and Environmental Microbiology, 59*, 695–700.

Myers, E. W., & Miller, W. (1988). Optimal alignments in linear space. *Computer Applications in the Biosciences, 4*(1), 11–17.

Nübel, U., Engelen, B., Felske, A., Snaidr, J., Wieshuber, A., Amann, R. I., et al. (1996). Sequence heterogeneities of genes encoding 16S rRNAs in *Paenibacillus polymyxa* detected by temperature gradient gel electrophoresis. *Journal of Bacteriology, 178*(19), 5636–5643.

Padmanabhan, R., Mishra, A. K., Raoult, D., & Fournier, P. E. (2013). Genomics and metagenomics in medical microbiology. *Journal of Microbiological Methods, 95*(3), 415–424.

Park, K. S., Ki, C. S., Kang, C. I., Kim, Y. J., Chung, D. R., Peck, K. R., et al. (2012). Evaluation of the GenBank, EzTaxon, and BIBI services for molecular identification of clinical blood culture isolates that were unidentifiable or misidentified by conventional methods. *Journal of Clinical Microbiology, 50*, 1792–1795.

Saitou, N., & Nei, M. (1987). The neighbor-joining method: A new method for reconstructing phylogenetic trees. *Molecular Biology and Evolution, 4*, 406–425.

Stackebrandt, E., & Ebers, J. (2006). Taxonomic parameters revisited: Tarnished gold standards. *Microbiology Today, 33*, 152–155.

Stackebrandt, E., & Goebel, B. M. (1994). Taxonomic note: A place for DNA-DNA reassociation and 16S rRNA sequence analysis in the present species definition in bacteriology. *International Journal of Systematic Bacteriology, 44*, 846–849.

Tamura, K., Peterson, D., Peterson, N., Stecher, G., Nei, M., & Kumar, S. (2011). MEGA5: Molecular evolutionary genetics analysis using maximum likelihood, evolutionary distance, and maximum parsimony methods. *Molecular Biology and Evolution, 28*(10), 2731–2739.

Tindall, B. J., Rossello-Mora, R., Busse, H. J., Ludwig, W., & Kampfer, P. (2010). Notes on the characterization of prokaryote strains for taxonomic purposes. *International Journal of Systematic and Evolutionary Microbiology, 60*, 249–266.

Wang, Y., & Qian, P.-Y. (2009). Conservative fragments in bacterial 16S rRNA genes and primer design for 16S ribosomal DNA amplicons in metagenomic studies. *PLoS One, 4*(10), e7401.

Wayne, L. G., Brenner, D. J., Colwell, R. R., Grimont, P. A. D., Kandler, O., Krichevsky, M. I., et al. (1987). Report of the *ad hoc* committee on reconciliation of approaches to bacterial systematics. *International Journal of Systematic Bacteriology, 37*, 463–464.

Woese, C. R. (1987). Bacterial evolution. *Microbiological Reviews, 51*(2), 221–271.

Woese, C. R., & Fox, G. E. (1977). Phylogenetic structure of the prokaryotic domain: The primary kingdoms. *Proceedings of the National Academy of Sciences of the United States of America, 74*, 5088–5090.

Revolutionizing Prokaryotic Systematics Through Next-Generation Sequencing

Vartul Sangal*, Leena Nieminen[†], Nicholas P. Tucker[†], Paul A. Hoskisson[†,1]

*Faculty of Health and Life Sciences, Northumbria University, Newcastle upon Tyne,
United Kingdom
[†]Strathclyde Institute of Pharmacy and Biomedical Sciences, University of Strathclyde, Glasgow,
United Kingdom
[1]Corresponding author: e-mail address: paul.hoskisson@strath.ac.uk

1 INTRODUCTION

Micro-organisms are ubiquitous in nature, inhabiting nearly every niche on the planet. Some have been exploited for their industrial potential, such as those involved in food production and the production of bioactive molecules (e.g. antibiotics), a relatively small number have been shown to be pathogenic to humans and animals, while the majority of organisms have vital ecological roles in terms of recycling nutrients and biogeochemical processes on a local and global stage. A unifying approach to understanding all of these organisms is to determine their phylogeny and, in this context, the reliable and stable classification and identification of bacteria are pivotal to discriminating between ecologically and phylogenetically distinct organisms.

Micro-organisms are primarily identified using biochemical and morphological properties followed by the application of genetic approaches. Genotyping approaches such as pulse field gel electrophoresis score whole genomic variation and are extensively used in epidemological studies and the characterization of outbreaks caused by pathogenic bacteria (Goering, 2010; Kaufmann, 1998). However, nucleotide sequence analyses of selectively neutral genes that are unlikely to be under positive selection, such as those involved in metabolic activities, are useful for inferring strain relatedness in global terms (Maiden et al., 1998). Multilocus sequence typing (MLST) is based on variation in fragments of seven housekeeping genes and has been widely used for studying evolution and population genetics in a variety of organisms (Maiden, 2006; chapter 'Multi-Locus Sequence Typing and the Gene-by-Gene Approach to Bacterial Classification Analysis of Population Variation' by Cody et al., and 'Multi-locus Sequence Analysis: Taking Prokaryotic Systematics to the Next Level' by Rong and Huang, in this volume). However,

Methods in Microbiology, Volume 41, ISSN 0580-9517, http://dx.doi.org/10.1016/bs.mim.2014.07.001

MLST suffers from low resolution for strain differentiation and lacks a proper inference of the evolution of virulence as only seven neutral loci are analysed.

The sequencing of entire genomes has the potential to answer most of the key questions related to evolution, global transmission and local adaptation in bacteria. The first bacterial genome sequenced was that of *Haemophilus influenzae* (Fleischmann et al., 1995), but sequencing genomes at that time was time and cost intensive. More recently, next-generation sequencing technology has revolutionized genome sequencing making it high throughput and economic, resulting in an exponential increase in the number of publicly available genomes (http://www.ncbi.nlm. nih.gov/genomes/MICROBES/microbial_taxtree.html; http://genomesonline.org). The development of advanced bioinformatics tools has helped further understanding of evolution and the local and global transmission of several pathogenic bacteria (Harris et al., 2010; Holt et al., 2012; chapter 'Whole-Genome Sequencing for Rapid and Accurate Identification of Bacterial Transmission Pathways' by Harris and Okoro, in this volume); these tools have also allowed the functional characterization of genome-wide variation. This chapter provides a brief summary of available next-generation sequencing platforms with an overview of bioinformatics tools available to handle the sequence data.

2 SEQUENCING APPROACHES

A number of sequencing platforms are available that use different chemistries and detection methods. Genome sequencing on most of the platforms involves library preparation, immobilization and sequencing. However, the read lengths, error rates, throughput and run times are highly variable (Loman et al., 2012). The most commonly used sequencing platforms are those of Roche (GS FLX and GS Junior), Illumina (HiSeq 2000/2500 and MiSeq) and Life Technologies (Ion Proton and Ion PGM). Other technologies focussed on single-molecule sequencing, including Pacific Biosciences (PacBio RS) and Oxford Nanopore Technologies, have been developed and offer great advances in read length. PacBio RS is available to users, but Oxford Nanopore Technologies is still under development in order to make it cost effective and high throughput in comparison to other available platforms.

For library preparation, DNA is extracted from bacterial cells and is fragmented using hydrodynamic shearing, enzymatic digestion, nebulization or sonication with different methods being favoured by different platforms (Loman et al., 2012; Myllykangas, Buenrostro, & Ji, 2012). The fragments are end-repaired to fill any gaps or nicks followed by adaptor ligation. The adaptors contain nucleotide sequences that are required for immobilization, amplification and sequencing of the library. Fragments of the preferred size are selected and used for the amplification and sequencing steps.

For Roche 454 genome sequencers, the library is immobilized on streptavidin coated beads and amplified in an oil–water emulsion PCR (Margulies et al., 2005). The beads are collected and loaded in the wells on a picotitre plate, one bead per well. Roche 454 genome sequencers follow the sequencing-by-synthesis approach using pyrosequencing chemistry for massively parallel sequencing. DNA polymerase incorporates nucleotides to the template library DNA on the bead that releases a pyrophosphate which is subsequently converted into visible light by ATP sulfurylase and luciferase enzymes (Margulies et al., 2005; Ronaghi, Uhlen, & Nyren, 1998). The nucleotides are sequentially flushed over the picotitre plate in each sequencing cycle, and the emitted light is recorded by a CCD camera and converted into the nucleotide sequence by capturing software. Although Roche 454 genome sequencers produce some of the longest sequence reads, they are relatively more cost intensive than other platforms; Roche recently announced a phase-out of their 454 sequencers and application support by mid-2016.

For Illumina sequencing, libraries are immobilized on the surface of a flow cell with an ultra-dense primer field on the inner surface (Bentley et al., 2008). These primers hybridize with the adaptor sequence of the immobilized library which acts as the template for amplification. Fluorescently labelled reversible terminator nucleotides (with protected 3′-OH groups) are added to the flow cell which extends the sequences and the fluorophores are imaged (Bentley et al., 2008; Loman et al., 2012; Myllykangas et al., 2012). The fluorophores are quenched and the reversible terminator moieties are chemically removed before the next sequencing cycle.

Life Technologies' Ion Torrent instruments work on a sequencing-by-synthesis approach using a semiconductor technology where the library is immobilized on beads called "ion spheres" and loaded onto a semiconductor chip with millions of sensors that detect hydrogen ions as they are released during nucleotide incorporation (Rothberg et al., 2011). A lack of sophisticated optics makes it much cheaper than other available platforms that rely upon optical signals to record sequences.

Life Technologies also provide another sequencing platform known as SOLiD, which sequences by oligonucleotide ligation and detection. The libraries of the appropriate size range are ligated to universal adaptors P1 and P2 that are immobilized onto P1-coated magnetic beads during emulsion PCR (Myllykangas et al., 2012; Valouev et al., 2008). The beads are enriched after PCR amplification and the 3′-end of the templates are modified for attachment to glass slides. The P1 adaptors are hybridized with a sequencing primer and a pool of labelled dinucleotide probes is applied. The complementary probes are ligated to the primers which are imaged, followed by the capping and cleavage of the fluorophores (Valouev et al., 2008). After several cycles of ligation, imaging and cleavage, the extension product is removed to reset the template. The latter is subsequently hybridized with a second template starting at the $n-1$ position for a second round of ligation. Five sequencing primers starting with n, $n-1$, $n-2$, $n-3$, $n-4$ and $n-5$ positions are used, and an approximately 35 bp insert is sequenced twice for >99% accuracy (Valouev et al., 2008). Life Technologies has recently released a newer version of the SOLiD genome sequencer (5500 W series) that has increased speed and economy. Moreover, libraries do not need immobilization onto magnetic beads and are directly amplified on the FlowChip.

Pacific Biosciences has recently developed a single-molecule real-time sequencing system (Eid et al., 2009). The library is ligated to adaptors that form an internal hairpin structure and are immobilized onto streptavidin coated zero-mode waveguide arrays (Levene et al., 2003). The sequencing is performed by synthesis where fluorescent nucleotides are incorporated on the growing strand by Φ29 DNA polymerase. The Φ29 DNA polymerase is a highly processive strand-displacing enzyme capable of rolling circle amplification and allows high-resolution imaging of the fluorescent nucleotide as it is being incorporated (Eid et al., 2009).

More recently, Oxford Nanopore Technologies developed an approach that exploits strand sequencing by passing DNA libraries through protein nanopores inserted into a synthetic polymer membrane (Loman et al., 2012; Myllykangas et al., 2012). DNA fragments form a complex with a processive enzyme that forces the DNA, a nucleotide at a time, from a single strand through the nanopore. A potential is applied to the membrane, and the disruption by the passing nucleotide is detected and decoded by the software. This technology is powerful and offers rapid bioinformatic analysis that may eventually supersede other sequencing platforms.

3 BIOINFORMATIC ANALYSES

Sequencing platforms generate reads that are generally filtered by their quality scores, using in-built software. Low-quality reads must be removed from downstream processing and analyses due to problems, such as wrongly called bases, which could introduce bias in the downstream analysis. Different platforms provide reads in different formats, for example, Roche 454 instruments store reads in a binary standard flowgram format (SFF format), whereas Illumina reads are generally in a text-based FASTQ format. Some programs accept all of these formats, but sometimes, additional bioinformatic scripts may be required for the format conversion for analysis by a specific program. Galaxy is a Web-based suite that is widely used for analysis of next-generation sequencing data (Blankenberg et al., 2010); several tools are present in the Galaxy tool shed that can interactively convert between various file formats for further processing. The primary reads from the sequencer also carry the adaptor sequences that were used for the sequencing reaction, and these need to be removed before processing; this feature is generally available in the programs used for assembly or mapping of the reads where a user can define the adaptor sequences to be trimmed during analysis. Next-generation sequencing produces enormous amounts of data that need to be handled and analysed using advanced bioinformatic tools. A wide range of programs have been developed by academic and commercial suppliers to make the analyses more user-friendly and more economic in terms of both time and computational resource. Here, we provide a brief description of some commonly used programs used for analyses of next-generation sequencing data. The programs and online resources described in this chapter have been summarized with links to their Web sites, as shown in Table 1.

3.1 *DE NOVO* ASSEMBLY AND MAPPING

A variety of programs are available to assemble reads from different platforms into larger contigs. Newbler (also known as GS *de novo* assembler) was developed by 454 Life Sciences to assemble long reads produced by the Roche 454 platform that allows assembly of reads in SFF, FASTA and FASTQ formats. Newbler runs on the Linux/

Table 1 A List of Major Software and Online Resources with Their Web Links

Software and Online Resources	Web site
Alien Hunter	http://www.sanger.ac.uk/resources/software/alien_hunter/
AMOS package	http://sourceforge.net/apps/mediawiki/amos/index.php?title=AMOS
ANNOVAR	http://www.openbioinformatics.org/annovar/
Artemis	http://www.sanger.ac.uk/resources/software/artemis/
Artemis comparison tool	http://www.sanger.ac.uk/resources/software/act/
BioCyc database	BioCyc.org
BLAST	http://blast.ncbi.nlm.nih.gov/Blast.cgi?
CARMEN	http://carmen.cebitec.uni-bielefeld.de/cgi-bin/index.cgi
CGView comparison tool	http://stothard.afns.ualberta.ca/downloads/CCT/
CLC Genomic Workbench (Qiagen)*	http://www.clcbio.com/products/clc-genomics-workbench/
ClonalFrame	http://www.xavierdidelot.xtreemhost.com/clonalframe.htm
COG database	http://www.ncbi.nlm.nih.gov/COG/
CRITICA	http://www.life.illinois.edu/gary/programs/CRITICA.html
EDGAR	https://edgar.computational.bio.uni-giessen.de
Galaxy	https://main.g2.bx.psu.edu/
GeneMark	http://opal.biology.gatech.edu/
GLIMMER	http://ccb.jhu.edu/software/glimmer/index.shtml
GS *de novo* Assembler/GS Reference Mapper*	http://454.com/products/analysis-software/index.asp
Genome-to-Genome Distance Calculator	http://ggdc.dsmz.de/
IS Finder	https://www-is.biotoul.fr//
Island Viewer	http://www.pathogenomics.sfu.ca/islandviewer/query.php
Jspecies	http://www.imedea.uib.es/jspecies/
KEGG database	http://www.genome.jp/kegg/
Lasergene Genomics Suite (DNASTAR)*	http://www.dnastar.com/t-products-dnastar-lasergene-genomics.aspx
Maq	http://maq.sourceforge.net/
Mauve	http://gel.ahabs.wisc.edu/mauve/
MEGA	http://www.megasoftware.net/
MetaCyc database	MetaCyc.org
MIRA	http://www.chevreux.org/projects_mira.html
MyTaxa	http://enve-omics.ce.gatech.edu/mytaxa/
NCBI Prokaryotic Genome Annotation Pipeline	https://www.ncbi.nlm.nih.gov/genome/annotation_prok/

Continued

Table 1 A List of Major Software and Online Resources with Their Web Links—cont'd

Software and Online Resources	Web site
NGS-SNP	http://stothard.afns.ualberta.ca/downloads/NGS-SNP/
Pan-seq	http://lfz.corefacility.ca/panseq/
Pfam database	http://pfam.sanger.ac.uk/
Phage Finder	http://phage-finder.sourceforge.net/
PHAST	http://phast.wishartlab.com/
PHYLIP	http://evolution.genetics.washington.edu/phylip.html
PhyloPhlAn	http://huttenhower.sph.harvard.edu/phylophlan
PRODIGAL	http://prodigal.ornl.gov/
RAST	http://rast.nmpdr.org/
REGANOR	https://www.cebitec.uni-bielefeld.de/groups/brf/software/reganor/cgi-bin/reganor_upload.cgi
SAMtools	http://samtools.sourceforge.net/
SEED Viewer	http://seed-viewer.theseed.org/
Sequencher (Gene Codes Corporation)*	http://genecodes.com/
Tendem Repeat Finder	http://tandem.bu.edu/trf/trf.html
TETRA	http://www.megx.net/tetra/index.html
TRAMS	http://dx.doi.org/10.6084/m9.figshare.782261
UniProt	http://www.uniprot.org/
VAT	http://vat.gersteinlab.org/index.php
Velvet	http://www.ebi.ac.uk/~zerbino/velvet/
xBASE	http://www.xbase.ac.uk/

Note: 1. All commercially available software are marked with a star (*). GS de novo Assembler/GS Reference Mapper may be obtained free from Roche.
2. Please see text for citation of these programs and online resources.

Unix platform with an interactive user interface. It generates contigs by aligning reads according to user-defined overlap length and minimum alignment percentage. The assemblies are produced in FASTA, QUAL, ACE, AGP and BAM formats which can be further analysed by a whole suite of freely available and commercial bioinformatics software. The assembly parameters can be adjusted for the shorter reads generated by other platforms such as Illumina and Ion Torrent. Additional programs such as MIRA (Chevreux, Wetter, & Suhai, 1999) and Velvet (Zerbino & Birney, 2008) are also available to assemble short sequences. These programs also run on Linux/Unix machines using command lines but do not have a graphical user interface.

Draft *de novo* assemblies often contain gaps that appear due to difficulties in assembling repeat regions and complications due to sequencing bias. In order to gain complete genomes, gap closure between adjacent contiguous genomic regions (contigs) may be performed either *in silico* or through experimental determination using PCR (Nagarajan et al., 2010). An example of *in silico* gap closure of a bacterial genome sequence from Ion Torrent PGM is the use of two different *de novo* assemblies of the same sequencing data set (Nieminen & Tucker, unpublished data). The first assembly was carried out using MIRA (Chevreux et al., 1999), the software recommended by Life Technologies to assemble Ion Torrent PGM reads, resulting in 891 contigs of 6,145,564 bp total size. The second assembly was performed using CLC Genomic Workbench (Qiagen) that resulted in 376 contigs of >1 kb size with an average of 60-fold coverage. The contigs with <40-fold or >80-fold coverage (average coverage ±33%) were excluded; the total size of this assembly was 5,502,571 bp. The two assemblies were then merged using the program Minimus2 assembler from the AMOS package (Sommer, Delcher, Salzberg, & Pop, 2007) which uses a nucmer-based overlap detector. The final assembly contained 202 contigs, the total length of the final assembly was 5,331,091 bp (Nieminen & Tucker, unpublished data). This approach provides a powerful and very rapid improvement of genome assembly, without extensive "wet" experiments, that can be time consuming.

GS Mapper (Roche) can align the reads obtained from a sequencing run to a reference sequence and generates contigs according to user-defined criteria. GS Mapper can interactively map a variety of sequences onto reference genomes. The input and output formats are the same as those described for the Newbler program. Reference mapping is really useful for resequencing projects for organisms where genomic rearrangements are limited. Another widely used program is Maq (Li, Ruan, & Durbin, 2008) which was developed to map Illumina reads onto a reference genome. Maq also extracts a list of variations including single nucleotide polymorphisms (SNPs) and insertions and deletions (in–dels) based on the reference genome. The mapped alignments can be converted into SAM and BAM formats using SAMtools (Li et al., 2009) that will allow manual inspection using other programs, including Artemis (Rutherford et al., 2000) which can also be used for manual annotation tasks (Figure 1).

A number of other software packages including CLC Genomic Workbench (Qiagen), Lasergene Genomics Suite (DNASTAR) and Sequencher (Gene Codes Corporation), are now commercially available that can perform *de novo* assembly and mapping to reference sequences on reads from all sequencing platforms.

3.2 ANNOTATION

The power of genomics lies in its comparative aspects, and a prerequisite for this is typically the annotation of all genes and DNA features in the assembled genomes/contigs. Annotation is a complex process where genomes are screened for mobile genetic elements and the complements of RNA species (nc-RNA, tRNA and rRNA) followed by the identification of potential open reading frames (ORFs) by a gene prediction program. Several different programs can be used for gene predictions, such as PRODIGAL (Prokaryotic Dynamic Programming Genefinding Algorithm;

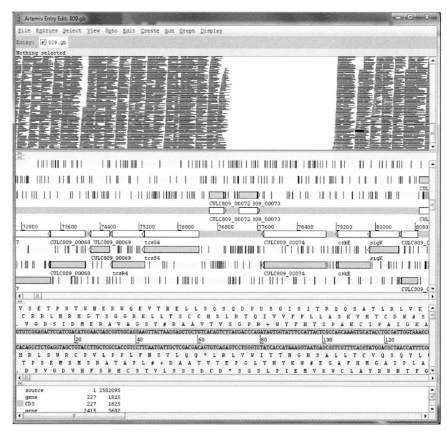

FIGURE 1

A screen snapshot of Artemis with Ion-Torrent reads mapped onto an annotated finished genome. (See the colour plate.)

Hyatt et al., 2010), REGANOR (Linke, McHardy, Neuweger, Krause, & Meyer, 2006), GLIMMER (Delcher, Harmon, Kasif, White, & Salzberg, 1999), CRITICA (Badger & Olsen, 1999) and GeneMark (Borodovsky & McIninch, 1993). The ORFs predicted by these programs can be further characterized for their potential functions using BLAST searches in the UniProt (Swiss-Prot and TrEMBL; http://www.uni prot.org/), Pfam (Finn et al., 2013), KEGG (Kanehisa, 2000) and COG databases (Tatusov et al., 2001). The gene annotations can also provide important insights about the metabolic pathways from information in the KEGG database using programs like CARMEN (Schneider et al., 2010) and SEED Viewer (Aziz et al., 2012). These predictions can be especially useful for defining potentially useful physiological assays on closely related strains. For example, these approaches have been used to develop media for growing organisms that were previously thought to be obligate intracellular parasites (Renesto et al., 2003). Different automated annotation pipelines are available that facilitate genome annotation using a specific or a combination of different gene prediction programs, for example, the NCBI Prokaryotic Genome Annotation Pipeline that primarily uses GeneMark for gene prediction. In contrast, GenDB combines several gene prediction software packages for identification, functional classification and annotation (Meyer et al., 2003). RAST is an extensively used annotation pipeline that assigns protein functions using a database of protein subsystems, FIGfam that was built using contributions from a number of scientists (Aziz et al., 2008).

The program xBASE provides a quick annotation of unfinished genomes using a similar/related finished genome as a reference sequence (Chaudhuri & Pallen, 2006). xBASE uses Glimmer for gene prediction, tRNAScan-SE to identify tRNA genes, RNAmmer to identify rRNA genes and Protein BLAST searches of translated coding sequences against the reference sequence to identify the best-hit according to the defined *E*-value cut-off for the transfer of annotation from the reference to the unfinished genome.

3.3 COMPARATIVE GENOMIC ANALYSIS

Different strains of a species can be expected to be functionally diverse in their metabolic properties. For example, *Corynebacterium diphtheriae* has historically been divided into four biovars based on biochemical differences (Sangal et al., 2014). In addition, some strains may undergo gain or loss of gene functions to adapt to a new environment or host. Bioinformatic tools to compare several genomes can be used to identify the differences in gene content that may correlate with variation in functional characteristics or any potential geographic, host or temporal association. Some annotation tools, such as RAST, also provide the opportunity to run comparative genomic analyses within known functional gene categories or whole genomic bidirectional protein BLAST searches (true orthologue searching) and provide a list of potential genomic differences. xBASE also has a useful function, allowing users to compare unfinished genomes with a specified reference genome (Chaudhuri & Pallen, 2006), producing a comparison file that can be viewed using the Artemis comparison tool (ACT; Carver et al., 2008).

Artemis is a free bioinformatics tool for visualization and manual annotation of genome sequences that has resulted from a variety of genome projects (Carver et al., 2008; Carver, Harris, Berriman, Parkhill, & McQuillan, 2012). ACT can be used to display pairwise comparisons between two or more DNA sequences and can provide a quick overview of the genomic synteny between different strains (Carver et al., 2008). The CGView comparison tool (CCT) is another freely available option that generates DNA and protein BLAST maps of a reference genome against a number of other genomes to allow visualization of the conservation at the nucleotide and protein level (Figure 2; Grant, Arantes, & Stothard, 2012) and has been used for comparing related strains for identification of variation in the gene content (Sangal et al., 2014; Sangal, Fineran, & Hoskisson, 2013; Sangal, Jones, Goodfellow, Sutcliffe, & Hoskisson, 2014).

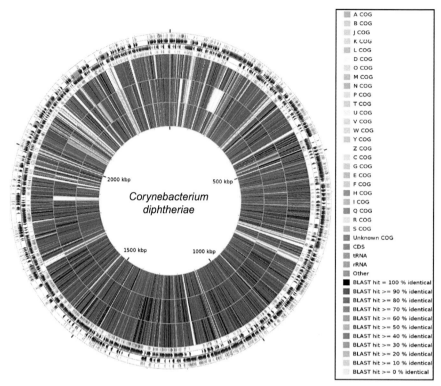

Legend:
- A COG
- B COG
- J COG
- K COG
- L COG
- D COG
- O COG
- M COG
- N COG
- P COG
- T COG
- U COG
- V COG
- W COG
- Y COG
- Z COG
- C COG
- G COG
- E COG
- F COG
- H COG
- I COG
- Q COG
- R COG
- S COG
- Unknown COG
- CDS
- tRNA
- rRNA
- Other
- BLAST hit = 100 % identical
- BLAST hit >= 90 % identical
- BLAST hit >= 80 % identical
- BLAST hit >= 70 % identical
- BLAST hit >= 60 % identical
- BLAST hit >= 50 % identical
- BLAST hit >= 40 % identical
- BLAST hit >= 30 % identical
- BLAST hit >= 20 % identical
- BLAST hit >= 10 % identical
- BLAST hit >= 0 % identical

Corynebacterium diphtheriae — 2000 kbp, 500 kbp, 1500 kbp, 1000 kbp

FIGURE 2

A CD BLAST map of *Corynebacterium diphtheriae* strain NCTC 13129 against NCTC 5011, INCA 402 and NCTC 3529 using the CG View Comparison Tool. (See the colour plate.)

Mauve can align a number of genomes and presents a visual representation of these as identity blocks with potential genomic rearrangements (Figure 3; Darling, Mau, Blattner, & Perna, 2004). Mauve can also generate a list of ortholo-gous genes that are present in all of the genomes and is able to extract an SNP table identifying all polymorphic nucleotides from the alignment. The contigs of unfin-ished genomes can also be reordered onto a closely related finished reference ge-nome using Mauve, thereby aiding in the assembly of genomes.

EDGAR is a powerful Web-based suite of software for performing comparative genomic analyses based on BLAST score ratios (Blom et al., 2009). The program calculates the core and pan genomes from genomic dataset and can also identify the set of genes that is specific to a particular strain or a particular group of strains within the data set. The program can create synteny plots and the sharing of genes can be visualized as a Venn diagram (Figure 4). EDGAR can also generate a phyloge-netic tree from the conserved core genome to study the evolutionary relationships between different strains. All these analyses are performed on a dedicated server and the user can view the results *via* a Web interface.

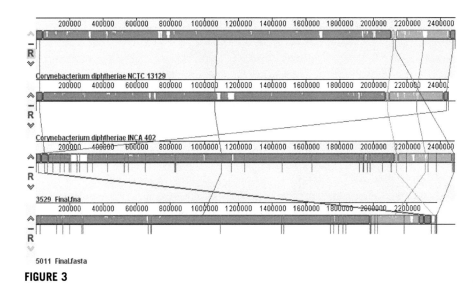

FIGURE 3

A Mauve alignment of *Corynebacterium diphtheriae* finished genomes of strains NCTC 13129 and INCA 402 and draft genomes of strains NCTC 3529 and NCTC 5011.

Pan-seq is another Web-based tool used to calculate core and accessory genomes (Laing et al., 2010). The program is able to extract the SNPs in the core genome and identifies the novel genomic regions that are specific to a strain or a group of strains and hence is of value in systematics.

As well as performing various analyses of NGS data, Galaxy provides a Web-based software suite for comparative and functional genomic analyses using powerful tools developed by several bioinformatics groups (Blankenberg et al., 2010). Galaxy provides a very user-friendly interface for scientists with limited computational background to analyse their genomic data; analysis details can be stored for future use as "pipelines" so that an almost automated approach to analysis can be undertaken, through the linking of common work-flow schemes. Some commercially available software, including CLC genomic workbench (Qiagen) and Lasergene Genomic Suite (DNASTAR), offer a similar capacity for performing comparative genomic analyses.

Another important task in genomic analyses is the identification of mobile genetic elements, such as transposable elements, insertion sequences (IS), repeat sequences and bacteriophage regions, as these elements are responsible for genetic rearrangements and gain/loss of gene function in genomes. These regions can be identified using programs such as IS Finder (Siguier, Perochon, Lestrade, Mahillon, & Chandler, 2006), PHAST (Zhou, Liang, Lynch, Dennis, & Wishart, 2011), Phage Finder (Fouts, 2006) and Tandem Repeat Finder (Benson, 1999). Island Viewer (Langille & Brinkman, 2009) and Alien Hunter (Vernikos & Parkhill, 2006) can be used to identify potentially horizontally acquired genomic islands.

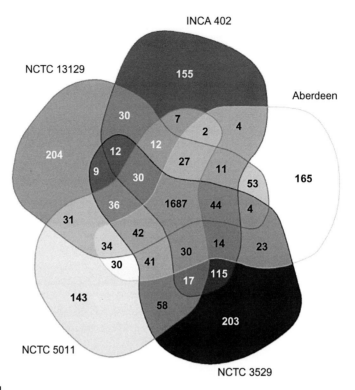

FIGURE 4

A Venn diagram of shared and strain specific genes between *Corynebacterium diphtheriae* strains NCTC 13129, NCTC 5011, INCA 402, NCTC 3529 and "Aberdeen". (See the colour plate.)

A number of comparative genomic tools have been developed recently for delineating species to assist prokaryotic systematists. Average nucleotide identity (see chapter 'Whole-Genome Analyses: Average Nucleotide Identity' by Arahal, in this volume) provides a measurement of genomic relatedness between different strains that can be calculated using the BLAST alignment (Goris et al., 2007) or a maximal unique match-based approach (Deloger, El Karoui, & Petit, 2009). These calculations can be performed using the work package Jspecies (Richter & Rosselló-Móra, 2009). Tetranucleotide frequencies can carry species-specific signatures and so can also be useful for taxonomic identification of prokaryotic species (Pride, Meinersmann, Wassenaar, & Blaser, 2003; van Passel, Kuramae, Luyf, Bart, & Boekhout, 2006). Genomic tetranucleotide usage patterns can be analysed using the program TETRA (Teeling, Waldmann, Lombardot, Bauer, & Glockner, 2004). The tetranucleotide signature correlation index has also been implemented in Jspecies (Richter & Rosselló-Móra, 2009) so that it is possible to perform pairwise comparisons by plotting tetranucleotide frequency for each strain and obtain a regression value. Digital DNA–DNA hybridization will also be helpful in genome sequence-based species delineation. These can be performed using the Web-based genome-to-genome distance calculator GGDC (Auch, von Jan, Klenk, & Göker, 2010; Meier-Kolthoff, Auch, Klenk, & Göker, 2013).

Average amino acid identity provides a robust measurement of strain relatedness and can be integrated into the genome-based taxonomy of prokaryotes (Konstantinidis & Tiedje, 2005b). The program MyTaxa can assist in taxonomic identification of genomic sequences using the genome-aggregate average amino acid identity concept (Luo, Rodriguez, & Konstantinidis, 2014).

3.4 SNP EXTRACTION AND FUNCTIONAL CHARACTERISTICS

The presence or absence of genes does not always correlate with the functional differences observed between bacterial strains. Some of these variations might be the result of SNPs altering functional amino acids in proteins (nonsynonymous SNPs) or the introduction of premature stop codons resulting in the loss of gene function (nonsense SNPs). Resequencing projects where sequencing reads are mapped onto a reference genome and the comparison of several genomes by sequence alignment produce a list of SNPs that can be annotated to assess the functional impact on the strains. Many programs are available for SNP annotation including ANNOVAR (Wang, Li, & Hakonarson, 2010), NGS-SNP (Grant, Arantes, Liao, & Stothard, 2011), VAT (Habegger et al., 2012) and TRAMS (Reumerman, Tucker, Herron, Hoskisson, & Sangal, 2013). Most of these programs were developed for analysing SNPs of human or eukaryotic genomes and need complex installation procedures or require command lines to run, which may limit their use by nonexpert NGS users with limited computational experience. TRAMS (Tool for Rapid Annotation of Microbial SNPs) is designed to annotate microbial SNPs and is available as a single file executable for Microsoft Windows users and as a python script for other computer platforms. TRAMS annotates the genome-wide SNPs as synonymous, nonsynonymous and nonsense, although the nonsynonymous SNPs in start and stop codons are separated as non-start and non-stop SNPs. TRAMS separately annotates SNPs in multiple overlapping features and merges multiple nucleotide polymorphisms within a codon before annotation. TRAMS only needs a simple input file in tab-delimited format including SNP locations, reference nucleotide and SNPs in variant strains along with a reference genome in GenBank or EMBL format. It is also available in the Galaxy tool shed and can be used in the Galaxy workflow for SNP analysis.

3.5 PHYLOGENETIC ANALYSES

Phylogenetic trees represent the evolutionary relatedness between different strains; several programs are available to infer phylogenetic relationships from whole genomic data. ClonalFrame is a program that infers evolutionary relatedness after accounting for recombination (Didelot & Falush, 2007) and accepts genome alignments that are produced by the program Mauve (Darling et al., 2004). EDGAR is also able to generate phylogenetic trees from the conserved core genome following the masking of nonmatching parts of the alignments (Blom et al., 2009). More recently, the program PhyloPhlAn was developed by the Huttenhower laboratory to identify taxonomic and evolutionary relationships between different strains using 400 protein sequences from microbial genomes (Segata, Bornigen, Morgan, & Huttenhower, 2013). Users need to decide which of these trees fit best to the purpose at hand. EDGAR trees show true evolutionary relatedness from the conserved core genome, whereas ClonalFrame uses the entire genomic alignment and highlights the impact of recombination on phylogenetic relationships. PhyloPhlAn derives phylogenies from conserved protein sequences and is designed to resolve taxonomic groupings for prokaryotes.

Pan-Seq provides SNPs in the core genome in phylip format that can be analysed by a phylogenetic package including PHYLIP (Retief, 2000). The genome alignments or SNP data can be formatted for a variety of phylogenetic packages by users. For example, MEGA (Tamura et al., 2011) can analyse nucleotide and protein sequence alignments by a variety of approaches.

4 APPLICATIONS OF NEXT-GENERATION SEQUENCING TECHNOLOGY

Next-generation sequencing has been widely applied in life science research. From the prokaryotic perspective, it has been very useful in studying the evolution, biogeographic and temporal associations of micro-organisms, particularly in relation to pathogen evolution. In this section, we summarize some of the important applications of next-generation sequencing in prokaryotic research.

4.1 PROKARYOTIC SYSTEMATICS

Prior to the introduction of 16S rRNA gene sequence analyses in the 1980s, the classification and identification of prokaryotes were based on phenotypic properties and DNA–DNA hybridization (Vandamme et al., 1996). The 70% DNA:DNA cut-off point recommended for assigning strains to the same species (Wayne et al., 1987) is now considered to correspond to 16S rRNA gene similarities of 98.7–99.0% (Kim, Oh, Park, & Chun, 2014; Stackebrandt & Ebers, 2006). These measures have provided key criteria for the delineation of prokaryotic species although well-known issues have been associated with them (Achtman & Wagner, 2008) which have led to calls for the concepts and practices used in prokaryotic systematics to be revisited (Sangal & Hoskisson, 2013; Sutcliffe, Trujillo, & Goodfellow, 2012; Vandamme & Peeters, 2014). Next-generation sequencing provides a way out of such dilemmas as genome sequences index the entire micro-variation and hence can help to identify the most reliable boundaries between species and other operational taxonomic units. It has been suggested that average nucleotide identity between genomes (Konstantinidis, Ramette, & Tiedje, 2006; Konstantinidis & Tiedje, 2005a) and average nucleotide identity between orthologous genes (Sentausa & Fournier, 2013) should become part of the current polyphasic classification system (Chun & Rainey, 2014). However, the pervasive nature of horizontal gene transfer and recombination between prokaryotes can introduce bias when using variation in gene contents for defining species and hence robust approaches are needed to counter these problems. To this end, Chen and his colleagues applied genomics to assign *Acinetobacter* strains into species (Chan, Halachev, Loman, Constantinidou, & Pallen, 2012) as 16S rRNA sequence data lacked sufficient resolution to do so. Goris et al. (2007) found that a combination of core genome phylogeny and average nucleotide identity are taxonomic metrics that have the potential to replace traditional and technically demanding DNA:DNA hybridization procedures.

"*Rhodococcus equi*" is another useful exemplar in this scenario. The taxonomic status of this species awaits clarification (Garrity, 2014; Kämpfer, Dott, Martin, & Glaeser, 2013; Tindall, 2014) although it has been proposed that it should be reclassified in the monospecific genus "*Prescottella*" as "*Prescottella equi*" (Jones, Sutcliffe, & Goodfellow, 2013a, 2013b). In the meantime, we have sequenced the whole genome of "*R. equi* C7^{T}" (Sangal, Jones, et al., 2014) and *R. defluvii* Ca11^{T} (Sangal et al., unpublished data). Our phylogenetic analyses of 400 universal protein sequences using PhyloPhlAn (Segata et al., 2013) and other genomic analyses including representative rhodococcal and nocardial type strains suggest that "*R. equi*" and *R. defluvii* together should be classified in a new genus (Sangal et al., unpublished data).

There has been some excellent work on why certain groups of organisms do not fall into clear and distinct clusters, taxa described as "fuzzy species" (Hanage, Fraser, & Spratt, 2005). Fuzzy species are a consequence of recombination that results in mosaic genotypes, creating taxonomic confusion (Hanage, 2013). It is clear that the need for better resolution in the definition of taxonomic groups will increase given a move from single gene (16S rRNA) to multi-gene (MLST or MLSA) and finally to whole genome phylogenies. However, genomic analyses can reinforce traditional polyphasic approaches in defining genetic boundaries between different OTUs and should become an integral part of prokaryotic systematics (Busarakam et al., 2014; Chun & Rainey, 2014; Vandamme & Peeters, 2014).

4.2 PATHOGEN EVOLUTION, TRANSMISSION AND ADAPTATION

An understanding of the evolutionary dynamics and routes of transmission of microbial pathogens is key to the development of effective preventive and intervention strategies (see chapter 'Whole-Genome Sequencing for Rapid and Accurate Identification of Bacterial Transmission Pathways' by Harris and Okoro, in this volume). Next-generation sequencing indexes entire genomic variation that can help in the accurate estimation of genome-wide mutation rates and identification of signatures of temporal and geographic association. These approaches have been applied to study the evolution, global transmission and geographic association of several pathogenic bacteria (Harris et al., 2010; Holt et al., 2008, 2012; Stewart et al., 2014). For example, *Shigella sonnei*, a human pathogen that descended from a common ancestor less than 500 years ago (Holt et al., 2012), had diversified in Europe before its global dissemination and adaptation in different geographical regions. A lack of frequent recombination led to the identification of six chromosomal SNPs that can precisely identify the lineage and major subgroups within the predominant global lineage of *S. sonnei* (Sangal, Holt, et al., 2013). Next-generation sequencing analyses, therefore, can help in identification of reliable diagnostic markers that can be used to develop rapid detection assays for specific bacterial pathogens.

4.3 GENETIC BASIS OF PHENOTYPIC CHARACTERISTICS

Strains belonging to the same species may exhibit differences in their phenotypic characteristics, including resistance to certain antibiotics or the ability to use certain carbon sources. Next-generation sequencing can be used to identify the genetic basis of these differences. In *S. sonnei*, for example, resistance to quinolones and reduced susceptibility to fluoroquinolones has been linked to a nonsynonymous SNP in the *gyrA* gene (Holt et al., 2012; Sangal, Holt, et al., 2013). Whole genomic sequence analyses can also help resolve inconsistencies between phenotypic characterization and genomic diversity. *C. diphtheriae* has been subdivided into four biovars based on the biochemical properties; however, this separation was not well supported by phylogenetic analyses of the genomic variation and the differences in gene contents in representative strains of each biovar (Sangal et al., 2014). Biovar belfanti was an exception as an insertion in the *narJ* gene has resulted in the inability of these strains to reduce nitrate. Similarly, Amaral et al. (2014) have linked metabolic pathways encoded in *Vibrio* genomes to diagnostic phenotypes, an approach which will undoubtedly prove of great utility, particularly through reference to metabolic pathway databases such as KEGG, MetaCyc and BioCyc (Caspi et al., 2014).

4.4 METAGENOMICS

Metagenomics is the genomic analysis of microbial communities using DNA directly extracted from environmental samples. Metagenomic studies reveal the extent of microbial diversity in natural habitats, including that of uncultured microbes, and help to improve our understanding of the biological functions within microbial communities. For example, analyses of the panda gut microbiome revealed the presence of micro-organisms that are responsible for lignin oxidation (Fang et al., 2012) and the degradation of cellulose present in the staple diet of bamboo (Zhu, Wu, Dai, Zhang, & Wei, 2011).

Another approach to genome sequencing of uncultivable prokaryotes is single-cell sequencing, which can allow a more meaningful interpretation of microbial diversity in complex ecosystems in combination with metagenomic data and can also facilitate discovery of novel metabolic features (Kamke et al., 2014; Kantor et al., 2013; Rinke et al., 2013). For example, analyses of 201 archaeal and bacterial single-cell genomes of 29 under-represented phyla identified two new superphyla and a novel use of the opal stop codon (Rinke et al., 2013). Metagenomic and single-cell sequencing analyses will have significant impact not only on microbial ecology and diversity studies but also on industrial biotechnology.

4.5 **TARGET RESEQUENCING**

Target resequencing is a cost-effective method designed to genotype large numbers of samples by selective multiplex amplification of certain genomic regions followed by analysis using next-generation sequencing procedures. Whole genome sequencing generates enormous amounts of data that are not easy to handle; sometimes, selected loci can offer sufficient resolution for strain typing (Amplicon sequencing). MLST has been used mainly for the evolutionary analysis of strains (Maiden, 2006; see chapter 'Multi-Locus Sequence Typing and the Gene-by-Gene Approach to Bacterial Classification and Analysis of Population Variation' by Cody et al., in this volume). However, when applied to a large collection of strains this method can be slow, costly and labour intensive. Targeted resequencing of MLST genes (MLST-seq) is potentially the way forward and has been applied to *Salmonella* (Singh, Foley, Nayak, & Kwon, 2013). Metagenomic analyses also involve target resequencing of 16S rRNA genes for identification of bacterial populations in environmental and clinical samples (Daniel, 2005; Salipante et al., 2013; Zhu, Wang, & Li, 2010). Targeted resequencing has the potential to replace traditional Sanger sequencing in clinical diagnostic settings (Sikkema-Raddatz et al., 2013).

4.6 **RNA-Seq AND TRANSCRIPTOMICS**

Transcriptomics aims to quantify all transcripts, including mRNAs, noncoding RNAs (ncRNAs) and small RNAs (sRNAs) at a given time within a cell that can reflect a specific developmental stage or physiological condition. RNA-seq (RNA sequencing) is a powerful tool for studying transcriptomics as it determines expression levels and can reveal post-transcriptional modification to a single-base resolution. It can also give insight into the transcriptional structure of genes, including start and end sites and directionality, and can be used to identify sequence variations such as SNPs in the transcribed regions (Wang, Gerstein, & Snyder, 2009). A combined approach using metagenomics together with metatranscriptomics has been used to identify a large number of putative novel and functional ncRNAs in both soil and marine microbial communities (Sorek & Cossart, 2010). Recently, RNA-seq was used to study *Staphylococcus aureus* responses to antimicrobials (Howden et al., 2013) and to phage-encoded transcription factor gp67 (Osmundson, Dewell, & Darst, 2013). It can also be used to investigate host–pathogen interactions (Westermann, Gorski, & Vogel, 2012) and protein–RNA interactions (Sittka et al., 2008).

The success of RNA-seq is dependent on the quality of RNA isolated from cells. A minimum of 1–2 µg of high-quality, pure, undegraded total RNA that represents a "snapshot" of all RNA species in natural proportions is needed for downstream processes. After either enzymatic or mechanical cell disruption, the total RNA is isolated using an organic extraction and/or adsorption onto a solid matrix. If spin column-based systems are used, care must be taken in sRNA analyses that the filters do not exclude the sRNA molecules. More than 95% of total RNA is rRNA and tRNA (Yi et al., 2011) so an enrichment step for mRNA and sRNA will dramatically increase the sequencing coverage of these RNA species. The fact that prokaryotic mRNAs lack a 3'-end poly(A) tail makes mRNA enrichment more challenging for these organisms. Several methods have been used in the past for mRNA enrichment: (i) hybridization capture of rRNAs by antisense oligonucleotides followed by pull down using magnetic beads, (ii) degradation of mature rRNAs and tRNAs by a 5'–3' exonuclease, (iii) polyadenylation of mRNAs followed by oligo(dT) purification, and (iv) antibody capture of RNAs interacting with a specific protein.

The rRNA-depleted total RNA sample is then converted into a cDNA library. Specific consideration should be given to selection of an appropriate cDNA library protocol since some protocols can result in the generation of strand-specific libraries. During library preparation, RNA (or in some protocols cDNA) molecules are fragmented by enzymatic or mechanical disruption, adapter ligated and reverse transcribed into cDNA. The cDNA library is then PCR amplified and sequenced using Illumina Genome Analyzer, Roche 454, Applied Biosystems SOLiD or Ion Torrent platforms (Mader, Nicolas, Richard, Bessieres, & Aymerich, 2011; Sorek & Cossart, 2010). Prokaryotic RNA-seq data analysis differs from eukaryotic RNA-seq data analysis as it does not need to take into account poly(A) tails or exon junctions from splicing events (Wang et al., 2009). However, challenges arise when trying to distinguish the start of one gene transcript from the end of another since bacterial genes often overlap. In addition to this, the presence of polycistronic messages, sRNAs and multiple promoters further complicates bacterial transcript assembly compared with analysis of eukaryotic systems. For these reasons, bioinformatics software designed specifically for prokaryotic RNA-seq data analyses are recommended to be used when possible. Good examples of prokaryotic specific software for RNA-seq include Rockhopper (McClure et al., 2013) and EDGE-pro (Magoc, Wood, & Salzberg, 2013).

The analysis of RNA-seq data starts with a quality check of the generated reads and then continues to map the reads either to a reference genome or to a reference transcriptome with, for example, Map with BWA in Galaxy (Li & Durbin, 2009), or by constructing a *de novo* assembly of the reads. The number of reads mapping to a feature (i.e. gene) can be quantified with, for example, htseq-count in Galaxy (HTSeq pyton module; Anders, Pyl, & Huber, 2014). The data are then normalized between samples. The appropriate normalization method depends on the study design and can include methods such as total count (TC), upper quartile (UQ), Median (Med), DESeq normalization using DESeq Bioconductor package, Trimmed Mean of M values (TMM) of edgeR Bioconductor package, Quantile (Q) and the fragments per kilobase per million mapped reads (FPKM) (Dillies et al., 2013). Several mathematical models/strategies such as Voom, Limma, Benjamini–Hochberg correction for multiple testing and false discovery rate have previously been adopted in order to assess the statistical significance of prokaryotic RNA-seq data. Recently, nonnegative matrix factorization was applied to discover patterns in RNA-seq data in order to gain comprehensive outcomes of these complex data (Howden et al., 2013; Osmundson et al., 2013). These data have great potential for investigating interstrain variation that may be the result of transcriptional variation, adding even finer resolution to data obtained from the analysis of whole genomes.

CONCLUSIONS

The ever-reducing costs and the availability of advanced bioinformatic tools have made next-generation sequencing suitable for a wide range of applications in biological research; hence, such studies are no longer the preserve of dedicated "sequencing labs". Next-generation sequencing generates huge amounts of data, the proper handling of which remains a challenge for most experimentalists with limited computational experience. However, easy-to-use software with interactive interfaces have started to make data handling more user-friendly. Next-generation sequencing is having a substantial impact on prokaryotic research in terms of establishing phylogenies, inferring evolution, exploring niche adaptation and realizing the true diversity of unculturable microbes in natural habitats. Now that bioinformatics tools are on the verge of being used to assemble complete genomes from metagenomic studies, it is clear that whole genome sequencing will revolutionise developments in clinical diagnostics, therapeutic strategies, as well as addressing some of the industrial needs of society; this will all need to be underpinned by a functional and theoretical understanding of systematics. It is surely time to embrace the genome and make it part of routine prokaryotic systematics.

ACKNOWLEDGEMENTS

V. S. is supported by an Anniversary Research Fellowship at Northumbria University, Newcastle upon Tyne. P. A. H. acknowledges support from Medical Research Scotland (422 FRG), The Leverhulme Trust, Scottish Universities Life Science Alliance (SULSA) and The University of Strathclyde.

REFERENCES

Achtman, M., & Wagner, M. (2008). Microbial diversity and the genetic nature of microbial species. *Nature Reviews. Microbiology*, *6*(6), 431–440. http://dx.doi.org/10.1038/nrmicro1872.

Amaral, G. R., Dias, G. M., Wellington-Oguri, M., Chimetto, L., Campeao, M. E., Thompson, F. L., et al. (2014). Genotype to phenotype: Identification of diagnostic *Vibrio* phenotypes using whole genome sequences. *International Journal of Systematic and Evolutionary Microbiology*, *64*(2), 357–365. http://dx.doi.org/10.1099/ijs.0.057927-0.

Anders, S., Pyl, P. T., & Huber, W. (2014). HTSeq; A Python framework to work with high-throughput sequencing data. *BioRxiv Preprint*. http://dx.doi.org/10.1101/002824.

Auch, A. F., von Jan, M., Klenk, H. P., & Göker, M. (2010). Digital DNA-DNA hybridization for microbial species delineation by means of genome-to-genome sequence comparison. *Standards in Genomic Sciences*, *2*(1), 117–134. http://dx.doi.org/10.4056/sigs.531120.

Aziz, R. K., Bartels, D., Best, A. A., DeJongh, M., Disz, T., Edwards, R. A., et al. (2008). The RAST Server: Rapid annotations using subsystems technology. *BMC Genomics*, *9*, 75. http://dx.doi.org/10.1186/1471-2164-9-75.

Aziz, R. K., Devoid, S., Disz, T., Edwards, R. A., Henry, C. S., Olsen, G. J., et al. (2012). SEED servers: High-performance access to the SEED genomes, annotations, and metabolic models. *PLoS One*, *7*(10), e48053. http://dx.doi.org/10.1371/journal.pone.0048053.

Badger, J. H., & Olsen, G. J. (1999). CRITICA: Coding region identification tool invoking comparative analysis. *Molecular Biology and Evolution*, *16*(4), 512–524.

Benson, G. (1999). Tandem repeats finder: A program to analyze DNA sequences. *Nucleic Acids Research*, *27*(2), 573–580, gkc131 [pii].

Bentley, D. R., Balasubramanian, S., Swerdlow, H. P., Smith, G. P., Milton, J., Brown, C. G., et al. (2008). Accurate whole human genome sequencing using reversible terminator chemistry. *Nature*, *456*(7218), 53–59. http://dx.doi.org/10.1038/nature07517.

Blankenberg, D., Von Kuster, G., Coraor, N., Ananda, G., Lazarus, R., Mangan, M., et al. (2010). Galaxy: A web-based genome analysis tool for experimentalists. *Current Protocols in Molecular Biology*. http://dx.doi.org/10.1002/0471142727.mb1910s89, Chapter 19, Unit 19 10 11–21.

Blom, J., Albaum, S. P., Doppmeier, D., Puhler, A., Vorholter, F. J., Zakrzewski, M., et al. (2009). EDGAR: A software framework for the comparative analysis of prokaryotic genomes. *BMC Bioinformatics*, *10*, 154. http://dx.doi.org/10.1186/1471-2105-10-154.

Borodovsky, M., & McIninch, J. (1993). GeneMark: Parallel gene recognition for both DNA strands. *Computers & Chemistry*, *17*(19), 123–133.

Busarakam, K., Bull, A. T., Girard, G., Labeda, D. P., van Wezel, G. P., & Goodfellow, M. (2014). *Streptomyces leeuwenhoekii* sp. nov., the producer of chaxalactins and chaxamycins, forms a distinct branch in *Streptomyces* gene trees. *Antonie Van Leeuwenhoek*, *105*(5), 849–861. http://dx.doi.org/10.1007/s10482-014-0139-y.

Carver, T., Berriman, M., Tivey, A., Patel, C., Bohme, U., Barrell, B. G., et al. (2008). Artemis and ACT: Viewing, annotating and comparing sequences stored in a relational database. *Bioinformatics*, *24*(23), 2672–2676. http://dx.doi.org/10.1093/bioinformatics/btn529.

Carver, T., Harris, S. R., Berriman, M., Parkhill, J., & McQuillan, J. A. (2012). Artemis: An integrated platform for visualization and analysis of high-throughput sequence-based experimental data. *Bioinformatics*, *28*(4), 464–469. http://dx.doi.org/10.1093/bioinformatics/btr703.

Caspi, R., Altman, T., Billington, R., Dreher, K., Foerster, H., Fulcher, C. A., et al. (2014). The MetaCyc database of metabolic pathways and enzymes and the BioCyc collection of Pathway/Genome Databases. *Nucleic Acids Research*, *42*(Database issue), D459–D471. http://dx.doi.org/10.1093/nar/gkt1103.

Chan, J. Z., Halachev, M. R., Loman, N. J., Constantinidou, C., & Pallen, M. J. (2012). Defining bacterial species in the genomic era: Insights from the genus *Acinetobacter*. *BMC Microbiology*, *12*, 302. http://dx.doi.org/10.1186/1471-2180-12-302.

Chaudhuri, R. R., & Pallen, M. J. (2006). xBASE, a collection of online databases for bacterial comparative genomics. *Nucleic Acids Research*, *34*(Database issue), D335–D337. http://dx.doi.org/10.1093/nar/gkj140.

Chevreux, B., Wetter, T., & Suhai, S. (1999). Genome sequence assembly using trace signals and additional sequence information. In *Proceedings of the German conference on bioinformatics, 99, Computer science and biology, Hannover, Germany* (pp. 45–56).

Chun, J., & Rainey, F. A. (2014). Integrating genomics into the taxonomy and systematics of the *Bacteria* and *Archaea*. *International Journal of Systematic and Evolutionary Microbiology*, *64*(2), 316–324. http://dx.doi.org/10.1099/ijs.0.054171-0.

Daniel, R. (2005). The metagenomics of soil. *Nature Reviews. Microbiology*, *3*(6), 470–478. http://dx.doi.org/10.1038/nrmicro1160.

Darling, A. C., Mau, B., Blattner, F. R., & Perna, N. T. (2004). Mauve: Multiple alignment of conserved genomic sequence with rearrangements. *Genome Research*, *14*(7), 1394–1403. http://dx.doi.org/10.1101/gr.2289704.

Delcher, A. L., Harmon, D., Kasif, S., White, O., & Salzberg, S. L. (1999). Improved microbial gene identification with GLIMMER. *Nucleic Acids Research*, *27*(23), 4636–4641.

Deloger, M., El Karoui, M., & Petit, M. A. (2009). A genomic distance based on MUM indicates discontinuity between most bacterial species and genera. *Journal of Bacteriology*, *191*(1), 91–99. http://dx.doi.org/10.1128/JB.01202-08.

Didelot, X., & Falush, D. (2007). Inference of bacterial microevolution using multilocus sequence data. *Genetics*, *175*(3), 1251–1266. http://dx.doi.org/10.1534/genetics.106.063305.

Dillies, M. A., Rau, A., Aubert, J., Hennequet-Antier, C., Jeanmougin, M., Servant, N., et al. (2013). A comprehensive evaluation of normalization methods for Illumina high-throughput RNA sequencing data analysis. *Briefings in Bioinformatics*, *14*(6), 671–683. http://dx.doi.org/10.1093/bib/bbs046.

Eid, J., Fehr, A., Gray, J., Luong, K., Lyle, J., Otto, G., et al. (2009). Real-time DNA sequencing from single polymerase molecules. *Science*, *323*(5910), 133–138. http://dx.doi.org/10.1126/science.1162986.

Fang, W., Fang, Z., Zhou, P., Chang, F., Hong, Y., Zhang, X., et al. (2012). Evidence for lignin oxidation by the giant panda fecal microbiome. *PLoS One*, *7*(11), e50312. http://dx.doi.org/10.1371/journal.pone.0050312.

Finn, R. D., Bateman, A., Clements, J., Coggill, P., Eberhardt, R. Y., Eddy, S. R., et al. (2013). Pfam: The protein families database. *Nucleic Acids Research*, *42*, D222–D230. http://dx.doi.org/10.1093/nar/gkt1223.

Fleischmann, R. D., Adams, M. D., White, O., Clayton, R. A., Kirkness, E. F., Kerlavage, A. R., et al. (1995). Whole-genome random sequencing and assembly of *Haemophilus influenzae* Rd. *Science*, *269*(5223), 496–512.

Fouts, D. E. (2006). Phage_Finder: Automated identification and classification of prophage regions in complete bacterial genome sequences. *Nucleic Acids Research*, *34*(20), 5839–5851. http://dx.doi.org/10.1093/nar/gkl732.

Garrity, G. M. (2014). Conservation of *Rhodococcus equi* (Magnusson 1923) Goodfellow and Alderson 1977 and rejection of *Corynebacterium hoagii* (Morse 1912) Eberson 1918. *International Journal of Systematic and Evolutionary Microbiology*, *64*(1), 311–312. http://dx.doi.org/10.1099/ijs.0.059741-0.

Goering, R. V. (2010). Pulsed field gel electrophoresis: A review of application and interpretation in the molecular epidemiology of infectious disease. *Infection, Genetics and Evolution*, *10*(7), 866–875. http://dx.doi.org/10.1016/j.meegid.2010.07.023.

Goris, J., Konstantinidis, K. T., Klappenbach, J. A., Coenye, T., Vandamme, P., & Tiedje, J. M. (2007). DNA-DNA hybridization values and their relationship to whole-genome sequence similarities. *International Journal of Systematic and Evolutionary Microbiology*, *57*(1), 81–91. http://dx.doi.org/10.1099/ijs.0.64483-0.

Grant, J. R., Arantes, A. S., Liao, X., & Stothard, P. (2011). In-depth annotation of SNPs arising from resequencing projects using NGS-SNP. *Bioinformatics*, *27*(16), 2300–2301. http://dx.doi.org/10.1093/bioinformatics/btr372.

Grant, J. R., Arantes, A. S., & Stothard, P. (2012). Comparing thousands of circular genomes using the CGView Comparison Tool. *BMC Genomics*, *13*(1), 202. http://dx.doi.org/10.1186/1471-2164-13-202.

Habegger, L., Balasubramanian, S., Chen, D. Z., Khurana, E., Sboner, A., Harmanci, A., et al. (2012). VAT: A computational framework to functionally annotate variants in personal genomes within a cloud-computing environment. *Bioinformatics*, *28*(17), 2267–2269. http://dx.doi.org/10.1093/bioinformatics/bts368.

Hanage, W. P. (2013). Fuzzy species revisited. *BMC Biololgy*, *11*, 41. http://dx.doi.org/10.1186/1741-7007-11-41.

Hanage, W. P., Fraser, C., & Spratt, B. G. (2005). Fuzzy species among recombinogenic bacteria. *BMC Biololgy*, *3*, 6. http://dx.doi.org/10.1186/1741-7007-3-6.

Harris, S. R., Feil, E. J., Holden, M. T., Quail, M. A., Nickerson, E. K., Chantratita, N., et al. (2010). Evolution of MRSA during hospital transmission and intercontinental spread. *Science*, *327*(5964), 469–474. http://dx.doi.org/10.1126/science.1182395.

Holt, K. E., Baker, S., Weill, F. X., Holmes, E. C., Kitchen, A., Yu, J., et al. (2012). *Shigella sonnei* genome sequencing and phylogenetic analysis indicate recent global dissemination from Europe. *Nature Genetics*, *44*(9), 1056–1059. http://dx.doi.org/10.1038/ng.2369.

Holt, K. E., Parkhill, J., Mazzoni, C. J., Roumagnac, P., Weill, F. X., Goodhead, I., et al. (2008). High-throughput sequencing provides insights into genome variation and evolution in *Salmonella typhi*. *Nature Genetics*, *40*(8), 987–993. http://dx.doi.org/10.1038/ng.195.

Howden, B. P., Beaume, M., Harrison, P. F., Hernandez, D., Schrenzel, J., Seemann, T., et al. (2013). Analysis of the small RNA transcriptional response in multidrug-resistant *Staphylococcus aureus* after antimicrobial exposure. *Antimicrobial Agents and Chemotherapy*, *57*(8), 3864–3874. http://dx.doi.org/10.1128/AAC.00263-13.

Hyatt, D., Chen, G. L., Locascio, P. F., Land, M. L., Larimer, F. W., & Hauser, L. J. (2010). Prodigal: Prokaryotic gene recognition and translation initiation site identification. *BMC Bioinformatics*, *11*, 119. http://dx.doi.org/10.1186/1471-2105-11-119.

Jones, A. L., Sutcliffe, I. C., & Goodfellow, M. (2013a). *Prescottia equi* gen. nov., comb. nov.: A new home for an old pathogen. *Antonie Van Leeuwenhoek*, *103*(3), 655–671. http://dx.doi.org/10.1007/s10482-012-9850-8.

Jones, A. L., Sutcliffe, I. C., & Goodfellow, M. (2013b). Proposal to replace the illegitimate genus name *Prescottia* Jones et al. 2013 with the genus name *Prescottella* gen. nov. and to replace the illegitimate combination *Prescottia equi* Jones et al. 2013 with *Prescottella equi* comb. nov. *Antonie Van Leeuwenhoek*, *103*(6), 1405–1407. http://dx.doi.org/10.1007/s10482-013-9924-2.

Kamke, J., Rinke, C., Schwientek, P., Mavromatis, K., Ivanova, N., Sczyrba, A., et al. (2014). The candidate phylum *Poribacteria* by single-cell genomics: New insights into phylogeny, cell-compartmentation, eukaryote-like repeat proteins, and other genomic features. *PLoS One*, *9*(1), e87353. http://dx.doi.org/10.1371/journal.pone.0087353.

Kämpfer, P., Dott, W., Martin, K., & Glaeser, S. P. (2013). *Rhodococcus defluvii* sp. nov. isolated from wastewater of a bioreactor and formal proposal to reclassify '*Corynebacterium*' *hoagii* and *Rhodococcus equi* as *Rhodococcus hoagii*. *International Journal of Systematic and Evolutionary Microbiology, 64*(Pt 3), 755–761. http://dx.doi.org/10.1099/ijs.0.053322-0.

Kanehisa, M. (2000). *Post-genome informatics*. Oxford, UK: Oxford University Press.

Kantor, R. S., Wrighton, K. C., Handley, K. M., Sharon, I., Hug, L. A., Castelle, C. J., et al. (2013). Small genomes and sparse metabolisms of sediment-associated bacteria from four candidate phyla. *mBio, 4*(5), e00708–e00713. http://dx.doi.org/10.1128/mBio.00708-13.

Kaufmann, M. E. (1998). Pulsed-field gel electrophoresis. *Methods in Molecular Medicine, 15*, 33–50. http://dx.doi.org/10.1385/0-89603-498-4:33.

Kim, M., Oh, H. S., Park, S. C., & Chun, J. (2014). Towards a taxonomic coherence between average nucleotide identity and 16S rRNA gene sequence similarity for species demarcation of prokaryotes. *International Journal of Systematic and Evolutionary Microbiology, 64*(2), 346–351. http://dx.doi.org/10.1099/ijs.0.059774-0.

Konstantinidis, K. T., Ramette, A., & Tiedje, J. M. (2006). The bacterial species definition in the genomic era. *Philosophical Transactions of the Royal Society B, 361*, 1929–1940.

Konstantinidis, K. T., & Tiedje, J. M. (2005a). Genomic insights that advance the species definition for prokaryotes. *Proceedings of the National Academy of Sciences of the United States of America, 102*(7), 2567–2572. http://dx.doi.org/10.1073/pnas.0409727102.

Konstantinidis, K. T., & Tiedje, J. M. (2005b). Towards a genome-based taxonomy for prokaryotes. *Journal of Bacteriology, 187*(18), 6258–6264. http://dx.doi.org/10.1128/JB.187.18.6258-6264.2005.

Laing, C., Buchanan, C., Taboada, E. N., Zhang, Y., Kropinski, A., Villegas, A., et al. (2010). Pan-genome sequence analysis using Panseq: An online tool for the rapid analysis of core and accessory genomic regions. *BMC Bioinformatics, 11*, 461. http://dx.doi.org/10.1186/1471-2105-11-461.

Langille, M. G., & Brinkman, F. S. (2009). IslandViewer: An integrated interface for computational identification and visualization of genomic islands. *Bioinformatics, 25*(5), 664–665. http://dx.doi.org/10.1093/bioinformatics/btp030.

Levene, M. J., Korlach, J., Turner, S. W., Foquet, M., Craighead, H. G., & Webb, W. W. (2003). Zero-mode waveguides for single-molecule analysis at high concentrations. *Science, 299*(5607), 682–686. http://dx.doi.org/10.1126/science.1079700.

Li, H., & Durbin, R. (2009). Fast and accurate short read alignment with Burrows-Wheeler transform. *Bioinformatics, 25*(14), 1754–1760. http://dx.doi.org/10.1093/bioinformatics/btp324.

Li, H., Handsaker, B., Wysoker, A., Fennell, T., Ruan, J., Homer, N., et al. (2009). The sequence alignment/map format and SAMtools. *Bioinformatics, 25*(16), 2078–2079. http://dx.doi.org/10.1093/bioinformatics/btp352.

Li, H., Ruan, J., & Durbin, R. (2008). Mapping short DNA sequencing reads and calling variants using mapping quality scores. *Genome Research, 18*(11), 1851–1858. http://dx.doi.org/10.1101/gr.078212.108.

Linke, B., McHardy, A. C., Neuweger, H., Krause, L., & Meyer, F. (2006). REGANOR: A gene prediction server for prokaryotic genomes and a database of high quality gene predictions for prokaryotes. *Applied Bioinformatics, 5*(3), 193–198.

Loman, N. J., Constantinidou, C., Chan, J. Z., Halachev, M., Sergeant, M., Penn, C. W., et al. (2012). High-throughput bacterial genome sequencing: An embarrassment of choice, a world of opportunity. *Nature Reviews. Microbiology, 10*(9), 599–606. http://dx.doi.org/10.1038/nrmicro2850.

Luo, C., Rodriguez, R. L., & Konstantinidis, K. T. (2014). MyTaxa: An advanced taxonomic classifier for genomic and metagenomic sequences. *Nucleic Acids Research, 42*(8), e73. http://dx.doi.org/10.1093/nar/gku169.

Mader, U., Nicolas, P., Richard, H., Bessieres, P., & Aymerich, S. (2011). Comprehensive identification and quantification of microbial transcriptomes by genome-wide unbiased methods. *Current Opinion in Biotechnology, 22*(1), 32–41. http://dx.doi.org/10.1016/j.copbio.2010.10.003.

Magoc, T., Wood, D., & Salzberg, S. L. (2013). EDGE-pro: Estimated degree of gene expression in prokaryotic genomes. *Evolutionary Bioinformatics Online, 9*, 127–136. http://dx.doi.org/10.4137/EBO.S11250.

Maiden, M. C. (2006). Multilocus sequence typing of bacteria. *Annual Review of Microbiology, 60*, 561–588. http://dx.doi.org/10.1146/annurev.micro.59.030804.121325.

Maiden, M. C., Bygraves, J. A., Feil, E., Morelli, G., Russell, J. E., Urwin, R., et al. (1998). Multilocus sequence typing: A portable approach to the identification of clones within populations of pathogenic microorganisms. *Proceedings of the National Academy of Sciences of the United States of America, 95*(6), 3140–3145.

Margulies, M., Egholm, M., Altman, W. E., Attiya, S., Bader, J. S., Bemben, L. A., et al. (2005). Genome sequencing in microfabricated high-density picolitre reactors. *Nature, 437*(7057), 376–380. http://dx.doi.org/10.1038/nature03959.

McClure, R., Balasubramanian, D., Sun, Y., Bobrovskyy, M., Sumby, P., Genco, C. A., et al. (2013). Computational analysis of bacterial RNA-Seq data. *Nucleic Acids Research, 41*(14), e140. http://dx.doi.org/10.1093/nar/gkt444.

Meier-Kolthoff, J. P., Auch, A. F., Klenk, H. P., & Göker, M. (2013). Genome sequence-based species delimitation with confidence intervals and improved distance functions. *BMC Bioinformatics, 14*, 60. http://dx.doi.org/10.1186/1471-2105-14-60.

Meyer, F., Goesmann, A., McHardy, A. C., Bartels, D., Bekel, T., Clausen, J., et al. (2003). GenDB – An open source genome annotation system for prokaryote genomes. *Nucleic Acids Research, 31*(8), 2187–2195.

Myllykangas, S., Buenrostro, J., & Ji, H. P. (2012). Overview of sequencing technology platforms. In N. Rodríguez-Ezpeleta, M. Hackenberg, & A. M. Aransay (Eds.), *Bioinformatics for high throughput sequencing* (pp. 11–25). New York: Springer.

Nagarajan, N., Cook, C., Di Bonaventura, M., Ge, H., Richards, A., Bishop-Lilly, K. A., et al. (2010). Finishing genomes with limited resources: Lessons from an ensemble of microbial genomes. *BMC Genomics, 11*, 242. http://dx.doi.org/10.1186/1471-2164-11-242.

Osmundson, J., Dewell, S., & Darst, S. A. (2013). RNA-Seq reveals differential gene expression in *Staphylococcus aureus* with single-nucleotide resolution. *PLoS One, 8*(10), e76572. http://dx.doi.org/10.1371/journal.pone.0076572.

Pride, D. T., Meinersmann, R. J., Wassenaar, T. M., & Blaser, M. J. (2003). Evolutionary implications of microbial genome tetranucleotide frequency biases. *Genome Research, 13*(2), 145–158. http://dx.doi.org/10.1101/gr.335003.

Renesto, P., Crapoulet, N., Ogata, H., La Scola, B., Vestris, G., Claverie, J. M., et al. (2003). Genome-based design of a cell-free culture medium for *Tropheryma whipplei. Lancet, 362*(9382), 447–449. http://dx.doi.org/10.1016/S0140-6736(03)14071-8.

Retief, J. D. (2000). Phylogenetic analysis using PHYLIP. *Methods in Molecular Biology, 132*, 243–258.

Reumerman, R. A., Tucker, N. P., Herron, P. R., Hoskisson, P. A., & Sangal, V. (2013). Tool for Rapid Annotation of Microbial SNPs (TRAMS): A simple program for rapid annotation of genomic variation in prokaryotes. *Antonie Van Leeuwenhoek, 104*(3), 431–434. http://dx.doi.org/10.1007/s10482-013-9953-x.

Richter, M., & Rosselló-Móra, R. (2009). Shifting the genomic gold standard for the pro-karyotic species definition. *Proceedings of the National Academy of Sciences of the United States of America*, *106*(45), 19126–19131. http://dx.doi.org/10.1073/pnas.0906412106.

Rinke, C., Schwientek, P., Sczyrba, A., Ivanova, N. N., Anderson, I. J., Cheng, J. F., et al. (2013). Insights into the phylogeny and coding potential of microbial dark matter. *Nature*, *499*(7459), 431–437. http://dx.doi.org/10.1038/nature12352.

Ronaghi, M., Uhlen, M., & Nyren, P. (1998). A sequencing method based on real-time pyro-phosphate. *Science*, *281*(5375), 363, 365.

Rothberg, J. M., Hinz, W., Rearick, T. M., Schultz, J., Mileski, W., Davey, M., et al. (2011). An integrated semiconductor device enabling non-optical genome sequencing. *Nature*, *475*(7356), 348–352. http://dx.doi.org/10.1038/nature10242.

Rutherford, K., Parkhill, J., Crook, J., Horsnell, T., Rice, P., Rajandream, M. A., et al. (2000). Artemis: Sequence visualization and annotation. *Bioinformatics*, *16*(10), 944–945.

Salipante, S. J., Sengupta, D. J., Rosenthal, C., Costa, G., Spangler, J., Sims, E. H., et al. (2013). Rapid 16S rRNA next-generation sequencing of polymicrobial clinical samples for diagnosis of complex bacterial infections. *PLoS One*, *8*(5), e65226. http://dx.doi.org/10.1371/journal.pone.0065226.

Sangal, V., Burkovski, A., Hunt, A. C., Edwards, B., Blom, J., & Hoskisson, P. A. (2014). A lack of genetic basis for biovar differentiation in clinically important *Corynebacterium diphtheriae* from whole genome sequencing. *Infection, Genetics and Evolution*, *21*, 54–57. http://dx.doi.org/10.1016/j.meegid.2013.10.019.

Sangal, V., Fineran, P. C., & Hoskisson, P. A. (2013). Novel configurations of type I and II CRISPR-Cas systems in *Corynebacterium diphtheriae*. *Microbiology*, *159*(10), 2118–2126. http://dx.doi.org/10.1099/mic.0.070235-0.

Sangal, V., Holt, K. E., Yuan, J., Brown, D. J., Filliol-Toutain, I., Weill, F. X., et al. (2013). Global phylogeny of *Shigella sonnei* strains from limited single nucleotide polymorphisms (SNPs) and development of a rapid and cost-effective SNP-typing scheme for strain iden-tification by high-resolution melting analysis. *Journal of Clinical Microbiology*, *51*(1), 303–305. http://dx.doi.org/10.1128/JCM.02238-12.

Sangal, V., & Hoskisson, P. A. (2013). Embrace the genome! *The Bulletin of BISMiS*, *4*(1), 15–17.

Sangal, V., Jones, A. L., Goodfellow, M., Sutcliffe, I. C., & Hoskisson, P. A. (2014). Com-parative genomic analyses reveal a lack of a substantial signature of host adaptation in *Rhodococcus equi* ("*Prescottella equi*"). *Pathogens and Disease*, *71*(3), 352–356. http://dx.doi.org/10.1111/2049-632X.12126.

Schneider, J., Vorholter, F. J., Trost, E., Blom, J., Musa, Y. R., Neuweger, H., et al. (2010). CARMEN – Comparative analysis and *in silico* reconstruction of organism-specific MEt-abolic networks. *Genetics and Molecular Research*, *9*(3), 1660–1672. http://dx.doi.org/10.4238/vol9-3gmr901.

Segata, N., Bornigen, D., Morgan, X. C., & Huttenhower, C. (2013). PhyloPhlAn is a new method for improved phylogenetic and taxonomic placement of microbes. *Nature Com-munications*, *4*, 2304. http://dx.doi.org/10.1038/ncomms3304.

Sentausa, E., & Fournier, P. E. (2013). Advantages and limitations of genomics in prokaryotic taxonomy. *Clinical Microbiology and Infection*, *19*(9), 790–795. http://dx.doi.org/10.1111/1469-0691.12181.

Siguier, P., Perochon, J., Lestrade, L., Mahillon, J., & Chandler, M. (2006). ISfinder: The ref-erence centre for bacterial insertion sequences. *Nucleic Acids Research*, *34*(Database issue), D32–36. http://dx.doi.org/10.1093/nar/gkj014.

Sikkema-Raddatz, B., Johansson, L. F., de Boer, E. N., Almomani, R., Boven, L. G., van den Berg, M. P., et al. (2013). Targeted next-generation sequencing can replace Sanger sequencing in clinical diagnostics. *Human Mutation, 34*(7), 1035–1042. http://dx.doi.org/10.1002/humu.22332.

Singh, P., Foley, S. L., Nayak, R., & Kwon, Y. M. (2013). Massively parallel sequencing of enriched target amplicons for high-resolution genotyping of *Salmonella* serovars. *Molecular and Cellular Probes, 27*(2), 80–85. http://dx.doi.org/10.1016/j.mcp.2012.11.004.

Sittka, A., Lucchini, S., Papenfort, K., Sharma, C. M., Rolle, K., Binnewies, T. T., et al. (2008). Deep sequencing analysis of small noncoding RNA and mRNA targets of the global post-transcriptional regulator, Hfq. *PLoS Genetics, 4*(8), e1000163. http://dx.doi.org/10.1371/journal.pgen.1000163.

Sommer, D. D., Delcher, A. L., Salzberg, S. L., & Pop, M. (2007). Minimus: A fast, light-weight genome assembler. *BMC Bioinformatics, 8*, 64. http://dx.doi.org/10.1186/1471-2105-8-64.

Sorek, R., & Cossart, P. (2010). Prokaryotic transcriptomics: A new view on regulation, physiology and pathogenicity. *Nature Reviews Genetics, 11*(1), 9–16. http://dx.doi.org/10.1038/nrg2695.

Stackebrandt, E., & Ebers, J. (2006). Taxonomic parameters revisited: Tarnished gold standards. *Microbiology Today, 33*, 152–155.

Stewart, L., Ford, A., Sangal, V., Jeukens, J., Boyle, B., Caim, S., et al. (2014). Draft genomes of twelve host adapted and environmental isolates of *Pseudomonas aeruginosa* and their position in the core genome phylogeny. *Pathogens and Disease, 71*(1), 20–25. http://dx.doi.org/10.1111/2049-632X.12107.

Sutcliffe, I. C., Trujillo, M. E., & Goodfellow, M. (2012). A call to arms for systematists: Revitalising the purpose and practises underpinning the description of novel microbial taxa. *Antonie Van Leeuwenhoek, 101*(1), 13–20. http://dx.doi.org/10.1007/s10482-011-9664-0.

Tamura, K., Peterson, D., Peterson, N., Stecher, G., Nei, M., & Kumar, S. (2011). MEGA5: Molecular evolutionary genetics analysis using maximum likelihood, evolutionary distance, and maximum parsimony methods. *Molecular Biology and Evolution, 28*(10), 2731–2739.

Tatusov, R. L., Natale, D. A., Garkavtsev, I. V., Tatusova, T. A., Shankavaram, U. T., Rao, B. S., et al. (2001). The COG database: New developments in phylogenetic classification of proteins from complete genomes. *Nucleic Acids Research, 29*(1), 22–28.

Teeling, H., Waldmann, J., Lombardot, T., Bauer, M., & Glockner, F. O. (2004). TETRA: A web-service and a stand-alone program for the analysis and comparison of tetranucleotide usage patterns in DNA sequences. *BMC Bioinformatics, 5*, 163. http://dx.doi.org/10.1186/1471-2105-5-163.

Tindall, B. J. (2014). The correct name of the taxon that contains the type strain of *Rhodococcus equi. International Journal of Systematic and Evolutionary Microbiology, 64*(1), 302–308. http://dx.doi.org/10.1099/ijs.0.059584-0.

Valouev, A., Ichikawa, J., Tonthat, T., Stuart, J., Ranade, S., Peckham, H., et al. (2008). A high-resolution, nucleosome position map of *C. elegans* reveals a lack of universal sequence-dictated positioning. *Genome Research, 18*(7), 1051–1063. http://dx.doi.org/10.1101/gr.076463.108.

Vandamme, P., & Peeters, C. (2014). Time to revisit polyphasic taxonomy. *Antonie Van Leeuwenhoek, 106*(1), 57–65. http://dx.doi.org/10.1007/s10482-014-0148-x.

Vandamme, P., Pot, B., Gillis, M., de Vos, P., Kersters, K., & Swings, J. (1996). Polyphasic taxonomy, a consensus approach to bacterial systematics. *Microbiological Reviews*, *60*(2), 407–438.

van Passel, M. W., Kuramae, E. E., Luyf, A. C., Bart, A., & Boekhout, T. (2006). The reach of the genome signature in prokaryotes. *BMC Evolutionary Biology*, *6*, 84. http://dx.doi.org/10.1186/1471-2148-6-84.

Vernikos, G. S., & Parkhill, J. (2006). Interpolated variable order motifs for identification of horizontally acquired DNA: Revisiting the *Salmonella* pathogenicity islands. *Bioinformatics*, *22*(18), 2196–2203. http://dx.doi.org/10.1093/bioinformatics/btl369.

Wang, K., Li, M., & Hakonarson, H. (2010). ANNOVAR: Functional annotation of genetic variants from high-throughput sequencing data. *Nucleic Acids Research*, *38*(16), e164. http://dx.doi.org/10.1093/nar/gkq603.

Wang, Z., Gerstein, M., & Snyder, M. (2009). RNA-Seq: A revolutionary tool for transcriptomics. *Nature Reviews Genetics*, *10*(1), 57–63. http://dx.doi.org/10.1038/nrg2484.

Wayne, L. G., Brenner, D. J., Colwell, R. R., Grimont, P. A. D., Kandler, O., Krichevsky, M. I., et al. (1987). Report of the *ad hoc* committee on reconciliation of approaches to bacterial systematics. *International Journal of Systematic Bacteriology*, *37*, 463–464.

Westermann, A. J., Gorski, S. A., & Vogel, J. (2012). Dual RNA-seq of pathogen and host. *Nature Reviews. Microbiology*, *10*(9), 618–630. http://dx.doi.org/10.1038/nrmicro2852.

Yi, H., Cho, Y. J., Won, S., Lee, J. E., Jin Yu, H., Kim, S., et al. (2011). Duplex-specific nuclease efficiently removes rRNA for prokaryotic RNA-seq. *Nucleic Acids Research*, *39*(20), e140. http://dx.doi.org/10.1093/nar/gkr617.

Zerbino, D. R., & Birney, E. (2008). Velvet: Algorithms for de novo short read assembly using de Bruijn graphs. *Genome Research*, *18*(5), 821–829. http://dx.doi.org/10.1101/gr.074492.107.

Zhou, Y., Liang, Y., Lynch, K. H., Dennis, J. J., & Wishart, D. S. (2011). PHAST: A fast phage search tool. *Nucleic Acids Research*, *39*(Web Server issue), W347–352. http://dx.doi.org/10.1093/nar/gkr485.

Zhu, B., Wang, X., & Li, L. (2010). Human gut microbiome: The second genome of human body. *Protein & Cell*, *1*(8), 718–725. http://dx.doi.org/10.1007/s13238-010-0093-z.

Zhu, L., Wu, Q., Dai, J., Zhang, S., & Wei, F. (2011). Evidence of cellulose metabolism by the giant panda gut microbiome. *Proceedings of the National Academy of Sciences of the United States of America*, *108*(43), 17714–17719. http://dx.doi.org/10.1073/pnas.1017956108.

Whole-Genome Analyses: Average Nucleotide Identity

6

David R. Arahal[1]

Colección Española de Cultivos Tipo (CECT) Parque Científico Universidad de Valencia,
Paterna, Spain
Departamento de Microbiología y Ecología, Universidad de Valencia, Burjassot,
Valencia, Spain
[1]*Corresponding author: e-mail address: david.ruiz@uv.es*

1 INTRODUCTION

It is obvious that genomics has revolutionized Life Sciences, opening up new lines of enquiry that yield valuable data and insights into biological processes. Prokaryotic systematics has contributed to such developments and to the explosion of knowledge, but it has also benefited from them. Even before the introduction of genomics, it was clear to bacterial systematists that "the complete deoxyribonucleic acid (DNA) sequence would be the reference standard to determine phylogeny and that phylogeny should determine taxonomy" (Wayne et al., 1987). At that time, DNA reassociation was considered to be the best applicable procedure for the circumscription of prokaryotic species, the basic unit of classification. More than a decade later, this method was still seen to be an integral part of the pragmatic species definition for prokaryotes (Stackebrandt et al., 2002), but at the same time, investigators were encouraged to propose new species based upon other genomic methods provided that there was a sufficient degree of congruence between data derived from the new method and DNA reassociation, or in other words, that new recommendations were compatible with existing classifications. Ideally, the ultimate answer was seen to be dependent on the direct comparison of the full genome sequences of the strains to be compared. However, at least three requisites had to be met to turn this approach into a workable option (i) that the balance of advantages and disadvantages justified the use of full genome sequences; (ii) that bioinformatic tools and models were available to handle the wealth of information present in genomes (including unannotated parts and sections with unknown function); and (iii) that the approach could be validated in light of corresponding DNA–DNA hybridization (DDH) data.

1.1 CALCULATION OF AVERAGE NUCLEOTIDE IDENTITY

By 2005, after a decade of microbial whole-genome sequencing, the number of available complete bacterial genomes approached 400. Over this period, the cost of sequencing per nucleotide had dropped about 10^2–10^3 fold, thereby making whole-genome sequencing an affordable option for many laboratories. Konstantinidis and Tiedje (2005), in their seminal paper, fully sequenced the genomes of 70 closely related strains, that is, ones showing >94% 16S rRNA gene similarity, to determine both the conserved predicted protein-coding genes between each pair of strains and the strain-specific genes. They went on to determine how these parameters correlated with evolutionary distances between the strains and their assignment to species. The article by Konstantinidis and Tiedje also explored other data and observations with respect to the definition of prokaryotic species. In contrast, this chapter is focussed on the use of average nucleotide identity (ANI) to delineate archaeal and bacterial species. The strengths of using ANI for this purpose are becoming increasingly evident. ANI is a simple, useful and overall descriptor of genetic relatedness, and in addition to widely distributed genes (typically >1000 genes in total), it is derived from lineage-specific genes, thereby increasing the robustness and resolution of extracted phylogenetic signals. Furthermore, since ANI is based on a large number of genes, it is a better measure of relatedness than data derived from a single gene, such as the 16S rRNA gene. Again, ANI values are not significantly affected by varied evolutionary rates or by horizontal gene transfer of single or a few genes as the effect of fast-evolving genes is mitigated by the slow evolution of other genes.

The ANI values recorded for the set of strains studied by Konstantinidis and Tiedje (2005) showed a strong linear correlation to experimental DDH values ($r^2 = 0.93$). However, in the majority of cases, the sequenced strains were not the same as the ones used in the DDH experiments, a problem that was overcome by using average DDH and ANI values based on several strains of the same species, a strategy that may have improved the correlation between both parameters. Finally, the data set examined was not only limited but biased towards human pathogenic species. Nevertheless, the study revealed some major trends in, and impressions about, the definition of prokaryotic species that were confirmed in subsequent studies: the 70% DDH standard seen as a pragmatic cut-off value for the delineation of species corresponds to $\approx$94% ANI.

In order to examine more accurately the relationship between DDH values and genomic sequence-derived parameters, such as ANI, Goris et al. (2007) determined 124 DDH values among 28 related strains for which whole-genome sequences were available. The results revealed a close relationship between DDH and ANI values. Regression analysis between DDH values and ANI data performed using linear, exponential, power and logarithmic models gave comparably high r^2 correlation values (0.94, 0.94, 0.95 and 0.94, respectively). The recommended cut-off point of 70% DDH for species delineation thereby corresponded to an ANI of $95 \pm 0.5\%$, depending on the specific regression model used. Consequently, it was concluded that ANI can be used with confidence to replace DDH values for strains for which genome sequences are available.

Deloger, El Karoui, and Petit (2009) developed a genomic-distance index based on DNA maximal unique matches shared by two genomes and not only found that it correlated well with ANI but performed better in terms of computational needs.

In this study, 638 pairs of bacterial genomes representing 68 bacterial species were examined.

Richter and Rosselló-Móra (2009) broadened considerably the validation of ANI to include 200 pairwise comparisons with DDH values, representing many more prokaryotic lineages than before. They also advanced its use for the species circumscription of uncultured organisms such as the endosymbionts *Buchnera aphidicola* and *Wolbachia* spp.

1.2 THEORETICAL BACKGROUND: BLAST/MUMmer SOFTWARE

In the study by Konstantinidis and Tiedje (2005), the conserved genes between each pair of genomes were determined using the BLAST program (version 2.2.5; Altschul et al., 1997). For this, all predicted protein-coding sequences from the query genome were searched against the genomic sequence of the reference genome and were considered conserved when they had a BLAST match that was at least 60% of the overall sequence identity (recalculated to an identity along the entire sequence) and an alignable region >70% of the length (nucleotide level) in the reference genome. Protein-coding sequences that had no match or a match below these cut-off values were considered to be genome-specific in the query genome. The BLAST search was run with the following settings: $x = 150$ (drop-off value for gapped alignment), $q = -1$ (penalty for nucleotide mismatch) and $F = F$ (filter for repeated sequences); the rest of the parameters were defaults. These settings give better sensitivity with more distantly related genomes, compared with default settings, because the default settings target more similar sequences. The evolutionary distance between a pair of strains was measured by the ANI of all of the conserved genes between the two genomes as computed by the BLAST local alignment algorithm. Duplicated genes within a genome were defined as genes that had a better match within their genome than in the reference genome during a pairwise whole-genome comparison, using, in all cases, a minimum cut-off for a match of 60% identity over at least 70% of the length of the query gene. Despite the use of rather stringent cut-offs in these comparisons, cases of independent acquisition of very similar genes (instead of gene duplication) cannot be excluded.

The settings mentioned above were refined by Konstantinidis, Ramette, and Tiedje (2006) who used BLASTn release 2.2.9 and a 50% identity cut-off (less stringent than the previous value of 60%) over at least 70% of the length of the gene in the reference genome to identify the reciprocally best-match-conserved (and presumably orthologous) gene set, in order to evaluate the performance and phylogenetic robustness of genetic markers within whole-genome sequences.

Goris et al. (2007) introduced a new method involving the segmentation of the query genome sequence into consecutive 1020 nt fragments to correspond with the fragmentation of the genomic DNA to approximately 1 kb fragments that are formed during experimental DDH reactions, performed using slight modifications of the microplate hybridization method of Ezaki, Hashimoto, and Yabuuchi (1989) or others. The 1020 nt long fragments of the query genome were then used to search against the whole-genomic sequence of the subject genome by using BLASTn with the settings mentioned above; the best match was saved for further analysis. The ANI between the query and the reference genome was calculated as the mean identity of all BLASTn matches that showed more than 30% overall sequence identity (recalculated to an identity along the entire sequence) over an alignable region of at least 70%. Reverse searching, i.e., in which the reference genome is used as the query, was also performed to provide reciprocal values. Although it might seem artificial and unnecessary, a direct implication of this fixed-length partitioning step is that it circumvents the need to use *in silico* ORF predictions and the search for orthologs (Goris et al., 2007).

The BLAST search algorithm follows a hashing approach that involves the generation of a large number of short sequence stretches and character comparisons, whereas MUMmer (Kurtz et al., 2004) makes use of suffix trees to find potential anchors for an alignment. A suffix tree is a data structure that represents all of the substrings of a string (a DNA sequence, a protein sequence, or plain text); the run-time of suffix-tree algorithms, unlike hashing algorithms, does not depend on the length of the maximal matches and thereby provides much faster searches at a genome scale.

The software package JSpecies (Richter & Rosselló-Móra, 2009) was developed as a user-friendly, biologist-oriented interface to analyse and compare species boundaries between genomes, draft genomes, or partial random genome sequences. It offers a graphical user interface to assist microbiologists in the exploration of intergenome similarities. The BLAST calculation of ANI values (called ANIb) in JSpecies was implemented as described by Goris et al. (2007). Once again, ANI values are also calculated by using the NUCmer program in the MUMmer software package (Kurtz et al., 2004) (called ANIm).

2 PREPARATION AND DNA SEQUENCING
2.1 STRAIN CULTIVATION

Media formulations and culture conditions depend on the strains set under study. Bacterial systematics is mainly based on isolated organisms (pure cultures) that can be maintained *ex situ*; hence, this section focusses on them. However, it should not be forgotten that genomic data obtained from enrichments (important for *Candidata* species), single cells and environmental samples (metagenome assemblies) can also be used in comparative studies.

One practical question that arises with respect to a set of strains considered to form a novel species is how many of them need to be included in whole-genome sequencing studies to obtain reliable average nucleotide identities; this situation also applies to other taxonomic studies. Extending analyses to whole strain sets is highly recommended at both the phenotypic and the genomic level in order to obtain a better assessment of intraspecies variability. Again, such comprehensive taxonomic studies may show that the tested strains represent more than one species, a situation easily overlooked if only one or a few strains had been studied.

In addition to confirming the internal coherence of strains that constitute the novel species, it is important to demonstrate the genomic distinctiveness of the taxon from members of related validly named species; at the very least, the type strains of such species should be examined. Currently, a 16S rRNA sequence similarity between strains above 98.65% can be taken as an indication of a close relatedness (Kim, Oh, Park, & Chun, 2014), although it is good practice to use other data to complement and assist in this decision. Once the strain list is at hand, it is important to check whether whole-genome sequences need to be determined for any of the reference strains. Type strains are reference material in taxonomic studies and should be obtained readily from service culture collections, some of which can also provide high-quality DNA, either from in-house DNA banks or upon request. The availability of high-quality DNA avoids the need to cultivate strains, which is especially advantageous with respect to fastidious strains and for strains that cannot be provided as viable cells due to legal or administrative restrictions. It is also an advantage to be able to avoid having to extract DNA from strains as that can be a very tricky step for many organisms.

As Richter and Rosselló-Móra (2009) pointed out, one of the major drawbacks that taxonomists may find in the current genome database (http://www.ncbi.nlm.nih.gov) is that it relies on the identification of the strains that have been sequenced. Errors in the strain designation and/or its identification are known to occur. At that time, only about 10% of the entries were tagged with one of the international strain collection numbers and less than 30% of the sequenced genomes belonged to the type strain of the species for which they were identified (Richter & Rosselló-Móra, 2009).

The availability of whole-genome data from bacterial and archaeal type strains into public databases has increased. But, so has the genome data of other strains, not always well identified, and in many cases not available for research purposes. For these reasons, the use of curated and optimized databases such as EzGenome (http://ezgenome.ezbiocloud.net) is very much recommended.

2.2 DNA EXTRACTION AND QUANTIFICATION

Many different procedures, including conventional methods and the use of commercial kits, can be used to extract sufficient high-quality DNA for whole-genome sequencing. However, since not all methods are equally suited for this purpose, it is necessary to check the quantity, integrity and purity of extracted DNA samples. It is, of course, better to know in advance the demands of the sequencing platform of the sequencing project to ensure that the required standard of DNA preparations has been reached.

DNA extraction protocols, developed for other taxonomic methods such as DDH (Rosselló-Móra, Urdain, & López-López, 2011) and estimation of G + C content (Mesbah, Whitman, & Mesbah, 2011), can be followed as they also require high-quality DNA, often in even larger amounts than are needed for WGS. Low-yield problems experienced with some commercial kits that work well for PCR amplification can be solved by scaling up, that is, by using several extraction columns. Low

DNA concentration and associated impurities can be resolved by using additional steps. However, highly sheared DNA cannot be used for WGS. Purification of DNA preparations can be achieved by using columns (e.g. Qiagen), AMPure Beads (Beckman Coulter), or by precipitation with ethanol. In the latter case, it is important to completely remove traces of ethanol so as not to interfere in the enzymatic steps involved in the preparation of DNA libraries.

In general, laboratories performing NGS sequencing will expect DNA samples to be free of any contaminant. Fluorometric methods such as Quant-IT (Qubit, Invitrogen) are preferred to spectrophotometric methods for accurate quantification of DNA samples, which are less reproducible and tend to overestimate concentrations by a factor of 2. An image of an agarose gel of the samples is frequently requested to confirm their integrity (genomic DNAs must be observed as single bands without smears).

The total amount of sample required is determined by the experimental strategy, including details of the sequencing platform, as well as on the type and number of DNA libraries to be constructed. The amount can be as low as 0.5–5 µg in 100 µl of nuclease-free water for shot-gun libraries, but in the paired-end strategy, more DNA is required as the size of the insert increases (>60 µg in 100 µl of nuclease-free water for paired-end 40 kb fragments).

It is wise to perform an authenticity check before processing samples or shipping them to a next-generation facility though this step is often overlooked. This step can involve partial sequencing of the 16S rRNA gene (first ~900 bp; Arahal, Sánchez, Macián, & Garay, 2008) or any other known gene, or by using a fingerprinting method. Authenticity checks do not always ensure that samples are of a sufficiently high standard, but they will detect contamination and whether samples have been mislabelled.

2.3 WHOLE-GENOME SEQUENCING

High-throughput DNA sequencing technology is evolving to such an extent that improvements in the quality, speed and cost of the method are making other recent cutting edge technologies obsolete. The advantages and details associated with each NGS system, including costs, components, error rates and applications, have been the subject of several excellent reviews (Liu et al., 2012; Loman et al., 2012; Quail et al., 2012). The impact of such developments on microbial genome assembly, annotation and analysis of data of taxonomic value has also been addressed (Borriss et al., 2011; Dark, 2013; Mavromatis et al., 2012). Benchtop instruments include 454 GS FLX Titanium/GS Junior (Roche), Genome Analyzer/HiSeq 2000/MiSeq (Illumina), SOLiD/Ion Torrent PGM (Life Technologies) and RS (Pacific Biosciences). Some of these produce low output and are suitable for individual laboratories. The availability of such instruments in biotech companies, hospitals, research institutes and universities allows laboratory staff to carry out whole-genome sequencing (preparation of libraries, emulsion PCR, sequencing) and initial post-sequencing (assembly, metrics) procedures. More advanced bioinformatic analyses of whole-genome sequences can usually be provided in house.

Inexperienced users (in massive sequencing) will also appreciate receiving assistance for the experimental design that best fits their needs and budget. For example, multiplexing with barcodes to tag two or more samples sounds tempting because it permits the cost of one GS Junior run among those samples to be split, lowering the cost per sample. However, the total cost of the run, compared to just one sample, is higher because of the additional library preparation and barcoding required. Moreover, the coverage per strain will of course be lower because the reads are shared, while the likely uneven distribution of reads may mean that not all samples reach the desired sequencing depth of coverage, thereby causing delays and additional costs. To avoid this, it is recommended to aim for a 25-fold coverage as this is usually enough to assemble correctly 95–99% of a good-quality sequence (Borriss et al., 2011); for most prokaryotes (genome sizes 2–8 Mb) this means 50–200 Mb length. The current tendency for prokaryotic genomic sequencing and assembly is to use a long-read, high-error platform (such as Pacific Biosciences and Oxford Nanopore Technologies) to get the scaffold of the sequence and then a short-read, low-error platform (such as Illumina or SOLiD) which can provide an extraordinarily high coverage of reads that are mapped onto the scaffold to provide a highly reliable genomic sequence.

ANI calculations are not seriously affected by low coverage as it has been demonstrated that randomly sequencing $\geq 20\%$ of the genome of query strains, or producing an alignment equivalent >4% of their genome sizes, can serve taxonomic purposes (Richter & Rosselló-Móra, 2009). We have tested this approach a number of times (Lucena, Ruvira, Arahal, Macián, & Pujalte, 2012; Lucena, Ruvira, Macián, Pujalte, & Arahal, 2013, 2014; Ruvira, Lucena, Pujalte, Arahal, & Macián, 2013; and unpublished results) and are convinced of its validity. However, at the same time, we believe that it is worth investing, at least with type strains, in the generation of good-quality permanent drafts or even finished genomes rather than doing low-depth *de novo* sequencing. Again, methodological progress makes this approach increasingly feasible.

3 ANI CALCULATIONS USING JSPECIES

3.1 INSTALLATION

JSpecies is written in the platform-independent, object-oriented programming language Java (Richter & Rosselló-Móra, 2009). It can be started using the Java Web Start technology, which automatically downloads and installs the software locally. This ensures the user will always obtain access to the latest available version. Alternatively, it can be downloaded and installed manually. To calculate species relationships, two additional software packages need to be locally installed: BLAST (Altschul et al., 1997) and MUMmer (Kurtz et al., 2004). JSpecies is freely available from the project Web site (http://www.imedea.uib.es/jspecies/), where further information and documentation about the tool and how to use it are provided. JSpecies runs on Linux or Microsoft Windows operating systems, but unfortunately, one of the components, the MUMmer software, is only available for Linux. In the case of BLAST, updates are frequent. The latest version, release 2.2.29+, can be found at ftp://ftp.ncbi.nih.gov/blast/executables/LATEST/ though it seems that BLAST+ cannot be used (personal observation) because JSpecies calls three subprograms (blastall, fastcmd and formatdb) that can be found in legacy BLAST, either the latest or previous consolidated releases at ftp://ftp.ncbi.nih.gov/blast/executables/release.

3.2 OPERATION

Some parameters need to be adjusted the first time JSpecies is used. The user has to specify the folder where all JSpecies data are to be stored (the workspace), and point to blastall, nucmer, formatdb and fastacmd executables indicating their paths. This can be done by opening the preferences dialogue box (Figure 1) through the menu item Edit. If sequences are going to be retrieved from a remote location, it is important to give its URL. These general settings can be changed at any time. Users can also make changes to the ANIb and ANIm parameters or just leave the default values since they have been evaluated and perform well for the intended use.

Once the program has started, the files (fasta format sequences) to be compared must be downloaded. The whole set is termed a group in JSpecies. So, initially a new group has to be created though in subsequent sessions it is also possible to open previously stored groups (these also retain the calculations performed on them). Groups, newly created or reopened, can be fed with sequences stored in the computer (import Fasta(s) from file(s)) and/or from remote sites (import Fasta(s) from www). The second option does, of course, require internet connection.

At this stage several menus will be shown. In the upper part, right below the main menu bar the downloaded files (sequences of the group) will be arranged in different

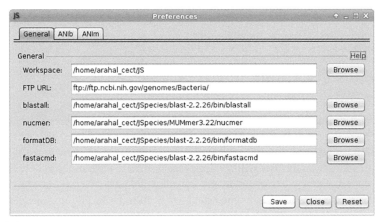

FIGURE 1

Screenshot of the general settings lash in the preferences dialogue box. Data of a real installation have been left to serve as an example.

lashes. By selecting the desired file, the information on the sequence will appear in the window below. A graphic will show the number of fragments (chromosomes/plasmids/contigs) that will be taken into account together with their lengths, G+C content and counts of each base. An example of a finished and draft genome can be seen in Figures 2 and 3, respectively.

Calculation, at the main menu bar, prompts a display of a new window where the different pairs of files to be compared and the different parameters to be calculated can be selected. Progress can be read at the bottom of the main navigation window which also registers the values already obtained (in a sort of self-navigation fashion). When all of the calculations are completed, which may take minutes to hours depending on the number of files, the bottom line displays Ready and the user can retrieve the matrices of results directly from the Result button at the main menu. This also permits them to be saved as text files. To aid interpretation of data, values which show that sequences belong to separate species are coloured in red, while members of the same genomic species are coloured in green.

Other interesting data can be retrieved by navigating the main window. This can be done with the buttons right below the downloaded files: Sequence shows the features of the fasta file that is active at that moment, while ANIb, Tetra and ANIm give

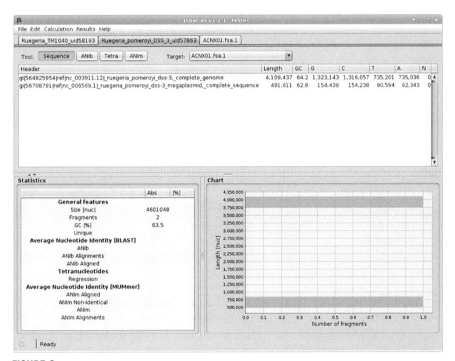

FIGURE 2

Screenshot of JSpecies main window displaying sequence data. The example corresponds to the closed genome of *Ruegeria pomeroyi* DSS-3^T (one chromosome and a megaplasmid).

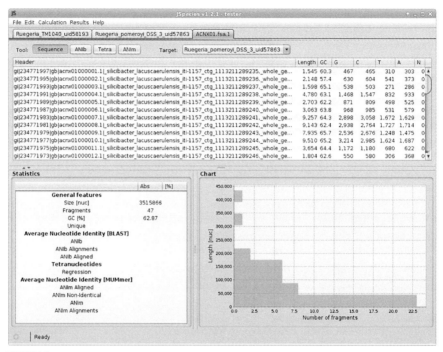

FIGURE 3

Screenshot of JSpecies main window displaying sequence data. The example corresponds to the draft genome of *Ruegeria lacuscaerulensis* ITI-1157[T] (47 contigs).

the results of the respective comparisons between the query and the displayed target genome. The strain lashes can be selected to complete the navigation.

The ANIb tool shows the different generated fragments (default size of 1020 nt): the length and position of the different fragments in the fasta file (Figure 4). The chart shows the number of compared fragments distributed according to the percentages of ANI identities. The statistics window shows the overall calculated results. ANIb alignments indicate the number of fragments used for comparisons, and the percentage corresponding to the total fragments generated. ANIb aligned indicates the total number of nucleotides compared and the percentage corresponding to the total nucleotides present in the query fasta file.

The ANIm tool shows the different fragments of the query fasta file used for calculation with their different lengths and positions within the file (Figure 5). The chart window shows the distribution and number of fragments used to calculate ANIm, the statistics window the overall results calculated and ANIm aligned the number of nucleotides used for calculation and its percentage corresponding to the complete query fasta file. ANIm Non-Identical shows the number of nucleotides not used for calculation and the percentage corresponding to the complete query fasta file and ANIm the total nucleotides used to calculate the parameter, and the ANIm value. ANIm alignments indicate the number of fragments used for calculation independently of their size.

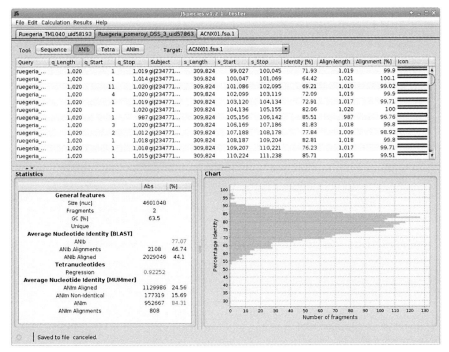

FIGURE 4

Screenshot of JSpecies main window displaying data related to an ANIb calculation. (See the colour plate.)

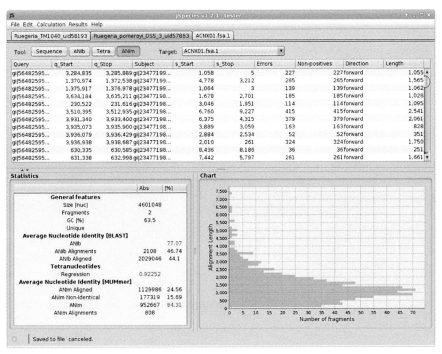

FIGURE 5

Screenshot of JSpecies main window displaying data related to an ANIm calculation. (See the colour plate.)

3.3 CALCULATIONS ON-LINE

During the writing of this chapter a JSpecies Web service was launched (http://jspecies.ribohost.com/jspeciesws/) making unnecessary the local installation described above. It contains several interesting functionalities, for instance, the possibility of sharing an analysis with colleagues by exchanging a generated session code. And among other useful tools, EzGenome (http://ezgenome.ezbiocloud.net) also allows calculating pairwise ANI for a query and a subject genome.

4 INTERPRETATION AND PUBLICATION OF RESULTS

The ease by which ANI results can be interpreted for genetic relatedness compared among strains does not mean that other microbiological practices or even scientific judgement have to be abandoned. On the contrary, the decision to include two or more strains in a new species should only be taken when underpinned by sound-independent genomic and phenotypic data (Tindall, Rosselló-Móra, Busse, Ludwig, & Kämpfer, 2010). ANI values of 95–96% correspond to the 70% DDH cut-off point recommended by Wayne et al. (1987) for the delineation of bacterial species. However, these values, while important, need to be in agreement with corresponding phenotypic and phylogenetic data.

Despite pressure to keep descriptions of new taxa short and to the point, authors should, as with other types of sequence data, provide key information to both reviewers and the scientific community. This includes identification of the sequences used in the study (by their accession numbers) and details of associated parameters (particularly if non-default settings have been used). Furthermore, any new sequences should be submitted to a public repository, even low-coverage assemblies and raw sequence data. The Sequence Read Archive (SRA) at NCBI (http://www.ncbi.nlm.nih.gov/Traces/sra/) stores raw sequence data and alignments of next-generation sequencing technologies including 454, Ion Torrent, Illumina, SOLiD, Helicos, Pacific Biosciences and Complete Genomics. Aligned sequences may be submitted in BAM format. This can also be done at the European Bioinformatics Institute through the ENA's Sequence Read Archive (SRA) Webin submission service (https://www.ebi.ac.uk/ena/submit/sra/#home) and at the DNA Database of Japan through the DDBJ Sequence Read Archive (DRA) (http://trace.ddbj.nig.ac.jp/dra/index_e.html). All three are members of the International Nucleotide Sequence Database Collaboration (INSDC); hence, data submitted to any of these organizations are shared between them.

5 APPLICATION TO PROKARYOTIC CLASSIFICATION: CASE STUDIES

The impact of ANI calculations in microbiology is enormous, as testified by the many citations to the papers that introduced and promoted the ANI methodology

(Goris et al., 2007; Konstantinidis & Tiedje, 2005; Richter & Rosselló-Móra, 2009). These articles spell out the relevance of ANI values to both pure and applied microbiology. To the best of my knowledge, the first article making use of ANI values in the description of new species was by Bakermans et al. (2006), who described *Psychrobacter cryohalolentis* and *Psychrobacter arcticus* by taking an advantage of the availability of whole-genome sequences of their type strains, isolates K5^T and 273-4^T; these strains shared an ANI value of 88%. This lead has been followed by many additional workers, as summarized in Table 1.

Table 1 Taxonomic Proposals Supported by ANI Values Arranged According to Publication Year Then According to Names of Senior Authors

References	Taxonomic Proposal	*n*	New WGS (Assembly Level)
Bakermans et al. (2006)	*Psychrobacter cryohalolentis* sp. nov., *Psychrobacter arcticus* sp. nov.	2	0
Vanlaere et al. (2009)	*Burkholderia contaminans* sp. nov., *Burkholderia lata* sp. nov.	14	0
Haley et al. (2010)	"*Vibrio metecus*" sp. nov., "*Vibrio parilis*" sp. nov.	28	2 (contig)
Hamilton-Brehm et al. (2010)	*Caldicellulosiruptor obsidiansis* sp. nov.	4	0
Prakash et al. (2010)	*Geobacter daltonii* sp. nov.	2	2 (gapless)
Blainey, Mosier, Potanina, Francis, and Quake (2011)	"*Candidatus* Nitrosoarchaeum limnia"	2	1 (contig)
Nemec et al. (2011)	*Acinetobacter pittii* sp. nov., *Acinetobacter nosocomialis* sp. nov.	7	0
Campbell et al. (2011)	"*Nitrosococcus watsonii*" sp. nov.	10	0
García-Aljaro et al. (2012)	"*Neoscardovia arbecensis*" gen. nov., sp. nov.	8	4 (contig)
Hoffmann et al. (2012)	*Vibrio caribbeanicus* sp. nov.	19	7 (contig)
Lü and Lu (2012)	*Methanocella conradii* sp. nov.	3	0
Lucena et al. (2012)	*Vibrio aestivus* sp. nov., *Vibrio quintilis* sp. nov.	5	5 (contig)
Nesbø et al. (2012)	*Mesotoga prima* gen. nov., sp. nov.	2	1 (contig)
Chi et al. (2013)	"*Gayadomonas joobiniege*" gen. nov, sp. nov.	27	0
Cho et al. (2013)	"*Mycobacterium abscessus* subsp. *massiliense*" subsp. nov.	3	1 (contig)

Continued

Table 1 Taxonomic Proposals Supported by ANI Values Arranged According to Publication Year Then According to Names of Senior Authors—cont'd

References	Taxonomic Proposal	n	New WGS (Assembly Level)
Delamuta et al. (2013)	*Bradyrhizobium diazoefficiens* sp. nov.	4	1 (contig)
den Bakker, Manuel, Fortes, Wiedmann, and Nightingale (2013)	*Listeria fleischmannii* subsp. *coloradonensis* subsp. nov.	9	2 (contig)
Guinebretière et al. (2013)	*Bacillus cytotoxicus* sp. nov.	8	0
Jiménez et al. (2013)	*Bacillus toyonensis* sp. nov.	41	1 (gapless)
Jores et al. (2013)	"*Mycoplasma feriruminatoris*" sp. nov.	10	1 (contig)
Lebedeva et al. (2013)	"*Candidatus* Nitrosotenuis uzonensis"	3	1 (contig)
Lee, Park, Yi, and Chun (2013)	*Flavobacterium limnosediminis* sp. nov.	4	4 (contig)
Löffler et al. (2013)	*Dehalococcoidia* classis nov., *Dehalococcoidales* ord. nov., *Dehalococcoidaceae* fam. nov., *Dehalococcoides mccartyi* gen. nov., sp. nov.	5	0
Lucena et al. (2013)	*Tropicibacter mediterraneus* sp. nov., *Tropicibacter litoreus* sp. nov.	5	5 (contig)
Matsumoto et al. (2013)	*Ilumatobacter nonamiense* sp. nov. *Ilumatobacter coccineum* sp. nov.	2	1 (gapless), 1 (contig)
Ruvira et al. (2013)	*Marinifilum flexuosum* sp. nov.	2	2 (contig)
Saw et al. (2013)	"*Gloeobacter kilaueensis*" sp. nov.	2	1 (gapless)
Tomás, Karakashev, and Angelidaki (2013)	*Thermoanaerobacter pentosaceus* sp. nov.	10	1 (contig)
Villeneuve, Martineau, Mauffrey, and Villemur (2013)	*Methylophaga nitratireducenticrescens* sp. nov., *Methylophaga frappieri* sp. nov.	2	2 (gapless)
Camelo-Castillo, Benítez-Páez, Belda-Ferre, Cabrera-Rubio, and Mira (2014)	*Streptococcus dentisani* sp. nov.	139	2 (contig)
van der Wolf et al. (2014)	*Dickeya solani* sp. nov.	12	0
Liu et al. (2014)	"*Bacillus bingmayongensis*" sp. nov.	7	0
Lucena et al. (2014)	"*Roseovarius albus*" sp. nov.	3	3 (contig)

n, the number of compared genomes is followed by the number of new genome sequences generated in the study together with an indication of the assembly level.

CONCLUDING REMARKS

It is evident from Table 1 that prokaryote taxonomists are increasingly making use of ANI values in the delineation of new species. Indeed, the limiting factor in the application of this approach to the description of new species is the lack of whole-genome sequences of type strains (Chun & Rainey, 2014), a problem that is increasingly being addressed given the reduction in costs of whole-genome sequences. The sum of genomic sequences in the 33 cases outlined in Table 1 is about 8 times higher than the number of contributed sequences (404 to 51) which illustrates the advantage of cumulative repositories. It is interesting that one-third of the case studies draw on sequences from public databases though sequences do have to be generated for novel isolates. It also needs to be stressed that additional valuable taxonomic information can be gleaned from whole-genome sequences (even from partial ones with reasonable random coverage), including the G + C content of the DNA, and sequences of genetic markers. Moreover, other analyses can be used to define the core genome, singleton genes , genomic islands and the pan genome, thereby providing other estimates of genomic relatedness (Ben Zakour, Beatson, van den Broek, Thoday, & Fitzgerald, 2012; Borriss et al., 2011; Chan, Halachev, Loman, Constantinidou, & Pallen, 2012; Flowers et al., 2013; Frank, Klockow, Richter, Glöckner, & Harder, 2013; Romano et al., 2013; Scortichini, Marcelletti, Ferrante, & Firrao, 2013; Sentausa & Fournier, 2013; Tamura et al., 2012; Yi, Chun, & Cha, 2014; Zhang, Tian, Sui, Chen, & Chen, 2012).

Finally, whole-genome sequences generated by taxonomists provide valuable data for biotechnology and bioprospecting, not least for data mining.

ACKNOWLEDGEMENTS

The author is grateful for support underpinning several projects (Grants CGL2005-02292/BOS and CGL2010-18134/BOS from the Spanish Government and grants ACOMP06/177 and PROMETEO2012/040 from Generalitat Valenciana, Spain). Thanks are also due to J. Chun, M. Goodfellow, A. Mira and R. Rosselló-Móra for critically reading the manuscript.

REFERENCES

Altschul, S. F., Madden, T. L., Schäffer, A. A., Zhang, J., Zhang, Z., Miller, W., et al. (1997). Gapped BLAST and PSI-BLAST: A new generation of protein database search programs. *Nucleic Acids Research, 25,* 3389–3402.

Arahal, D. R., Sánchez, E., Macián, M. C., & Garay, E. (2008). Value of *recN* sequences for species identification and as a phylogenetic marker within the family "*Leuconostocaceae*" *International Microbiology, 11,* 33–39.

Bakermans, C., Ayala-del-Río, H. L., Ponder, M. A., Vishnivetskaya, T., Gilichinsky, D., Thomashow, M. F., et al. (2006). *Psychrobacter cryohalolentis* sp. nov. and *Psychrobacter arcticus* sp. nov., isolated from Siberian permafrost. *International Journal of Systematic and Evolutionary Microbiology, 56,* 1285–1291.

Ben Zakour, N. L., Beatson, S. A., van den Broek, A. H., Thoday, K. L., & Fitzgerald, J. R. (2012). Comparative genomics of the *Staphylococcus intermedius* group of animal pathogens. *Frontiers in Cellular and Infection Microbiology*, *2*, 44.

Blainey, P. C., Mosier, A. C., Potanina, A., Francis, C. A., & Quake, S. R. (2011). Genome of a low-salinity ammonia-oxidizing archaeon determined by single-cell and metagenomic analysis. *PLoS One*, *6*, e16626.

Borriss, R., Rueckert, C., Blom, J., Bezuidt, O., Reva, O., & Klenk, H.-P. (2011). Whole genome sequence comparisons in taxonomy. In F. Rainey & A. Oren (Eds.), *Methods in microbiology: Vol. 38*, (pp. 409–436). London: Academic Press.

Camelo-Castillo, A., Benítez-Páez, A., Belda-Ferre, P., Cabrera-Rubio, R., & Mira, A. (2014). *Streptococcus dentisani* sp. nov. a novel member of the Mitis group. *International Journal of Systematic and Evolutionary Microbiology*, *64*, 60–65.

Campbell, M. A., Chain, P. S., Dang, H., El Sheikh, A. F., Norton, J. M., Ward, N. L., et al. (2011). *Nitrosococcus watsonii* sp. nov., a new species of marine obligate ammonia-oxidizing bacteria that is not omnipresent in the world's oceans: Calls to validate the names '*Nitrosococcus halophilus*' and '*Nitrosomonas mobilis*'. *FEMS Microbiology Ecolology*, *76*, 39–48.

Chan, J. Z., Halachev, M. R., Loman, N. J., Constantinidou, C., & Pallen, M. J. (2012). Defining bacterial species in the genomic era: Insights from the genus *Acinetobacter*. *BMC Microbiology*, *12*, 302.

Chi, W. J., Park, J. S., Kwak, M. J., Kim, J. F., Chang, Y. K., & Hong, S. K. (2013). Isolation and characterization of a novel agar-degrading marine bacterium, *Gayadomonas joobiniege* gen. nov, sp. nov., from the Southern Sea, Korea. *Journal of Microbiology and Biotechnology*, *23*, 1509–1518.

Cho, Y. J., Yi, H., Chun, J., Cho, S. N., Daley, C. L., Koh, W. J., et al. (2013). The genome sequence of '*Mycobacterium massiliense*' strain CIP 108297 suggests the independent taxonomic status of the *Mycobacterium abscessus* complex at the subspecies level. *PLoS One*, *8*, e81560.

Chun, J., & Rainey, F. A. (2014). Integrating genomics into the taxonomy and systematics of the *Bacteria* and *Archaea*. *International Journal of Systematic and Evolutionary Microbiology*, *64*, 316–324.

Dark, M. J. (2013). Whole-genome sequencing in bacteriology: State of the art. *Infection and Drug Resistance*, *6*, 115–123.

Delamuta, J. R., Ribeiro, R. A., Ormeño-Orrillo, E., Melo, I. S., Martínez-Romero, E., & Hungria, M. (2013). Polyphasic evidence supporting the reclassification of *Bradyrhizobium japonicum* group Ia strains as *Bradyrhizobium diazoefficiens* sp. nov. *International Journal of Systematic and Evolutionary Microbiology*, *63*, 3342–3351.

Deloger, M., El Karoui, M., & Petit, M. A. (2009). A genomic distance based on MUM indicates discontinuity between most bacterial species and genera. *Journal of Bacteriology*, *191*, 91–99.

den Bakker, H. C., Manuel, C. S., Fortes, E. D., Wiedmann, M., & Nightingale, K. K. (2013). Genome sequencing identifies *Listeria fleischmannii* subsp. *coloradonensis* subsp. nov., isolated from a ranch. *International Journal of Systematic and Evolutionary Microbiology*, *63*, 3257–3268.

Ezaki, T., Hashimoto, Y., & Yabuuchi, E. (1989). Fluorometric deoxyribonucleic acid-deoxyribonucleic acid hybridization in microdilution wells as an alternative to membrane-filter hybridization in which radioisotopes are used to determine genetic relatedness among bacterial strains. *International Journal of Systematic Bacteriology*, *39*, 224–229.

Flowers, J. J., He, S., Malfatti, S., Del Rio, T. G., Tringe, S. G., Hugenholtz, P., et al. (2013). Comparative genomics of two 'Candidatus Accumulibacter' clades performing biological phosphorus removal. *ISME Journal, 7*, 2301–2314.

Frank, C. S., Klockow, C., Richter, M., Glöckner, F. O., & Harder, J. (2013). Genetic diversity of *Rhodopirellula* strains. *Antonie Van Leeuwenhoek, 104*, 547–550.

García-Aljaro, C., Ballesté, E., Rosselló-Móra, R., Cifuentes, A., Richter, M., & Blanch, A. R. (2012). *Neoscardovia arbecensis* gen. nov., sp. nov., isolated from porcine slurries. *Systematic and Applied Microbiology, 35*, 374–379.

Goris, J., Konstantinidis, K. T., Klappenbach, J. A., Coenye, T., Vandamme, P., & Tiedje, J. M. (2007). DNA-DNA hybridization values and their relationship to whole-genome sequence similarities. *International Journal of Systematic and Evolutionary Microbiology, 57*, 81–91.

Guinebretière, M. H., Auger, S., Galleron, N., Contzen, M., De Sarrau, B., De Buyser, M. L., et al. (2013). *Bacillus cytotoxicus* sp. nov. is a novel thermotolerant species of the *Bacillus cereus* group occasionally associated with food poisoning. *International Journal of Systematic and Evolutionary Microbiology, 63*, 31–40.

Haley, B. J., Grim, C. J., Hasan, N. A., Choi, S. Y., Chun, J., Brettin, T. S., et al. (2010). Comparative genomic analysis reveals evidence of two novel *Vibrio* species closely related to *V. cholerae*. *BMC Microbiology, 10*, 154.

Hamilton-Brehm, S. D., Mosher, J. J., Vishnivetskaya, T., Podar, M., Carroll, S., Allman, S., et al. (2010). *Caldicellulosiruptor obsidiansis* sp. nov., an anaerobic, extremely thermophilic, cellulolytic bacterium isolated from Obsidian Pool, Yellowstone National Park. *Applied and Environmental Microbiology, 76*, 1014–1020.

Hoffmann, M., Monday, S. R., Allard, M. W., Strain, E. A., Whittaker, P., Naum, M., et al. (2012). *Vibrio caribbeanicus* sp. nov., isolated from the marine sponge *Scleritoderma cyanea*. *International Journal of Systematic and Evolutionary Microbiology, 62*, 1736–1743.

Jiménez, G., Urdiain, M., Cifuentes, A., López-López, A., Blanch, A. R., Tamames, J., et al. (2013). Description of *Bacillus toyonensis* sp. nov., a novel species of the *Bacillus cereus* group, and pairwise genome comparisons of the species of the group by means of ANI calculations. *Systematic and Applied Microbiology, 36*, 383–391.

Jores, J., Fischer, A., Sirand-Pugnet, P., Thomann, A., Liebler-Tenorio, E. M., Schnee, C., et al. (2013). *Mycoplasma feriruminatoris* sp. nov., a fast growing *Mycoplasma* species isolated from wild *Caprinae*. *Systematic and Applied Microbiology, 36*, 533–538.

Kim, M., Oh, H. S., Park, S. C., & Chun, J. (2014). Towards a taxonomic coherence between average nucleotide identity and 16S rRNA gene sequence similarity for species demarcation of prokaryotes. *International Journal of Systematic and Evolutionary Microbiology, 64*, 346–351.

Konstantinidis, K. T., Ramette, A., & Tiedje, J. M. (2006). Toward a more robust assessment of intraspecies diversity, using fewer genetic markers. *Applied and Environmental Microbiology, 72*, 7286–7293.

Konstantinidis, K. T., & Tiedje, J. M. (2005). Genomic insights that advance the species definition for prokaryotes. *Proceedings of the National Academy of Sciences of the United States of America, 102*, 2567–2572.

Kurtz, S., Phillippy, A., Delcher, A. L., Smoot, M., Shumway, M., Antonescu, C., et al. (2004). Versatile and open software for comparing large genomes. *Genome Biology, 5*, R12.

Lebedeva, E. V., Hatzenpichler, R., Pelletier, E., Schuster, N., Hauzmayer, S., Bulaev, A., et al. (2013). Enrichment and genome sequence of the group i.1a ammonia-oxidizing archaeon "Ca. Nitrosotenuis uzonensis" representing a clade globally distributed in thermal habitats. *PLoS One, 8*, e80835.

Lee, K., Park, S. C., Yi, H., & Chun, J. (2013). *Flavobacterium limnosediminis* sp. nov., isolated from sediment of a freshwater lake. *International Journal of Systematic and Evolutionary Microbiology, 63,* 4784–4789.

Liu, L., Li, Y., Li, S., Hu, N., He, Y., Pong, R., et al. (2012). Comparison of next-generation sequencing systems. *Journal of Biomedicine and Biotechnology, 2012,* 251364.

Liu, B., Liu, G. H., Hu, G. P., Sengonca, C., Lin, N. Q., Tang, J. Y., et al. (2014). *Bacillus bingmayongensis* sp. nov., isolated from the pit soil of Emperor Qin's Terra-cotta warriors in China. *Antonie Van Leeuwenhoek, 105,* 501–510.

Löffler, F. E., Yan, J., Ritalahti, K. M., Adrian, L., Edwards, E. A., Konstantinidis, K. T., et al. (2013). *Dehalococcoides mccartyi* gen. nov., sp. nov., obligately organohalide-respiring anaerobic bacteria relevant to halogen cycling and bioremediation, belong to a novel bacterial class, *Dehalococcoidia* classis nov., order *Dehalococcoidales* ord. nov. and family *Dehalococcoidaceae* fam. nov., within the phylum *Chloroflexi. International Journal of Systematic and Evolutionary Microbiology, 63,* 625–635.

Loman, N. J., Misra, R. V., Dallman, T. J., Constantinidou, C., Gharbia, S. E., Wain, J., et al. (2012). Performance comparison of benchtop high-throughput sequencing platforms. *Nature Biotechnology, 30,* 434–439.

Lü, Z., & Lu, Y. (2012). *Methanocella conradii* sp. nov., a thermophilic, obligate hydrogenotrophic methanogen, isolated from Chinese rice field soil. *PLoS One, 7,* e35279.

Lucena, T., Ruvira, M. A., Arahal, D. R., Macián, M. C., & Pujalte, M. J. (2012). *Vibrio aestivus* sp. nov. and *Vibrio quintilis* sp. nov., related to Marisflavi and Gazogenes clades, respectively. *Systematic and Applied Microbiology, 35,* 427–431.

Lucena, T., Ruvira, M. A., Macián, M. C., Pujalte, M. J., & Arahal, D. R. (2013). Description of *Tropicibacter mediterraneus* sp. nov. and *Tropicibacter litoreus* sp. nov. *Systematic and Applied Microbiology, 36,* 325–329.

Lucena, T., Ruvira, M. A., Macián, M. C., Pujalte, M. J., & Arahal, D. R. (2014). *Roseovarius albus* sp. nov., a new Alphaproteobacterium isolated from the Mediterranean Sea. *Antonie Van Leeuwenhoek, 105,* 671–678.

Matsumoto, A., Kasai, H., Matsuo, Y., Shizuri, Y., Ichikawa, N., Fujita, N., et al. (2013). *Ilumatobacter nonamiense* sp. nov. and *Ilumatobacter coccineum* sp. nov., isolated from seashore sand. *International Journal of Systematic and Evolutionary Microbiology, 63,* 3404–3408.

Mavromatis, K., Land, M. L., Brettin, T. S., Quest, D. J., Copeland, A., Clum, A., et al. (2012). The fast changing landscape of sequencing technologies and their impact on microbial genome assemblies and annotation. *PLoS One, 7,* e48837.

Mesbah, N. M., Whitman, W. B., & Mesbah, M. (2011). Determination of the G+C content of prokaryotes. In F. Rainey & A. Oren (Eds.), *Methods in microbiology: Vol. 38,* (pp. 299–324). London: Academic Press.

Nemec, A., Krizova, L., Maixnerova, M., van der Reijden, T. J., Deschaght, P., Passet, V., et al. (2011). Genotypic and phenotypic characterization of the *Acinetobacter calcoaceticus-Acinetobacter baumannii* complex with the proposal of *Acinetobacter pittii* sp. nov. (formerly *Acinetobacter* genomic species 3) and *Acinetobacter nosocomialis* sp. nov. (formerly *Acinetobacter* genomic species 13TU). *Research in Microbiology, 162,* 393–404.

Nesbø, C. L., Bradnan, D. M., Adebusuyi, A., Dlutek, M., Petrus, A. K., Foght, J., et al. (2012). *Mesotoga prima* gen. nov., sp. nov., the first described mesophilic species of the *Thermotogales. Extremophiles, 16,* 387–393.

Prakash, O., Gihring, T. M., Dalton, D. D., Chin, K. J., Green, S. J., Akob, D. M., et al. (2010). *Geobacter daltonii* sp. nov., an Fe(III)- and uranium(VI)-reducing bacterium isolated from a shallow subsurface exposed to mixed heavy metal and hydrocarbon contamination. *International Journal of Systematic and Evolutionary Microbiology, 60*, 546–553.

Quail, M. A., Smith, M., Coupland, P., Otto, T. D., Harris, S. R., Connor, T. R., et al. (2012). A tale of three next generation sequencing platforms: Comparison of Ion Torrent, Pacific Biosciences and Illumina MiSeq sequencers. *BMC Genomics, 13*, 341.

Richter, M., & Rosselló-Móra, R. (2009). Shifting the genomic gold standard for the prokaryotic species definition. *Proceedings of the National Academy of Sciences of the United States of America, 106*, 19126–19131.

Romano, C., D'Imperio, S., Woyke, T., Mavromatis, K., Lasken, R., Shock, E. L., et al. (2013). Comparative genomic analysis of phylogenetically closely related *Hydrogenobaculum* sp. isolates from Yellowstone National Park. *Applied and Environmental Microbiology, 79*, 2932–2943.

Rosselló-Móra, R., Urdiain, M., & López-López, A. (2011). DNA-DNA hybridization. In F. Rainey & A. Oren (Eds.), *Methods in microbiology: Vol. 38*, (pp. 325–347). London: Academic Press.

Ruvira, M. A., Lucena, T., Pujalte, M. J., Arahal, D. R., & Macián, M. C. (2013). *Marinifilum flexuosum* sp. nov., a new *Bacteroidetes* isolated from coastal Mediterranean Sea water and emended description of the genus *Marinifilum* Na et al., 2009. *Systematic and Applied Microbiology, 36*, 155–159.

Saw, J. H., Schatz, M., Brown, M. V., Kunkel, D. D., Foster, J. S., Shick, H., et al. (2013). Cultivation and complete genome sequencing of *Gloeobacter kilaueensis* sp. nov., from a lava cave in Kīlauea Caldera, Hawai'i. *PLoS One, 8*, e76376.

Scortichini, M., Marcelletti, S., Ferrante, P., & Firrao, G. (2013). A genomic redefinition of *Pseudomonas avellanae* species. *PLoS One, 8*, e75794.

Sentausa, E., & Fournier, P. E. (2013). Advantages and limitations of genomics in prokaryotic taxonomy. *Clinical Microbiology and Infection, 19*, 790–795.

Stackebrandt, E., Frederiksen, W., Garrity, G. M., Grimont, P. A., Kämpfer, P., Maiden, M. C., et al. (2002). Report of the ad hoc committee for the re-evaluation of the species definition in bacteriology. *International Journal of Systematic and Evolutionary Microbiology, 52*, 1043–1047.

Tamura, T., Matsuzawa, T., Oji, S., Ichikawa, N., Hosoyama, A., Katsumata, H., et al. (2012). A genome sequence-based approach to taxonomy of the genus *Nocardia*. *Antonie Van Leeuwenhoek, 102*, 481–491.

Tindall, B. J., Rosselló-Móra, R., Busse, H. J., Ludwig, W., & Kämpfer, P. (2010). Notes on the characterization of prokaryote strains for taxonomic purposes. *International Journal of Systematic and Evolutionary Microbiology, 60*, 249–266.

Tomás, A. F., Karakashev, D., & Angelidaki, I. (2013). *Thermoanaerobacter pentosaceus* sp. nov., an anaerobic, extremely thermophilic, high ethanol-yielding bacterium isolated from household waste. *International Journal of Systematic and Evolutionary Microbiology, 63*, 2396–2404.

van der Wolf, J. M., Nijhuis, E. H., Kowalewska, M. J., Saddler, G. S., Parkinson, N., Elphinstone, J. G., et al. (2014). *Dickeya solani* sp. nov., a pectinolytic plant pathogenic bacterium isolated from potato (*Solanum tuberosum*). *International Journal of Systematic and Evolutionary Microbiology, 64*, 768–774.

Vanlaere, E., Baldwin, A., Gevers, D., Henry, D., De Brandt, E., LiPuma, J. J., et al. (2009). Taxon K, a complex within the *Burkholderia cepacia* complex, comprises at least two novel species. Burkholderia contaminans sp. nov. and Burkholderia lata sp. nov. *International Journal of Systematic and Evolutionary Microbiology, 59*, 102–111.

Villeneuve, C., Martineau, C., Mauffrey, F., Villemur, R., Villeneuve, C., Martineau, C., et al. (2013). *Methylophaga nitratireducenticrescens* sp. nov. and *Methylophaga frappieri* sp. nov., isolated from the biofilm of the methanol-fed denitrification system treating the seawater at the Montreal Biodome. *International Journal of Systematic and Evolutionary Microbiology, 63*, 2216–2222.

Wayne, L. G., Brenner, D. J., Colwell, R. R., Grimont, P. A. D., Kandler, O., Krichevsky, M. I., et al. (1987). International Committee on Systematic Bacteriology. Report of the ad hoc committee on reconciliation of approaches to bacterial systematics. *International Journal of Systematic Bacteriology, 37*, 463–464.

Yi, H., Chun, J., & Cha, C. J. (2014). Genomic insights into the taxonomic status of the three subspecies of *Bacillus subtilis*. *Systematic and Applied Microbiology, 37*, 95–99.

Zhang, Y. M., Tian, C. F., Sui, X. H., Chen, W. F., & Chen, W. X. (2012). Robust markers reflecting phylogeny and taxonomy of rhizobia. *PLoS One, 7*, e44936.

Whole-Genome Sequencing for Rapid and Accurate Identification of Bacterial Transmission Pathways

Simon R. Harris[1], Chinyere K. Okoro[1]

Pathogen Genomics, Wellcome Trust Sanger Institute, Cambridge, United Kingdom
[1]Corresponding authors: e-mail address: simon.harris@sanger.ac.uk;
chinyere.okoro@sanger.ac.uk

1 INTRODUCTION

The goal of molecular epidemiology is to combine epidemiological data with information about the relatedness of the pathogen causing disease in order to understand the process of transmission. An ideal method for assessing relatedness of pathogenic bacteria would at the same time be specific enough to identify clusters of related isolates and sensitive enough to differentiate recent from historical transmission. Most traditional molecular typing methods have been based on the sequencing of a small number of genomic loci or on patterns of DNA fragment lengths resulting from digestion with restriction enzymes. Unfortunately, bacterial pathogens do not all accumulate variation at the same rate, so these typing techniques have had to be tailored to the pathogen in question and the level of resolution required.

For species with relatively high mutation rates, multilocus sequence typing (MLST) (Maiden et al., 1998) can be used to provide some resolution within the species. MLST involves sequencing regions of a number (often seven) of housekeeping genes that are thought to be conserved, under low selective pressures and distributed around the entire genome. Every time a novel sequence is found for one of the genes, it is allocated a unique identification number. The combination of sequence numbers for the MLST loci is the sequence type (ST) of the sequenced isolate. Each novel ST is also uniquely numbered. However, even for species such as *Staphylococcus aureus*, which mutate relatively quickly, MLST does not provide the resolution required to consistently identify variants involved in outbreaks or resolve transmission dynamics. The majority of hospital-associated methicillin-resistant *S. aureus* (MRSA) in the UK, for example, belong to ST22. More sensitive typing methods often involve the sequencing of repetitive genomic regions in which the number of repeat sequences varies at a faster rate than mutations in MLST genes. An example

Methods in Microbiology, Volume 41, ISSN 0580-9517, http://dx.doi.org/10.1016/bs.mim.2014.07.003

of a typing scheme based on this premise is the variable number tandem repeat (VNTR) typing scheme. For example, differentiation within MRSA MLST types can be obtained by spa-typing, which is based on the number of copies of a 24 base VNTR in the 3′-region of the protein A (*spa*) gene, or by comparing pulse-field gel electrophoresis patterns. Although an improvement over MLST, these methods still provide limited resolution power and can also suffer from spurious predictions of association that may not be necessarily based on horizontally inherited characteristics. For slowly evolving pathogens, such as *Mycobacterium tuberculosis*, which causes tuberculosis, discrimination between isolates is only possible using a combination of larger numbers of VNTR loci, by comparing IS6110 restriction fragment length polymorphism patterns, or by spoligotyping (spacer oligonucleotide typing). The advent of second-generation, high-throughput sequencing technology has opened up the opportunity of employing whole-genome sequencing in clinical settings, providing unprecedented resolution, specificity, and reproducibility which has been lacking in the approaches employed up to now in clinical epidemiology.

2 THE SEQUENCING REVOLUTION

In 1977, the first DNA genome, that of the bacteriophage, φX174, was determined using a new sequencing method (Sanger, Nicklen, & Coulson, 1977). This method, now commonly referred to as the Sanger, dideoxy sequencing or chain termination method, is based on the use of dideoxynucleotides (ddNTPs) in addition to the normal deoxynucleotides (dNTPs) found in DNA. In dideoxynucleotides, a hydrogen group is incorporated on the 3′-carbon instead of a hydroxyl group (OH). Separate reactions are run, each containing a different radioactively or fluorescently labelled dideoxy terminator corresponding to one of the bases. During each reaction, a ddNTP is randomly incorporated into a strand synthesised from the template and, at that point, prevents further extension of the strand. Termination occurs because a phosphodiester bond cannot be established between the dideoxynucleotide and the next incoming nucleotide. The products of the four reactions are separated by electrophoresis on the basis of their sizes. Based on the patterns of the band, the sequence of bases in the template is thus determined (Shendure et al., 2011).

Improvements in the Sanger method have been based on the automation of the sequencing process. The resulting automated technology is often referred to as capillary sequencing or automated Sanger sequencing (Hutchison, 2007). Automated Sanger sequencing incorporates one or more additional techniques including: bacterial cloning (for *de novo* sequencing) (Shendure & Ji, 2008) or PCR (for resequencing) (Mardis, 2008); a template purification step; fluorescently labelling the chain terminating ddNTPs (McCombie, Heiner, Kelley, Fitzgerald, & Gocayne, 1992; Prober et al., 1987); capillary electrophoresis and fluorescent detection that provides the four-colour plots that reveal the DNA sequence (Behr, Matzig, Levin, Eickhoff, & Heller, 1999); and improvement in the read length (Metzker, 2010). After about three decades of improvement, the Sanger method can now produce sequences of read length ~1 kb, at raw accuracies of >99% with sequencing costs of about or less than \$0.50/kb (Shendure & Ji, 2008). However, for sequencing to become possible in routine clinical practice an exponential increase in throughput and decrease in cost, which could only be provided by new sequencing technologies that are not based on the Sanger method. These new technologies have been collectively referred to as next-generation sequencing (NGS), or second-generation sequencing (SGS) technologies (Mardis, 2008; Shendure & Ji, 2008).

2.1 SECOND-GENERATION SEQUENCING TECHNOLOGIES

The automated Sanger method is regarded as 'first-generation' technology and other methods not based on Sanger sequencing which have been developed within the last few years are often referred to as NGS (Metzker, 2010) or SGS technologies. The common feature of these technologies is that they are extensively parallel, which means that generated reads are much greater than the 96 obtained with automated Sanger sequencers (Hutchison, 2007). However, the reads generated with NGS/SGS technologies are usually shorter and possess lower raw accuracy. Second-generation technologies are primarily based on cyclic-array sequencing, in which an array of target DNA fragments to be sequenced are fixed at constant locations on a solid substrate. During each sequencing cycle, the identity of a single base is determined for each DNA fragment by imaging of the array. At the end of the sequencing cycle, image data for each fragment is collected for all sequencing cycles and the data are then used to determine the sequence of the fragment (Shendure & Ji, 2008; Shendure et al., 2011). Since 2004, a number of second-generation technologies have been commercially available and in widespread use. These include 454 sequencing (LifeSciences/Roche), Solexa (now, Illumina) technology (currently available as Genome Analyser, Hiseq® and Miseq® systems from Illumina), the Ion Torrent personal genome machine (PGM) and Proton (Life Technologies). Although these platforms differ in their sequencing biochemistry and the arrays generated, the basic sequencing steps are similar, and will most often include random fragmentation of template DNA followed by ligation with common adaptor sequences; a PCR reaction is carried out *in situ* or on beads to generate clonally clustered amplicons; followed by alternating cycles of enzyme-based sequencing chemistry using either 'sequencing by synthesis or sequencing by ligation' (Shendure & Ji, 2008). Using these approaches, the primer templates are extended by either a ligase or a polymerase reaction. Imaging the full array at each cycle enables sequential data acquisition. Index tagging allows the massive throughput of the technologies. Initially designed for sequencing human genomes, index tagging on these platforms are now being utilised for sequencing large numbers of smaller bacterial and parasite genomes. Brief descriptions of the three main NGS platforms are outlined in the next sections.

2.1.1 454 pyrosequencing

454, the first NGS platform became commercially available in 2004 (Mardis, 2008; Margulies et al., 2005), deployed a novel sequencing technology known as pyrosequencing. Each nucleotide incorporated by the polymerase stimulates the release of pyrophosphate, which initiates a series of downstream reactions leading to the production of light by the enzyme luciferase. The amount of light produced is proportional to the number of nucleotides incorporated and recurrent steps continue until the detector is saturated. A mixture of DNA fragments on agarose beads are covalently linked to complementary oligonucleotide adapters at the fragments' ends. This mixture is encapsulated in an oil and water emulsion containing PCR reactants, usually referred to as micelles. PCR amplification within these micelles produces about 1 million copies of the original single-stranded fragments attached to the beads. This provides sufficient signal strength during pyrosequencing to detect and record incorporation of the nucleotides (Mardis, 2008).

A major drawback of the 454 platform is that the reads contain a high rate of insertion or deletion errors concentrated in homopolymeric tracts, i.e., stretches of the same base. This is a consequence of the addition of a singe nucleotide at a time, with no termination. A fundamental impact of this is that assemblies based on the 454 platform alone are prone to false insertion and deletion errors. However, the rate of substitution errors is comparatively low. The 454 platform, available in the form of the GS FLX and GS Junior systems, is currently capable of producing long reads that are 200–1000 bp long. This feature makes the 454 platform suitable for *de novo* assemblies of larger more complex genomes and metagenomics analyses, as longer read lengths are an advantage in these processes (Metzker, 2010).

2.1.2 Illumina sequencing technology

The Illumina sequencing platform (formerly known as Solexa) is based on the 'cyclic reversible termination' method described in Bentley et al. (2008). DNA libraries prepared by random fragmentation and linked to adaptors are immobilised on a solid surface called a 'flowcell'. Flowcells are made up of eight channels, or lanes, which are densely covered with a 'lawn' of covalently bound primers complementary to the 5' and 3'-adapters. Each lane can be loaded with a separate library, either made up from a single sample, or more commonly for bacteria, multiplexed libraries containing up to 96 index-tagged libraries. DNA fragments ligate to the primers on the flow cell, where they are amplified *in situ* by the process of solid-phase PCR, also known as bridge amplification, which creates identical copies of each single template molecule in close proximity (clusters). This process can produce up to 10 million single-molecule-based clusters in each channel of the flowcell (Mardis, 2008; Shendure & Ji, 2008), all of which are then sequenced simultaneously.

During each sequencing cycle, a single labelled deoxynucleoside triphosphate (dNTP) is added to the nucleic acid chain. The nucleotide also serves as a terminator for polymerization, so after each round of dNTP incorporation, the fluorescent dye is imaged to identify the base and then cleaved along with the termination modification to allow incorporation of the next nucleotide (Bentley et al., 2008). Since all four reversible terminator-bound dNTPs (A, C, T, G) are present as single, separate molecules, natural competition decreases the chances of one nucleotide being incorporated more than the others. Base calls are made directly from signal intensity measurements during each cycle. After each cycle, the labels are removed and this allows the next level of incorporation (Turcatti, Romieu, Fedurco, & Tairi, 2008). Illumina sequencing has become the most commercially successful platform (Metzker, 2010; Shendure et al., 2011), producing reads of up to 125 bp on the Hiseq machines using the SBS v4 chemistry, or 300 bp on the Miseq, depending on the number of cycles. The substitution error rate is typically about 1% or lower, with few false indels.

Multiplexing of 96 samples per lane or 768 samples per flow cell has led to increased throughput for bacterial sequences. Library preparation methods can be modified to perform paired-end sequencing, i.e., the generation of sequence reads from both ends of the template DNA fragments using two sets of sequencing primers (Roach, Boysen, Wang, & Hood, 1995). The Illumina Hiseq 2500 generates up to 1000 Gb of bases per run, with a maximum read number of 4000 M and maximum read length of 2×125 bp. The Miseq, a lower throughput, faster turnaround instrument targeted for smaller laboratories and clinical diagnostics was also released in 2011. The Miseq has a maximum output of 15 Gb, read number of 25 M and read length of up to 2×300 bp. In 2014, Illumina released the Nextseq 500, which is a desktop sequencer with the capabilities of a Hiseq or Miseq. The Nextseq 500 is currently capable of delivering up to 120 Gb in total output, 400 M read number and 2×150 bp reads (http://www.illumina.com/systems.ilmn).

2.1.3 Ion Torrent

There is the constant need to continue to lower the cost of sequencing, reduce sample-handling time and provide more portable bench top devices. This requirement has fuelled the development of technologies that are primarily based on non-optical sequencing on integrated circuits and aimed at bypassing labelling and imaging-based detection steps during the sequencing process (Rothberg et al., 2011). An example of such a platform is the Ion Torrent PGM. The Ion Torrent technology is based on the use of semi-conductor technology to detect protons released as nucleotides incorporated as the DNA chain is synthesised (Quail et al., 2012; Rothberg et al., 2011). The central processes include the linking of DNA fragments to specific adapter sequences and subsequent clonal amplification by emulsion PCR on the surface of so-called ion sphere particles (Quail et al., 2012). The template beads are then loaded into silicon-lined wells. Sequencing is initiated from a specific location on the adapter sequence and proceeds as each base is sequentially introduced into the growing DNA chain. Incorporation of bases causes the release of protons, which can be detected by an ion-sensitive field-effect transistor located below the well. Similar to 454 pyrosequencing, when the DNA being sequenced includes homopolymeric tracts (strings of the same base next to one another), multiple bases can be incorporated in one sequencing cycle. This leads to increased proton release, so that the intensity of the detected signal increases proportionally to the number of bases incorporated. Unfortunately, however, interpreting the signal strength is difficult, leading to escalating indel error rates as the length of homopolymers increases.

3 BACTERIAL TYPING WITH NEXT-GENERATION SEQUENCING

The release of second-generation technologies provided the ability to sequence bacterial genomes on an unprecedented scale, allowing analyses to move from single genomes and small-scale comparative genomics to large-scale, population-based studies (Brinkman & Parkhill, 2008). This high-throughput sequencing has enabled the study of variation within highly clonal bacterial pathogens at a much higher resolution compared to that achieved by pre-existing typing schemes such as MLST and VNTR. The methods employed have allowed investigation of evolution, transmission and genome-driven epidemiology in a way that has not previously been possible. As a test of the utility of whole-genome sequencing for assessing bacterial evolution at varying scales, Harris et al. analysed 63 MRSA genomes from a single ST, including 43 temporally and geographically spread isolates and a further 20 isolates from a single hospital in Thailand (Harris et al., 2010). Using phylogenetic techniques, historical transmissions between continents could be reconstructed, but they also noted that no two isolates exhibited identical patterns of whole-genome single-nucleotide polymorphisms (SNPs). The ability to differentiate isolates taken only a few days apart in the same hospital ward highlighted the potential for using the technology for clinical epidemiology. A year later, Gardy and her colleagues showed how whole-genome data could be used alongside traditional epidemiology to trace chains of transmission of *M. tuberculosis* (Gardy et al., 2011), and in 2012, two contemporaneous studies proved that similar techniques could be used to differentiate isolates from clinical outbreaks of the hospital 'superbugs' MRSA and *Clostridium difficile* from sporadic cases imported into the same setting (Eyre et al., 2012; Koser et al., 2012).

In other cases, whole-genome sequencing has been used to investigate the variation within bacteria implicated in clinical infectious diseases and for elucidating transmission within and between clinical and community settings. One such study was conducted in 2012 by researchers based in Malawi and the United Kingdom. They investigated the nature of recurrent invasive non-typhoidal *Salmonella* disease in a cohort of 14 patients from the Queen Elizabeth hospital in Blantyre between 2002 and 2004. High-resolution SNP-typing and phylogenetic methods were used to distinguish between very closely related isolates to a degree not achievable by widely employed sub-genomic typing tools, and showed that recurrent disease may be the result of recrudescence, where the index isolate and the recurrent isolate are either genetically indistinguishable or plausibly phylogenetically derived one from the other, or as a result of re-infection, where the subsequent recurrent isolate is phylogenetically distinct from the index isolate (Okoro et al., 2012) (Figure 1). Importantly, it was possible to distinguish between these two recurrent syndrome episodes with extremely closely (<10 SNPs) related strains within the same individual. In another example, a recent study conducted with a cohort of 31 cystic fibrosis (CF) patients attending a lung infection clinic in Cambridge, UK, utilised a combination of the SNP-based phylogenetics, phylogeographical analyses, epidemiological and ecological analyses (Bryant et al., 2013) to change our understanding of the epidemiology of *Mycobacterium abscessus*, an environmental bacterium which can cause

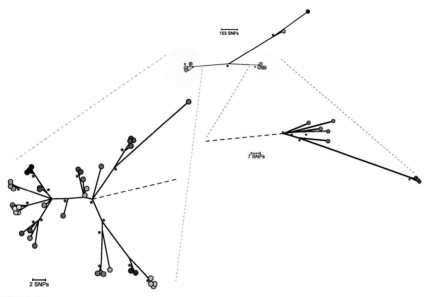

FIGURE 1

Maximum-likelihood phylogenetic tree based on 1463 chromosomal SNP loci of 51 human invasive *S. typhimurium* isolates from 14 patients with re-current invasive disease monitored over 486 days. Zoomed views of phylogenetic clusters bound by coloured ellipses are shown. Isolates are coloured based on the patient from which they were isolated. Grey coloured isolates represent control isolates from a different cohort. Phylogenetic positioning and SNP cut-offs were used to determine isolates representing recrudescence (recurrent isolates derived from the index isolate and borne on the same phylogenetic branch, e.g., pink isolates) and re-infection (recurrent isolates borne on different phylogenetic branches from index isolates, e.g., light blue isolates). Scale bars represent the number of SNPs that have occurred on each branch. (See the colour plate.)

Okoro et al. (2012). Copyright © Oxford University Press, 2012. Reprinted with permission.

chronic lung infections in immune suppressed patients. Contrary to conventional wisdom, which assumed that all infections were from the environment, the authors were able to highlight a strong possibility of human-to-human transmission of genomically identical isolates between patients infected with one subspecies of *M. abscessus*. Such findings highlight the potential of whole-genome sequencing for enriching our understanding of infectious diseases.

4 IDENTIFYING TRANSMISSION PATHWAYS USING WHOLE-GENOME SEQUENCING – THE TOOLKIT

The ability to reconstruct transmission networks from whole genome data has required innovations beyond physical DNA sequencing alone. The introduction of SGS technology has necessarily led to a proliferation of *in silico* data analysis applications designed to deal with the data they produce. These platforms generate millions of short (typically up to 500 base-pair) reads, which are often presented to the end user in FASTQ format. This format emerged as a consensus sequence format for sequencing read exchange between different sequencing technologies and associated bioinformatics-based platforms. It provides an efficient means of representing sequencing reads along with a numeric quality score associated with each nucleotide in a sequence (Cock, Fields, Goto, Heuer, & Rice, 2010). Although there are a number of variations in the way quality scores are represented, the standardised Sanger version of the FASTQ format has found widespread acceptance and is supported by many read mapping and bioinformatic algorithms (Deorowicz & Grabowski, 2011). The Sanger FASTQ format stores base qualities scores as Phred scores from 0 to 93 using the ASCII character encoding scheme characters 33–126. Phred quality scores were devised as a measure of the quality scores of each sequenced base in automated Sanger sequencing (Ewing, Hillier, Wendl, & Green, 1998). Simply, a Phred score of 10 means there is a 1/10 probability of an error in the base call, a score of 100 means a 1/100 probability and so on.

To transform raw sequencing reads into a measure of genome variation that can be used for assessing transmissions can be accomplished in a number of ways. The two main approaches that have been taken in the majority of studies are aligning (or mapping) the reads against a reference genome and assembling them into contigs *de novo*, using only information stored within the sequence reads themselves. Both methods have their strengths and weaknesses. Figure 2 illustrates example workflows for using each of these methods to produce multiple genome alignments. Mapping to a reference genome provides extra information from the high-quality reference sequence to simplify the assembly problem, and, if the same reference is used for mapping a set of new samples, multiple alignments of all samples is simple. However, mapping to a reference limits identification of variation to DNA sequence present within that reference genome. Deletions and short insertions (less that the length of a sequencing read) can be identified, but longer insertions, or novel extra-chromosomal DNA, such as plasmids will be missed. *De novo* assembly, on the other hand, should provide a more comprehensive representation of the gene content of a sample, meaning that novel mobile elements, such as bacteriophage, pathogenicity islands and plasmids should be present in the output. However, a *de novo* assembly approach also presents a number of challenges. Assembly itself is not trivial from short reads, particularly if the genome of the species of interest contains repeat sequences or regions of low complexity. Furthermore, once a set of genomes has been assembled into contigs, they must be aligned against each other in order for meaningful variation information necessary for identifying potential transmissions to be identified. Multiple alignments of many complete genomes are a highly complex problem for which new algorithms are still required. For these reasons, it is currently common to take a combined approach when analysing clinical sequencing data. Mapping is often used to quickly identify variation in the core genome (the portion of the genome shared by all sequences of interest), while *de novo* assembly is

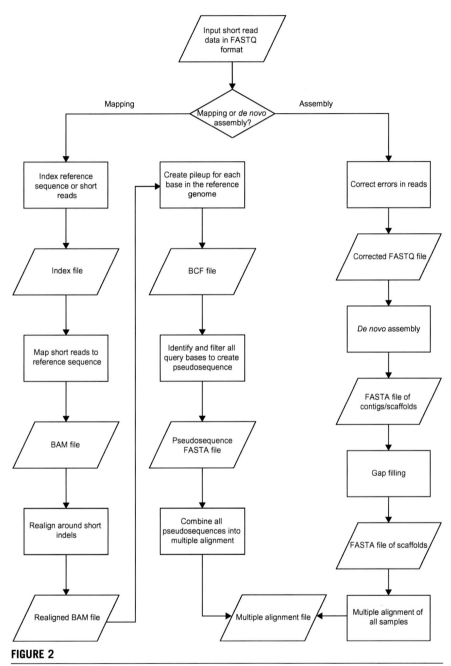

FIGURE 2

Schematic flow diagram of the two most common methods for identifying genomic variation between bacterial isolates.

used to identify accessory regions that often harbour genes associated with virulence or antibiotic resistance.

4.1 MAPPING AND ALIGNMENT OF WHOLE GENOMES

Identification of genome variation using a mapping approach requires a number of steps. A primary pre-requisite of this is availability of a high-quality genome sequence of a bacterial isolate that is suitably similar to the newly sequenced data. Usually, a reference within a few hundred or thousand SNPs will suffice, but if available, it is sensible to map against a number of prospective references to find the best candidate. To speed up this process, a small proportion of the sequenced reads can be used in this initial assessment. If no suitable reference is available, an assembly of one of the samples of interest can be used.

Once a suitable reference has been found, the short reads from the query genomes must be aligned against it efficiently, so that each read can be placed in the appropriate position in the reference genome while taking into account that sequencing error and genetic variation will mean that matches will not always be exact. Since the introduction of SGS technologies, many programs have been developed to tackle the mapping problem, with both speed and accuracy consistently being improved, along with modifications to exploit sequencing developments, such as the introduction of paired-end reads. Because of the frequent release of new programs and improved updates of existing programs, we will describe some of the common methods rather than suggesting the use of particular software for short-read mapping.

4.1.1 Indexing

Short-read mapping usually first requires the reference sequence or the sequence reads to be indexed. Indexing allows fast access to data within large files without the need to read every character in that file every time a search is made. The cost is in the disk space required to store the index file. The indexing process makes mapping millions of reads against multi-megabase reference genomes tractable. The two most common methods for indexing are hash tables and the Burrows–Wheeler transform.

4.1.1.1 Hash tables

Hash tables are data structures comprising keys that are each associated with a particular value or set of values. A simple analogy is a dictionary, where each word would be the key and the value would be the definitions of that word. In short-read mapping, either the reference sequence or the individual short reads can be hashed. In either case, each key would be a subsequence, and the associated values would be the locations in the file in which that subsequence occurs. For example, the mapping program MAQ (Li, Ruan, & Durbin, 2008) works by first constructing a hash table of the short reads, and then scanning along the length of the reference genome to search for matches in the table. SMALT (http://www.sanger.ac.uk/resources/software/smalt/), Novoalign (http://www.novocraft.com/main/page.php?s=novoalign) and many others, on the other hand, create their hash tables by splitting the reference genome into overlapping kmers (often in the region of 1–15 bases). Reads can then mapped to the hash table using various alignment algorithms.

4.1.1.2 The Burrows–Wheeler transform

The Burrows–Wheeler Transform (Burrows & Wheeler, 1994) is a data compression algorithm which has been adapted for short-read mapping in several programs, including BWA (Li, Handsaker, et al., 2009), SOAP2 (Li, Yu, et al., 2009), SOAP3 (Liu et al., 2012) and Bowtie (Langmead, Trapnell, Pop, & Salzberg, 2009). The method works by rearranging strings of characters so that similar characters are adjacent to each other. This allows the data to be compressed more efficiently, but importantly, the transformation is reversible, allowing data to be decompressed simply. An extension of the BWT, known as an FM-index after its creators Paolo Ferragina and Giovanni Manzini (Ferragina & Manzini, 2000), allows efficient searching for substrings, facilitating fast, memory efficient identification of exact matches between short reads and reference genomes. However, further modifications were necessary to allow inexact matches to be made. Mapping algorithms based on the BWT tend to be more memory efficient that those using hash tables, but often require more time to create the index.

4.1.2 Realigning indels

The biggest obstacles for most sequence alignment algorithms are insertions or deletions (indels) in either of the two sequences being aligned. Particular problems for short-read mapping are indels of short repeats, for which there may be multiple optimal alignments. Many read mapping software algorithms treat the alignment of each read independently, which can result in reads which span the same indel being aligned differently relative to that indel. Such inconsistencies compound the tricky process of variant identification, so after the initial mapping, it is advisable to realign reads around indels using a multiple-alignment process, such as that included in the Genome Analyses ToolKit (DePristo et al., 2011; McKenna et al., 2010).

4.1.3 The SAM format

Given the large variety of software available for mapping short read data to a reference genome, it became clear that a unified file format for representing aligned read data was necessary to facilitate data exchange and continuity of downstream analysis. Although not universally supported, many read mappers now produce output in the sequence alignment map (SAM) format (Li, Handsaker, et al., 2009). The SAM format was designed to be an efficient storage method, both in terms of file size and lookup speed, yet also be flexible enough to allow individual programs to store their own specific information (Li, Handsaker, et al., 2009). The format is tab-delimited and comprises an optional header and a number of alignment lines. The header can contain information such as the names and lengths of reference sequences that have been mapped to, and the software and parameters used to create the alignments, facilitating replication of the analysis that has created the file. The alignment lines contain 11 required fields followed by any number of additional fields, which may be program-specific. The required fields include the name of the read, its sequence and Phred-scaled quality scores for each base in ASCII format, and the reference contig and location to which the read has been mapped. To further reduce file size, a binary version of the SAM format, known as the BAM format was also developed, allowing improved compression.

4.1.4 Identifying variation from mapped reads

Once individual short reads have been successfully aligned to the reference genome, each site in the alignment can be assessed to identify variation from the reference sequence in the query sample. Together with the description of the SAM format, SAMtools, a toolkit including utilities for post-processing alignments in SAM format, was released (Li, Handsaker, et al., 2009). SAMtools provides the options to sort, index and filter alignments, as well as a pileup function (now superseded by mpileup). The pileup functions summarise the information from all bases aligned to each particular base in the reference genome. The most straightforward form of pileup is a simple count of the number of As, Cs, Gs and Ts mapped to each reference position. SNPs and short indels (those short enough to be included within individual reads) can be identified directly from the pileup file, but should be subject to a filtering process to reduce the risk of error. All bases, the same as, or a variant of, the reference, must be filtered to ensure accurate base calls. Filtering may include requiring a minimum number and proportion of reads mapping to the base to support the call, or more complex parameters can be utilised to avoid systematic errors.

4.2 *DE NOVO* ASSEMBLY AND GENOME ALIGNMENT

An alternative method for identifying genomic variation between samples is to first assemble *de novo* the short reads of each sample into contigs and then align these contigs, or regions of them, from all samples in order to create a multiple-genome alignment. Assuming accurate assembly, this method is advantageous for identifying larger variations in genomic structure, such as large insertions or deletions, the gain of novel accessory elements, such as plasmids, and genomic rearrangements including translocations or inversions. As with read mapping, there are a surfeit of assembly programs available, so here we will discuss the most common algorithms employed for assembling genomes from short reads.

4.2.1 Read correction

Read assembly algorithms build up contigs by tiling overlapping reads or subreads (fragments of individual reads). In the same way that read mapping is complicated by having to distinguish true genome variation from sequencing errors, these same errors lead to imperfect subread matches which obfuscate the identification of overlaps during *de novo* assembly. One method for reducing the impact of sequencing errors is to reduce the length of the target overlaps. For this reason, many assemblers search for subread (sometimes also called kmer) matches rather than requiring long read overlaps. An alternative method for simplifying the assembly problem is to attempt to reduce sequencing errors in the read data before assembly. This is particularly important for data generated by technologies with higher error rates, such as 454 or Ion Torrent, which usually contain high levels of erroneous short indels. However, read correction can improve assembly of data from any second-generation technology, and can also reduce both assembly times and memory requirements.

4.2.2 Assemblers

Most assemblers fall into two categories: overlap-layout-consensus methods and de Bruijn graph base methods. Both approaches are based on the concept of graphs, which are structures comprising a set of nodes joined by edges. In the case of assembly algorithms, the nodes represent sequences or subsequences, and the edges represent overlaps between them.

4.2.2.1 Overlap-layout-consensus

Overlap-layout-consensus, or OLC, methods first search for pairs of reads which overlap with each other. Overlaps are often required to be longer than a minimum length cutoff and/or may only contain a maximum number of mismatches. Read overlaps meeting these criteria are joined as edges on a graph, which can then be searched to find the most likely path through the data in order to contiguate the reads. OLC methods were commonly used to assemble reads produced by Sanger sequencing, which are longer and more accurate, but far smaller in number than those provided by second-generation technologies. Due to the enormous complexity of the overlap graphs produced during genome assembly, OLC methods can be extremely computationally expensive. Some programs, such as the Celera assembler, use string graphs to reduce the complexity of the overlap graph by removing redundant edges (i.e. edges correlating to perfect repeats in the genome sequence) (Myers, 2005). By counting the number of entries and exits into these repeats in the graph, it is possible to estimate the number of copies of each perfect repeat in the genome. This copy number can be used to expand the string graph to allow all possible paths through the graph to be reconstructed, one of which should be the correct assembly. Despite the enhanced speed provided by string graphs, OLC assembly algorithms lost favour after the introduction of SGS, being widely replaced by kmer-based approaches which scale much better with the increasing number of reads. However, recently OLC assembly has been revisited with the release of the program SGA (Simpson & Durbin, 2012) which introduced fast, efficient overlap read correction and string graph assembly algorithms based on the FM-index.

4.2.2.2 de Bruijn graphs

Most short-read assemblers, including the popular Velvet (Zerbino & Birney, 2008), Abyss (Simpson et al., 2009), SOAPdenovo2 (Luo et al., 2012) and SPAdes (Bankevich et al., 2012) make use of a type of de Bruijn graph structure known as a kmer graph. As with OLC assembly, the first step in creating a kmer graph is to identify overlaps between reads. However, rather than doing this for entire reads, de Bruijn graph assemblers search for matches between kmer subreads (subreads of length k). The kmer graph is produced by joining kmers with exact overlap matches of length k-1 and can be subsequently filtered to remove any patterns that are not present in at least one read. The requirement for exact overlaps makes the choice of kmer crucial for assembly using de Bruijn graphs. Using longer kmers reduces the number of incorrect overlaps due to short genomic repeats, thus simplifying the graph. However, as the kmer length is increased, the probability of the kmer containing sequencing errors also increases, leading to a reduction in the number of exact matches. To optimise this trade-off between sensitivity and specificity, assemblies are often run using a range of kmers and the best assembly chosen based on statistics such as the N50. However, some algorithms, such as that used in the program SPAdes, now allow de Bruijn graphs to be built from multiple kmers simultaneously, with small kmers being used to reduce fragmentation in low-coverage regions, and longer kmers being employed in regions with higher coverage to increase accuracy.

4.2.2.3 Platform-specific assemblers

Due to the nuances of the data generated by some sequencing technologies, it is necessary to use assemblers that have been specifically designed to deal with the error profiles of those technologies. Due to the inconsistencies in homopolymer lengths in reads produced by the 454 sequencer, Roche provided the Newbler assembler (http://454.com/products/analysis-software/index.asp), which can also be applied to Ion Torrent data which suffer from similar homopolymer errors.

4.2.3 Scaffolding and gap filling

Contiguation of short reads using OLC or kmer graphs will fail to produce finished assemblies in most cases, due to uncertainties introduced by genomic repeat regions that cannot be resolved using a single read or kmer, regions of very low or missing read coverage, or regions of increased error rate. However, the paired-end nature of the data produced by many SGS technologies can be used to scaffold contigs that are separated by only short gaps and can also allow those gaps to be filled. Scaffolding is the process of ordering and orienting contigs, and takes advantage of the expected insert sizes of read pairs produced by paired-end sequencing. If a read is incorporated into one contig and its mate into a second contig, it suggests that the DNA fragment from which those reads were sequenced may span the gap between the contigs. By ordering and orienting the contigs based on the read-pair information, the length of missing sequence between the contigs can be estimated based on the insert size of the reads. For short-insert libraries, the insert size is generally in the order of a few hundred bases, so only short contig breaks can be spanned, often resulting in little improvement over the original assembly. However, using mate-pair libraries (libraries that are made by circularising multi-kilobase-length DNA fragments with biotinylated ends, then fragmenting and capturing the biotinylated fragments before proceeding with a normal library preparation), gaps in the order of a few kilobases can be spanned, allowing scaffolding across many repeat regions, such as insertion sequences which are common in bacterial genomes. Many assembly programs produce scaffolds as part of the assembly process. However, there are also a vast array of stand-alone scaffolding programs that can be run on a set of pre-assembled contigs.

The process of scaffolding provides ordered contigs separated by gaps, but does not attempt to use the paired-end information in the reads to fill in the missing sequence. A number of algorithms have been implemented for this process, including a GapCloser option in the SOAPdenovo assembler (Li et al., 2010), and the stand-alone program, IMAGE which both use de Bruijn graphs to create local assemblies of reads around the gap (Tsai, Otto, & Berriman, 2010). These assemblies are added to the contigs and the process repeated until no more improvements can be made or the gap is closed. An alternative algorithm in the GapFiller program identifies read-pairs where one read maps to a contig, but the other overlaps the gap (Boetzer & Pirovano, 2012). Kmers are produced from the reads that overlap the gap and are used to eat into the gap from each end using kmers that overlap the end of the contig by k-1. The process is iterated until the gap is filled, or no more overlaps can be found. As an extra sanity-check, GapFiller only joins contigs when the inserted sequence corresponds to the size of the gap calculated in the scaffolding process.

4.2.4 Identifying variation using co-assembly

In 2012, Iqbal et al. introduced a method for co-assembling multiple genomes using coloured de Bruijn graphs (Iqbal, Caccamo, Turner, Flicek, & McVean, 2012). The method extends the concept of the de Bruin graph by colouring nodes and vertices in the graph based on the sample or samples within which they occur. Reference genomes can also be added to the graph. Using this method, it is possible to identify variation between the input samples *de novo*.

4.3 IDENTIFYING VARIATION FROM WHOLE-GENOME ASSEMBLIES

Comparing genome assemblies, requires alignment of the whole, or parts of each genome in large multiple alignments. Multiple alignment, even of individual genes is a difficult problem, and large genomes, which may include rearrangements as well as mutations and indels, compound this problem. In addition to this, deducing variation using whole genomes calls for inferences to be made on different aspects of homology. This means that variation observed in different segments of the aligned genomes can be either orthologous, paralogous or horizontally acquired genomic sequences. Several concepts have been developed to take these issues into consideration when using whole-genome alignments in variation analyses. In the following segment, we discuss a few of the whole genome aligners that have been developed to tackle some of these challenges.

4.3.1 Whole-genome alignment

The algorithms underpinning many whole genome aligners borrow from the concepts of either or both of Needleman–Wunsch and Smith–Waterman algorithms. Needleman–Wunsch alignment is based on evaluating the collinearity of two amino acid or nucleotide sequences; it considers a 'global' alignment of the sequences and works optimally when comparing highly similar sequences. The Smith–Waterman algorithm considers local alignments of similar regions that fall within what may be dissimilar sets of sequences. Both algorithms can be too simplistic for whole-genome sequences because of the large sizes and intrinsic non-collinearity of whole-genome sequences. An exhaustive discussion on methods and algorithms that underpinned many whole-genome assembly (WGA) methods has been included in an earlier volume of *Methods in Microbiology*. In chapter eight of volume 855 of *Methods in Molecular Biology*, Dewey split WGA methods into two main categories. The first approach, termed the hierarchical approach splits WGA into a set of multiple global alignments, identifies collinear and homologous segments (blocks) of the aligned genomes and subsequently produces a nucleotide-level alignment from the identified blocks. On the other hand, the second approach, the local approach will depend on finding the collinear segments between and within the compared genomes, usually by nucleotide-level pairwise comparisons. These local alignments are then combined into a whole-genome alignment by filtering out the non-collinear segments. The resulting filtered pairwise alignments are subsequently merged into a multiple alignment (Dewey, 2012). We describe in the following segments some of the WGA alignment methods that utilise the concepts described above in order to produce alignments that can be used for variation analyses.

4.3.1.1 BLAT

BLAT is a BLAST-like alignment tool based on the basic local alignment search tool (BLAST) which conducts alignment of two sequences by using maximal segment pair score as a measurement of local similarity (Altschul, Gish, Miller, Myers, & Lipman, 1990). BLAT is faster and more sensitive than BLAST and is based on the 'local' approach. The algorithm looks for kmers shared by both the reference and query genomes. An index is formed from all non-overlapping kmers in the genome. This index is then used to find homologous sequences shared between the reference and query genomes and establish local alignments across the length of the genomes. These local alignments are then merged in a larger whole-genome alignment. An additional improvement step allows for gaps to be closed and produces a more refined alignment (Kent, 2002).

4.3.1.2 MUMmer

MUMmer, is a package used to compare one entire genome against another, in either a draft or complete form. It relies on a suffix tree data structure for efficient pattern matching in contrast to using the hashing approach, which produces kmers. A suffix tree is a data structure for representing all the substrings of a string, which can be DNA sequence, protein sequence or plain text. This allows MUMmer to find all 20 bp maximal exact matches between two bacterial genomes and these exact matches are used to generate pairwise alignments similar to BLAST output (Kurtz et al., 2004).

4.3.1.3 Mauve

Mauve attempts to combat the problem of recombination in whole genome sequences. The developers present it as a method for identifying evolutionary conserved sequences within genomes that are highly recombinogenic or contain a high level of re-arrangements (Darling, Mau, Blattner, & Perna, 2004). The alignment algorithm in Mauve incorporates five basic steps which are summarised below. First, local alignments, or the so-called multi-MUMs are identified. This step uses a hashing method to find matching segments that are present in all sequences under consideration as well as those present only in subsets of the genomes to be aligned. Subsequently, the identified multi-MUMs are used to calculate a phylogenetic guide tree in the second step. Thirdly, a subset of the identified multi-MUMs are then segregated into local colinear blocks (known as locally collinear blocks or LCBs) of sequences conserved among all the genomes under consideration. The LCBs are required to meet the minimum weight criteria specified by the user, which evaluate the degree of confidence that an LCB contains only conserved homologous (orthologs) sequences that are not impacted by recombination or rearrangements, or that are paralogous or only found in a subset of sequences. The LCBs therefore serve as anchors to guide the rest of the steps. A further recursive anchoring step identifies additional anchors or multi-MUMs within the LCB and in the regions bounding the LCB. Finally, with the identified sets of anchors, a progressive alignment is subsequently made using the generated guide tree. As a tool, progressive mauve ameliorates one of the WGA problems, the aligner will ignore large regions >10 kb that are rearranged or acquired horizontally (Darling, Mau, & Perna, 2010). Improvements to the Mauve algorithm have been implemented in progressiveMauve, which is able to accurately align larger numbers of genomes, even in the presence of recombination (Darling, Mau, & Perna, 2010).

4.3.1.4 Mugsy

The Mugsy alignment tool performs whole-genome alignments of draft or finished genomes in single or multiple contiguous sequences. This makes it suitable for use in *de novo* sequencing projects where a suitable or more-closely related reference is not available (Angiuoli & Salzberg, 2011). The Mugsy algorithm encompasses the following key steps. Firstly, pairwise alignment of all candidate genomes against each other is performed. This is based on the Nucmer algorithm, a part of the MUMmer package The Nucmer algorithm allows for a pairwise alignment of an entire genome against another, in either a draft or complete form (Kurtz et al., 2004). Each pairwise alignment is checked to identify orthologous matches using a modified version 'delta filter' utility in the Nucmer package. This filtering step excludes matches found in repetitive and duplicated sequences. In a second step, the filtered pairwise alignments are used to form multiple alignments by building an alignment graph based on an approach called SeqAn::T-Coffee (Rausch et al., 2008). Finally, the alignment graph is processed to determine regions that are homologous, collinear and free of genomic rearrangements. As in Mauve these segments are also referred to as LCBs. Further identification of micro-rearrangements and duplication events are determined by several LCB refinement procedures before a multiple alignment for each LCB is carried out using the SeqAn::TCoffee algorithm. Mugsy requires low computing memory requirements, for example, >12 GB of RAM was required for the alignment of 57 *Escherichia coli* whole genomes. This makes it attractive for small-scale genome variation projects. However, more divergent genomes will require larger amounts of memory (Angiuoli, White, Matalka, White, & Fricke, 2011).

4.4 IDENTIFYING TRANSMISSIONS USING WHOLE-GENOME VARIATION

The discriminatory power of whole-genome variation means that interpretation of epidemiological links must be approached differently than for less discriminatory data. Unlike many molecular typing methods, where the evidence supporting links between cases would often be the classification of two isolates as the same discreet type, whole genome variation is a continuous measure, which requires determination of 'how close is close enough?'

4.4.1 SNP distances – Defining a cutoff

One method for using whole-genome SNPs in molecular epidemiology is to define a SNP cutoff differentiating pairs of isolates that are close enough to be potentially involved in a transmission from those for which the observed variation discounts direct transmission. Defining such cutoffs requires knowledge of the expected mutation rate of the bacterium and should always err on the side of caution. This invariably means that both SNP cutoff definitions and the accompanying epidemiological or clinical information must be considered together during data interpretation.

Walker et al. (2013) showed how SNP cutoffs could be used in practice. By comparing TB isolates taken longitudinally from the same patient, those within families and within groups with known epidemiological links, they showed that within groups that are likely to be part of the same transmission chains, most isolates differed by fewer than six SNPs. They defined this number as the cutoff for evidence of membership of a transmission chain, and a cutoff of greater than 12 SNPs as evidence against membership of a transmission chain, with between 6 and 12 SNPs being indeterminate. By applying these cutoffs to prospectively assess 11 community clusters previously defined based on MIRU-VNTR, they showed that WGS could both identify cryptic outbreaks, where contact tracing had been difficult, and to rule out transmissions in cases where MIRU-VNTR clustered isolates that were genomically diverse. Such overclustering can lead to wasted time and money attempting to epidemiologically link patients, which Walker and his colleagues suggest could be reduced using WGS data (Figure 3A).

Care must be taken in using SNP cutoffs, as mechanisms which quickly introduce large numbers of SNPs, such as homologous recombination, lateral transfer of mobile genetic elements or hypermutation, could lead to erroneous exclusions of isolates from transmission chains. Similarly, latency or variability in mutation rates, such as those observed in TB and *Clostridium difficile* infections, can complicate interpretation of SNP-count based epidemiology.

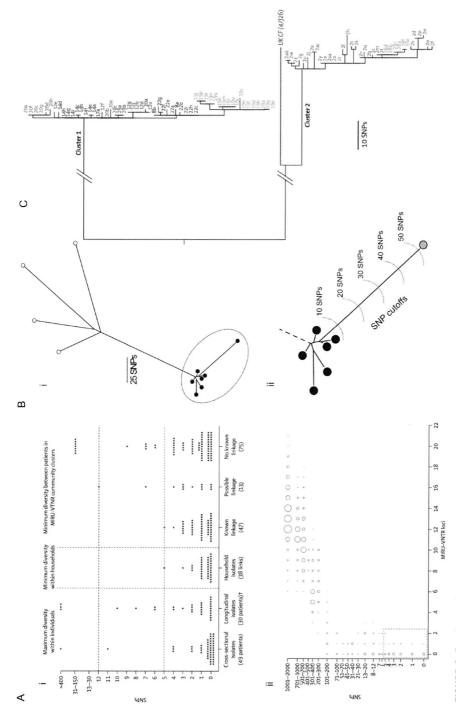

FIGURE 3 See legend on next page.

FIGURE 3

Identifying bacterial transmission using whole-genome sequencing. (A) Identifying transmission clusters using SNP cutoffs. (i) Pairwise genetic distances in SNPs between *M. tuberculosis* isolates within patients, within households and within isolates with known, possible and no known linkage. Top: horizontal dashed line indicates the threshold above which direct transmission can be judged to be unlikely; bottom horizontal dashed line indicates the threshold below which transmission should be investigated. (ii) Comparison of pairwise SNP distances and 24-locus MIRU-VNTR profile differences for all *M. tuberculosis* isolates from a frozen archive. SNP distances are plotted on a log scale. Circle sizes are proportionate to the number of pairs diverging by a specific number of loci and SNPs. Dashed red box includes isolates that differ by five or fewer SNPs. (B) (i) A phylogenetic tree of 11 MRSA isolates from Addenbrooke's hospital, Cambridge, UK. The seven outbreak isolates (black circles) cluster separately from isolates from patients not thought to be part of the outbreak (white circles). (ii) One of the outbreak isolates (grey circle) possessed a hypermutator phenotype, which led to an increased mutation rate. Using SNP cutoffs of up to 50 SNPs to define the outbreak would have excluded this isolate. However, including phylogenetic evidence showed that the isolate was clearly a part of the outbreak. Red dotted lines indicate possible SNP cutoff levels along the branch to the hypermutator isolate showing that even using a cutoff of 50 SNPs the hypermutator would be excluded from the outbreak using a SNP cutoff alone. (C) Maximum likelihood phylogenetic trees of two clusters of *M. abscessus* subspecies *massiliense* showing likely transmission between patients. The branches between the clusters are 95 and 90 SNPs in length but have been shortened for illustration purposes. Scale bar indicates number of SNPs on the branches of the trees. Isolates are coloured and numbered based on the patient from which they were isolated. Cluster 1 includes isolates from many patients separated by only a small number of SNPs, suggesting recent transmission or acquisition from a common source. In cluster two, all isolates from patient 28 cluster within the diversity of isolates from patient 2, providing strong evidence of direct or indirect transmission from patient 2 to patient 28. (See the colour plate.)

4.4.2 Phylogenetic evidence

Although SNP cutoffs can prove useful in identifying transmission clusters, the hierarchical nature of SNP data allows phylogenetic methods (see chapter "An Introduction to Phylogenetics and the Tree of Life" by Williams and Heaps, in this volume) to be used to reconstruct the relatedness of isolates. Again, availability of sequences from background isolates, or isolates not associated with the outbreak under investigation, is essential for determining outbreaks or transmission chains. Isolates from outbreaks would be expected to form tight clades, to the exclusion of non-outbreak isolates. Koser et al. (2012) illustrated the strength of the phylogenetic approach in whole-genome outbreak analysis by retrospectively sequencing isolates from an MRSA outbreak on a neonatal intensive care unit at a hospital in Cambridgeshire, UK. They found that all outbreak isolates formed a tight clade, while isolates from the same ward from the weeks prior to admission of the index case and isolates from other wards isolated contemporaneously with the outbreak samples were far more divergent. Furthermore, the phylogenetic approach was not confounded by one of the outbreak isolates developing a hypermutator phenotype, leading to it being 51 SNPs different from any other outbreak isolate (Figure 3B).

One important distinction that must be made is between phylogenetic reconstructions of the relationships between bacterial isolates and the transmission tree itself, with the latter representing the network of transmission events between hosts (Pybus & Rambaut, 2009). In the phylogenetic analysis, each sequenced isolate is a tip on the tree, with all internal nodes being assumed to be ancestral isolates that are not sampled. However, in a densely sampled outbreak, some isolates may be from hosts which are intermediates in the transmission chain, so these hosts would represent an internal node in the transmission tree. For this reason, Jombart, Eggo, Dodd, and Balloux (2011) introduced a novel graph-based approach for reconstructing transmission trees from densely sampled outbreaks, in which isolates can be reconstructed as both nodes and tips of the tree.

A second important consideration when employing cutoff or phylogenetic evidence for pathogen relatedness is intra-host variation. Often single bacterial colonies are purified for sequencing, providing the genome sequence of only a small subset of the bacteria within the host. In cases where multiple isolates have been sequenced from the same host (e.g. Golubchik et al., 2013; Harris et al., 2013), it has been shown that there can be considerable genome variation within the bacterial population, which may overlap the diversity of isolates from other hosts. Bryant et al. (2013) used this within-host diversity to improve their evidence for transmission between CF patients infected with *M. abscessus*. By sampling multiple isolates from each patient over time, they showed that in some cases the complete diversity of isolates from one patient formed a clade within the diversity of a second patient, providing strong support for transmission, either direct or indirect, between the two (Figure 3C).

4.4.3 Phylodynamics

Grenfell et al. (2004) coined the term phylodynamics to describe how the amalgamation of immunology, epidemiology and evolutionary biology shape the phylogenies of different pathogens. They showed, using viral examples, that the shapes of pathogen phylogenetic trees are affected by changes in the size of the pathogen population over time, immune evasion and the population structure of the host. Most phylodynamic analyses are based on Bayesian phylogenetic methods, as they allow complex evolutionary and population genetic models to be combined. To integrate epidemiological information the methods are often based on coalescent theory which includes an explicit model of the pathogen population. One common application of Bayesian coalescent models is to use isolation dates for samples to model the rate of molecular evolution across the tree and provide estimates for the dates of coalescent events. The software package BEAST (Bayesian Evolutionary Analysis by Sampling Trees) (Drummond & Rambaut, 2007; Drummond, Suchard, Xie, & Rambaut, 2012) includes a wide range of evolutionary, population genetic and phylogeographic models which build upon such a framework and provide a powerful tool for exploring pathogen population dynamics.

5 COMBINING GENOMIC AND EPIDEMIOLOGICAL EVIDENCE

Genomic data is only one line of evidence in the epidemiological toolkit, and even the increased differentiating power it provides is not alone sufficient to accurately determine bacterial transmission dynamics. Genome similarity and phylogenetic relatedness can provide powerful evidence to include or exclude an isolate in an outbreak, but only combined with detailed, traditional epidemiological data can it be used to best effect. Methods have been available for a number of years for using epidemiological data to attempt to reconstruct transmission chains, including timing of symptoms, movement records and tracing of contacts between patients. Combining such approaches with the added information provided by bacterial genome sequencing should improve our ability to correctly reconstruct transmission networks.

Early analytical approaches for combining genomic and epidemiological evidence have been designed primarily with viruses in mind, but the techniques are equally applicable to many bacteria. Cottam et al. (2008) used integrated data from the timings of infections alongside sequence data to determine transmission events between farms during the 2001 foot-and-mouth virus epidemic in the UK. Their method first employed a phylogenetic approach to create a set of equally parsimonious transmission trees from the sequence data, and then used likelihoods of transmission events calculated from the times of infection on the farms to select the most likely tree from this set. In 2012, both Ypma et al. (2012) and Morelli et al. (2012) produced likelihood functions that integrated genetic distance and epidemiological models in a Bayesian framework for estimating transmission trees, and showed that this combined approach led to improved estimation and identification of novel links. However, these approaches, like the method developed by Jombart et al. (2011), associated gene sequences with the host, rather than a bacterium infecting the host. Ypma, van Ballegooijen, and Wallinga (2013) showed that this assumption could lead to erroneous reconstruction of the transmission tree, and described a method for joint estimation of the phylogenetic tree of the infecting pathogen and the transmission tree of hosts to overcome this issue.

Recently, two software tools for reconstructing transmission trees using phylogeny and epidemiological data have been described. Unlike the methods described above, the outbreaker R package, released by Jombart et al. (2014) allows multiple introductions of the pathogen into the population, and unobserved cases. The method couples estimation of the epidemiological likelihood of a transmission tree, based on isolation dates and a predefined distribution representing the duration of the infectious period, with a simple model of sequence evolution. Using a Bayesian framework outbreaker reconstructs transmission chains for each introduction event, with posterior probabilities placed onto every predicted transmission.

A second approach, devised by Didelot, Gardy, and Colijn (2014), requires a dated phylogenetic tree, such as those generated by BEAST or Clonalframe (Didelot & Falush, 2007), as input. From this, transphylo reconstructs the transmission tree based on a within-host pathogen population genetic model. The result is a coloured phylogenetic reconstruction where the branch colours represent the inferred host of the ancestral genotype on each branch of the tree and changes in colour indicate transmissions between hosts. Unlike outbreaker, transphylo assumes complete coverage of an outbreak from a single source, but the modelling of within-host pathogen diversity mitigates errors that can arise from assuming that transmission events coincide with coalescent events in the phylogenetic tree. This is particularly problematic in pathogenic species with long carriage periods or that cause latent infections, such as *C. difficile* or *M. tuberculosis*. In such cases, it is possible for two isolates from hosts who have independently been infected from the same source to share a more recent common ancestor than they do to the source. By accounting for within-host diversity in the model, transphylo can potentially still reconstruct the correct transmission tree in such situations.

6 FUTURE DIRECTIONS

Whole-genome sequencing technologies and the sequencing process itself have become and will continue to be cheaper, faster and have a higher-throughput. It is now more pertinent than ever to overcome the technical challenges associated with these and develop streamlined 'sample-sequence-analyses' pipelines that are readily applicable in a standard laboratory. With the constantly changing landscape of genome sequencing, these challenges are not insurmountable.

An example is the drive to improve *de novo* genome assemblies. Here, sequencing using platforms that generate longer sequence reads >250 such as PacBio enable single contiguous sequences that can be combined with short reads to improve assembly (Quail et al., 2012). PacBio (Pacific Biosciences) is based on a single-molecule real-time (SMRT) sequencing technology (Eid et al., 2009; Ferrarini et al., 2013; Quail et al., 2012). DNA polymerase molecules are bound to a DNA template. The subsequent polymerase-bound DNA is attached to the bottom of the so-called zero-mode waveguides (ZMW), which are 50 nm-wide array plates with wells made from nanophotonic materials/structure. Second-strand DNA synthesis is carried out by each polymerase in the presence of gamma-phosphate fluorescent-labelled nucleotides. The width of the ZMW inhibits light propagation but encourages energy penetration at a short distance, which then excite the fluorophores attached to the nucleotides within the vicinity of polymerase attached to the bottom of the well. Base incorporation is subsequently detected in real time by monitoring/measuring fluorescence (Quail et al., 2012).

Due to the much higher error rate of individual reads from the PacBioRS II, Pacific Biosciences have packaged the HGAP (hierarchical Genome Assembly Process) algorithm into their SMRT Analysis software. HGAP first maps shorter reads against the longest reads produced during the sequencing run to produce consensus sequences for these long reads. After trimming using quality scores, an OCL assembly technique is employed to contiguate the long sequences. Finally, the Quiver algorithm is used to correct base call and indel errors in the assembly. Sequencing and assembly on this platform is capable of producing long reads of up to 5 Mb in size. This means that bacterial chromosomes and associated plasmids can be obtained in their entirety in a single sequencing reaction and assembly run. Furthermore, some of the quirks found in many bacterial genomes such as the high GC regions of some bacteria, such as *M. tuberculosis*, can be problematic with Illumina sequencing and using available references may introduce unnecessary complexities to the analyses. PacBio appears to be able to ameliorate these issues (Galagan, 2014).

Further, the challenges inherent in obtaining enough DNA from clinical samples can be overcome with novel pre-processing steps tailored towards maximising the genetic information that can be mined from the samples. Bacterial isolates come from samples that can often be precious, for example, from the very sick, young or elderly. Some samples may be difficult to culture or may be found mixed with other bacteria in culture, or may still be contaminated with host cells. These challenges can be overcome by using recently described protocols such as the immunomagnetic separation coupled with multiple displacement amplification, IMS-MDA (Seth-Smith et al., 2013). This method enriches for the targeted bacteria in low burden samples and includes an amplification of the target DNA in sufficient quantities suitable for whole-genome sequencing. The method has the advantages of being non-culture based, highly portable and very well suited for *de novo* sequence assembly (Seth-Smith et al., 2013) The IMS-MDA has been successfully implemented with respect to the obligate intracellular pathogen *Chlamydia trachomatis* (Seth-Smith et al., 2013; Seth-Smith & Thomson, 2013).

As whole-genome sequencing analyses are applied in studying bacterial transmission and evolution, an increased depth and breath in the knowledge of global bacterial genomic diversity is obtained. These will undoubtedly lead to accumulation of important genetic information such as conserved and variable genes in different lineages in a way that has never been achieved previously. Comparison of variable and conserved genomic regions from these populations will provide information to aid control measures for many human bacteria pathogens. For example, focus could then be on a selection of representative epitopes that will make for successful global vaccination campaigns. Whole-genome sequencing has helped in advancing our understanding of the extent and pattern of dissemination of mobile genetic elements that are often the vehicle for drug resistance acquisition in bacteria. Various studies lend credibility to the utility of WGS in monitoring drug resistance in bacteria and several successful interventions including changes in treatment regimens have been implemented following such studies. In the future, information from bacterial transmission studies based on whole-genome sequencing may very well become the mainstay of rational antibiotic prescription and monitoring.

In all aspects, it is clearly evident that the application of whole-genome sequencing in transmission studies is a powerful tool. Indeed, it has changed the way we study and understand almost every facet of biology. Microbiology perhaps has been the veritable beneficiary of the power of whole-genome sequencing in the past three decades. Here, we have described the ways, albeit non-exhaustive, that a single technology can replace many existing technologies and thereby offer great advantages in terms of speed, cost and data reproducibility.

REFERENCES

Altschul, S. F., Gish, W., Miller, W., Myers, E. W., & Lipman, D. J. (1990). Basic local alignment search tool. *Journal of Molecular Biology*, *215*(3), 403–410. http://dx.doi.org/10.1016/S0022-2836(05)80360-2.

Angiuoli, S. V., & Salzberg, S. L. (2011). Mugsy: Fast multiple alignment of closely related whole genomes. *Bioinformatics*, *27*(3), 334–342. http://dx.doi.org/10.1093/bioinformatics/btq665.

Angiuoli, S. V., White, J. R., Matalka, M., White, O., & Fricke, W. F. (2011). Resources and costs for microbial sequence analysis evaluated using virtual machines and cloud computing. *PLoS One*, *6*(10), e26624. http://dx.doi.org/10.1371/journal.pone.0026624.

Bankevich, A., Nurk, S., Antipov, D., Gurevich, A. A., Dvorkin, M., Kulikov, A. S., et al. (2012). SPAdes: A new genome assembly algorithm and its applications to single-cell sequencing. *Journal of Computational Biology*, *19*(5), 455–477. http://dx.doi.org/10.1089/cmb.2012.0021.

Behr, S., Matzig, M., Levin, A., Eickhoff, H., & Heller, C. (1999). A fully automated multicapillary electrophoresis device for DNA analysis. *Electrophoresis*, *20*(7), 1492–1507. http://dx.doi.org/10.1002/(SICI)1522-2683(19990601)20:7<1492::AID-ELPS1492>3.0.CO;2-V.

Bentley, D. R., Balasubramanian, S., Swerdlow, H. P., Smith, G. P., Milton, J., Brown, C. G., et al. (2008). Accurate whole human genome sequencing using reversible terminator chemistry. *Nature*, *456*(7218), 53–59. http://dx.doi.org/10.1038/nature07517.

Boetzer, M., & Pirovano, W. (2012). Toward almost closed genomes with GapFiller. *Genome Biology*, *13*(6), R56. http://dx.doi.org/10.1186/gb-2012-13-6-r56.

Brinkman, F. S., & Parkhill, J. (2008). Population genomics: Modeling the new and a renaissance of the old. *Current Opinion in Microbiology*, *11*(5), 439–441. http://dx.doi.org/10.1016/j.mib.2008.09.001.

Bryant, J. M., Grogono, D. M., Greaves, D., Foweraker, J., Roddick, I., Inns, T., et al. (2013). Whole-genome sequencing to identify transmission of *Mycobacterium abscessus* between patients with cystic fibrosis: A retrospective cohort study. *Lancet*, *381*(9877), 1551–1560. http://dx.doi.org/10.1016/S0140-6736(13)60632-7.

Burrows, M., & Wheeler, D. J. (1994). A block sorting lossless data compression algorithm. *Technical report 124.* Digital Equipment Corporation.

Cock, P. J., Fields, C. J., Goto, N., Heuer, M. L., & Rice, P. M. (2010). The Sanger FASTQ file format for sequences with quality scores, and the Solexa/Illumina FASTQ variants. *Nucleic Acids Research*, *38*(6), 1767–1771. http://dx.doi.org/10.1093/nar/gkp1137.

Cottam, E. M., Thebaud, G., Wadsworth, J., Gloster, J., Mansley, L., Paton, D. J., et al. (2008). Integrating genetic and epidemiological data to determine transmission pathways of foot-and-mouth disease virus. *Proceedings of the Royal Society B: Biological Sciences*, *275*(1637), 887–895. http://dx.doi.org/10.1098/rspb.2007.1442.

Darling, A. C., Mau, B., Blattner, F. R., & Perna, N. T. (2004). Mauve: Multiple alignment of conserved genomic sequence with rearrangements. *Genome Research*, *14*(7), 1394–1403. http://dx.doi.org/10.1101/gr.2289704.

Darling, A. E., Mau, B., & Perna, N. T. (2010). ProgressiveMauve: Multiple genome alignment with gene gain, loss and rearrangement. *PLoS One*, *5*(6), e11147. http://dx.doi.org/10.1371/journal.pone.0011147.

Deorowicz, S., & Grabowski, S. (2011). Compression of DNA sequence reads in FASTQ format. *Bioinformatics*, *27*(6), 860–862. http://dx.doi.org/10.1093/bioinformatics/btr014.

DePristo, M. A., Banks, E., Poplin, R., Garimella, K. V., Maguire, J. R., Hartl, C., et al. (2011). A framework for variation discovery and genotyping using next-generation DNA sequencing data. *Nature Genetics*, *43*(5), 491–498. http://dx.doi.org/10.1038/ng.806.

Dewey, C. N. (2012). Whole-genome alignment. *Methods in Molecular Biology*, *855*, 237–257. http://dx.doi.org/10.1007/978-1-61779-582-4_8.

Didelot, X., & Falush, D. (2007). Inference of bacterial microevolution using multilocus sequence data. *Genetics*, *175*(3), 1251–1266. http://dx.doi.org/10.1534/genetics.106.063305.

Didelot, X., Gardy, J., & Colijn, C. (2014). Bayesian inference of infectious disease transmission from whole-genome sequence data. *Molecular Biology and Evolution*, *31*(7), 1869–1879. http://dx.doi.org/10.1093/molbev/msu121.

Drummond, A. J., & Rambaut, A. (2007). BEAST: Bayesian evolutionary analysis by sampling trees. *BMC Evolutionary Biology*, *7*, 214. http://dx.doi.org/10.1186/1471-2148-7-214.

Drummond, A. J., Suchard, M. A., Xie, D., & Rambaut, A. (2012). Bayesian phylogenetics with BEAUti and the BEAST 1.7. *Molecular Biology and Evolution*, *29*(8), 1969–1973. http://dx.doi.org/10.1093/molbev/mss075.

Eid, J., Fehr, A., Gray, J., Luong, K., Lyle, J., Otto, G., et al. (2009). Real-time DNA sequencing from single polymerase molecules. *Science*, *323*(5910), 133–138. http://dx.doi.org/10.1126/science.1162986.

Ewing, B., Hillier, L., Wendl, M. C., & Green, P. (1998). Base-calling of automated sequencer traces using phred. I. Accuracy assessment. *Genome Research*, *8*(3), 175–185.

Eyre, D. W., Golubchik, T., Gordon, N. C., Bowden, R., Piazza, P., Batty, E. M., et al. (2012). A pilot study of rapid benchtop sequencing of *Staphylococcus aureus* and *Clostridium difficile* for outbreak detection and surveillance. *BMJ Open:2*(3). http://dx.doi.org/10.1136/bmjopen-2012-001124.

Ferragina, P., & Manzini, G. (2000). Opportunistic Data Structures with Applications. In *Paper presented at the 41st IEEE Symposium on Foundations of Computer Science Redondo Beach, CA*.

Ferrarini, M., Moretto, M., Ward, J. A., Surbanovski, N., Stevanovic, V., Giongo, L., et al. (2013). An evaluation of the PacBio RS platform for sequencing and *de novo* assembly of a chloroplast genome. *BMC Genomics*, *14*, 670. http://dx.doi.org/10.1186/1471-2164-14-670.

Galagan, J. E. (2014). Genomic insights into tuberculosis. *Nature Reviews Genetics*, *15*(5), 307–320. http://dx.doi.org/10.1038/nrg3664.

Gardy, J. L., Johnston, J. C., Ho Sui, S. J., Cook, V. J., Shah, L., Brodkin, E., et al. (2011). Whole-genome sequencing and social-network analysis of a tuberculosis outbreak. *New England Journal of Medicine*, *364*(8), 730–739. http://dx.doi.org/10.1056/NEJMoa1003176.

Golubchik, T., Batty, E. M., Miller, R. R., Farr, H., Young, B. C., Larner-Svensson, H., et al. (2013). Within-host evolution of *Staphylococcus aureus* during asymptomatic carriage. *PLoS One*, *8*(5), e61319. http://dx.doi.org/10.1371/journal.pone.0061319.

Grenfell, B. T., Pybus, O. G., Gog, J. R., Wood, J. L., Daly, J. M., Mumford, J. A., et al. (2004). Unifying the epidemiological and evolutionary dynamics of pathogens. *Science*, *303*(5656), 327–332. http://dx.doi.org/10.1126/science.1090727.

Harris, S. R., Cartwright, E. J., Torok, M. E., Holden, M. T., Brown, N. M., Ogilvy-Stuart, A. L., et al. (2013). Whole-genome sequencing for analysis of an outbreak of meticillin-resistant *Staphylococcus aureus*: A descriptive study. *The Lancet Infectious Diseases*, *13*(2), 130–136. http://dx.doi.org/10.1016/S1473-3099(12)70268-2.

Harris, S. R., Feil, E. J., Holden, M. T., Quail, M. A., Nickerson, E. K., Chantratita, N., et al. (2010). Evolution of MRSA during hospital transmission and intercontinental spread. *Science*, *327*(5964), 469. http://dx.doi.org/10.1126/science.1182395.

Hutchison, Clyde A. (2007). DNA sequencing: Bench to bedside and beyond. *Nucleic Acids Research*, *35*(18), 6227–6237. http://dx.doi.org/10.1093/nar/gkm688.

Iqbal, Z., Caccamo, M., Turner, I., Flicek, P., & McVean, G. (2012). De novo assembly and genotyping of variants using colored de Bruijn graphs. *Nature Genetics*, *44*(2), 226–232. http://dx.doi.org/10.1038/ng.1028.

Jombart, T., Cori, A., Didelot, X., Cauchemez, S., Fraser, C., & Ferguson, N. (2014). Bayesian reconstruction of disease outbreaks by combining epidemiologic and genomic data. *PLoS Computational Biology*, *10*(1), e1003457. http://dx.doi.org/10.1371/journal.pcbi.1003457.

Jombart, T., Eggo, R. M., Dodd, P. J., & Balloux, F. (2011). Reconstructing disease outbreaks from genetic data: A graph approach. *Heredity (Edinburgh)*, *106*(2), 383–390. http://dx.doi.org/10.1038/hdy.2010.78.

Kent, W. J. (2002). BLAT – The BLAST-like alignment tool. *Genome Research*, *12*(4), 656–664. http://dx.doi.org/10.1101/gr.229202, Article published online before March 2002.

Koser, C. U., Holden, M. T., Ellington, M. J., Cartwright, E. J., Brown, N. M., Ogilvy-Stuart, A. L., et al. (2012). Rapid whole-genome sequencing for investigation of a neonatal MRSA outbreak. *New England Journal of Medicine, 366*(24), 2267–2275. http://dx.doi.org/10.1056/NEJMoa1109910.

Kurtz, S., Phillippy, A., Delcher, A. L., Smoot, M., Shumway, M., Antonescu, C., et al. (2004). Versatile and open software for comparing large genomes. *Genome Biology, 5*(2), R12. http://dx.doi.org/10.1186/gb-2004-5-2-r12.

Langmead, B., Trapnell, C., Pop, M., & Salzberg, S. L. (2009). Ultrafast and memory-efficient alignment of short DNA sequences to the human genome. *Genome Biology, 10*(3), R25. http://dx.doi.org/10.1186/gb-2009-10-3-r25.

Li, H., Handsaker, B., Wysoker, A., Fennell, T., Ruan, J., Homer, N., et al. (2009). The sequence alignment/map format and SAMtools. *Bioinformatics, 25*(16), 2078–2079. http://dx.doi.org/10.1093/bioinformatics/btp352.

Li, H., Ruan, J., & Durbin, R. (2008). Mapping short DNA sequencing reads and calling variants using mapping quality scores. *Genome Research, 18*(11), 1851–1858. http://dx.doi.org/10.1101/gr.078212.108.

Li, R., Yu, C., Li, Y., Lam, T. W., Yiu, S. M., Kristiansen, K., et al. (2009). SOAP2: An improved ultrafast tool for short read alignment. *Bioinformatics, 25*(15), 1966–1967. http://dx.doi.org/10.1093/bioinformatics/btp336.

Li, R., Zhu, H., Ruan, J., Qian, W., Fang, X., Shi, Z., et al. (2010). De novo assembly of human genomes with massively parallel short read sequencing. *Genome Research, 20*(2), 265–272. http://dx.doi.org/10.1101/gr.097261.109.

Liu, C. M., Wong, T., Wu, E., Luo, R., Yiu, S. M., Li, Y., et al. (2012). SOAP3: Ultra-fast GPU-based parallel alignment tool for short reads. *Bioinformatics, 28*(6), 878–879. http://dx.doi.org/10.1093/bioinformatics/bts061.

Luo, R., Liu, B., Xie, Y., Li, Z., Huang, W., Yuan, J., et al. (2012). SOAPdenovo2: An empirically improved memory-efficient short-read *de novo* assembler. *GigaScience, 1*(1), 18. http://dx.doi.org/10.1186/2047-217X-1-18.

Maiden, M. C., Bygraves, J. A., Feil, E., Morelli, G., Russell, J. E., Urwin, R., et al. (1998). Multilocus sequence typing: A portable approach to the identification of clones within populations of pathogenic microorganisms. *Proceedings of the National Academy of Sciences of the United States of America, 95*(6), 3140–3145.

Mardis, E. R. (2008). Next-generation DNA sequencing methods. *Annual Review of Genomics and Human Genetics, 9,* 387–402. http://dx.doi.org/10.1146/annurev.genom.9.081307.164359.

Margulies, M., Egholm, M., Altman, W. E., Attiya, S., Bader, J. S., Bemben, L. A., et al. (2005). Genome sequencing in microfabricated high-density picolitre reactors. *Nature, 437*(7057), 376–380. http://dx.doi.org/10.1038/nature03959.

McCombie, W. R., Heiner, C., Kelley, J. M., Fitzgerald, M. G., & Gocayne, J. D. (1992). Rapid and reliable fluorescent cycle sequencing of double-stranded templates. *DNA Sequence, 2*(5), 289–296.

McKenna, A., Hanna, M., Banks, E., Sivachenko, A., Cibulskis, K., Kernytsky, A., et al. (2010). The Genome Analysis Toolkit: A MapReduce framework for analyzing next-generation DNA sequencing data. *Genome Research, 20*(9), 1297–1303. http://dx.doi.org/10.1101/gr.107524.110.

Metzker, M. L. (2010). Sequencing technologies – The next generation. *Nature Reviews Genetics, 11*(1), 31–46. http://dx.doi.org/10.1038/nrg2626.

Morelli, M. J., Thebaud, G., Chadoeuf, J., King, D. P., Haydon, D. T., & Soubeyrand, S. (2012). A Bayesian inference framework to reconstruct transmission trees using epidemiological and genetic data. *PLoS Computational Biology*, *8*(11), e1002768. http://dx.doi.org/10.1371/journal.pcbi.1002768.

Myers, E. W. (2005). The fragment assembly string graph. *Bioinformatics*, *21*(Suppl. 2), 79–85. http://dx.doi.org/10.1093/bioinformatics/bti1114.

Okoro, C. K., Kingsley, R. A., Quail, M. A., Kankwatira, A. M., Feasey, N. A., Parkhill, J., et al. (2012). High-resolution single nucleotide polymorphism analysis distinguishes recrudescence and reinfection in recurrent invasive nontyphoidal *Salmonella typhimurium* disease. *Clinical Infectious Diseases*, *54*(7), 955–963. http://dx.doi.org/10.1093/cid/cir1032.

Prober, J. M., Trainor, G. L., Dam, R. J., Hobbs, F. W., Robertson, C. W., Zagursky, R. J., et al. (1987). A system for rapid DNA sequencing with fluorescent chain-terminating dideoxynucleotides. *Science*, *238*(4825), 336–341.

Pybus, O. G., & Rambaut, A. (2009). Evolutionary analysis of the dynamics of viral infectious disease. *Nature Reviews Genetics*, *10*(8), 540–550. http://dx.doi.org/10.1038/nrg2583.

Quail, M. A., Smith, M., Coupland, P., Otto, T. D., Harris, S. R., Connor, T. R., et al. (2012). A tale of three next generation sequencing platforms: Comparison of Ion Torrent, Pacific Biosciences and Illumina MiSeq sequencers. *BMC Genomics*, *13*, 341. http://dx.doi.org/10.1186/1471-2164-13-341.

Rausch, T., Emde, A. K., Weese, D., Doring, A., Notredame, C., & Reinert, K. (2008). Segment-based multiple sequence alignment. *Bioinformatics*, *24*(16), i187–i192. http://dx.doi.org/10.1093/bioinformatics/btn281.

Roach, J. C., Boysen, C., Wang, K., & Hood, L. (1995). Pairwise end sequencing: A unified approach to genomic mapping and sequencing. *Genomics*, *26*(2), 345–353.

Rothberg, J. M., Hinz, W., Rearick, T. M., Schultz, J., Mileski, W., Davey, M., et al. (2011). An integrated semiconductor device enabling non-optical genome sequencing. *Nature*, *475*(7356), 348–352. http://dx.doi.org/10.1038/nature10242.

Sanger, F., Nicklen, S., & Coulson, A. R. (1977). DNA sequencing with chain-terminating inhibitors. *Proceedings of the National Academy of Sciences of the United States of America*, *74*(12), 5463–5467.

Seth-Smith, H. M., Harris, S. R., Scott, P., Parmar, S., Marsh, P., Unemo, M., et al. (2013). Generating whole bacterial genome sequences of low-abundance species from complex samples with IMS-MDA. *Nature Protocols*, *8*(12), 2404–2412. http://dx.doi.org/10.1038/nprot.2013.147.

Seth-Smith, H. M., & Thomson, N. R. (2013). Whole-genome sequencing of bacterial sexually transmitted infections: Implications for clinicians. *Current Opinion in Infectious Diseases*, *26*(1), 90–98. http://dx.doi.org/10.1097/QCO.0b013e32835c2159.

Shendure, J., & Ji, H. (2008). Next-generation DNA sequencing. *Nature Biotechnology*, *26*(10), 1135–1145. http://dx.doi.org/10.1038/nbt1486.

Shendure, J. A., Porreca, G. J., Church, G. M., Gardner, A. F., Hendrickson, C. L., Kieleczawa, J., et al. (2011). Overview of DNA sequencing strategies. *Current Protocols in Molecular Biology*. http://dx.doi.org/10.1002/0471142727.mb0701s96, Ausubel, F. M., et al. (Eds.), Chapter 7, Unit 7.1.

Simpson, J. T., & Durbin, R. (2012). Efficient de novo assembly of large genomes using compressed data structures. *Genome Research*, *22*(3), 549–556. http://dx.doi.org/10.1101/gr.126953.111.

Simpson, J. T., Wong, K., Jackman, S. D., Schein, J. E., Jones, S. J., & Birol, I. (2009). ABySS: A parallel assembler for short read sequence data. *Genome Research*, *19*(6), 1117–1123. http://dx.doi.org/10.1101/gr.089532.108.

Tsai, I. J., Otto, T. D., & Berriman, M. (2010). Improving draft assemblies by iterative mapping and assembly of short reads to eliminate gaps. *Genome Biology*, *11*(4), R41. http://dx. doi.org/10.1186/gb-2010-11-4-r41.

Turcatti, G., Romieu, A., Fedurco, M., & Tairi, A. P. (2008). A new class of cleavable fluorescent nucleotides: Synthesis and optimization as reversible terminators for DNA sequencing by synthesis. *Nucleic Acids Research*, *36*(4), e25. http://dx.doi.org/10.1093/ nar/gkn021.

Walker, T. M., Ip, C. L., Harrell, R. H., Evans, J. T., Kapatai, G., Dedicoat, M. J., et al. (2013). Whole-genome sequencing to delineate *Mycobacterium tuberculosis* outbreaks: A retrospective observational study. *Lancet Infectious Diseases*, *13*(2), 137–146. http:// dx.doi.org/10.1016/S1473-3099(12)70277-3.

Ypma, R. J., Bataille, A. M., Stegeman, A., Koch, G., Wallinga, J., & van Ballegooijen, W. M. (2012). Unravelling transmission trees of infectious diseases by combining genetic and epidemiological data. *Proceedings of the Royal Society B: Biological Sciences*, *279*(1728), 444–450. http://dx.doi.org/10.1098/rspb.2011.0913.

Ypma, R. J., van Ballegooijen, W. M., & Wallinga, J. (2013). Relating phylogenetic trees to transmission trees of infectious disease outbreaks. *Genetics*, *195*(3), 1055–1062. http://dx. doi.org/10.1534/genetics.113.154856.

Zerbino, D. R., & Birney, E. (2008). Velvet: Algorithms for de novo short read assembly using de Bruijn graphs. *Genome Research*, *18*(5), 821–829. http://dx.doi.org/10.1101/ gr.074492.107.

Identification of Conserved Indels that are Useful for Classification and Evolutionary Studies

Radhey S. Gupta[1]

Department of Biochemistry and Biomedical Sciences, McMaster University, Hamilton, Ontario, Canada

[1]Corresponding author: e-mail address: gupta@mcmaster.ca

1 LIMITATIONS OF THE PHYLOGENETIC TREES FOR UNDERSTANDING MICROBIAL CLASSIFICATION

The application of 16S rRNA sequences to study the evolutionary relationships among organisms has greatly advanced our understanding in these regards and it has been instrumental in revealing the tremendous diversity that exists within the prokaryotic world (Cole et al., 2009; Ludwig & Klenk, 2005; Pace, 2009; Woese, 1987, 1992; Yarza et al., 2010). Phylogenetic trees and other sequence similarity-based approaches using 16S rRNA and other gene/protein sequences have now largely replaced the earlier classification schemes based on resemblances in morphology and physiology (Brenner, Staley, & Krieg, 2005; Garrity, Bell, & Lilburn, 2005; Oren, 2010). The latter characteristics are generally plastic in nature and most of them have proven unsuitable for developing a reliable classification of prokaryotic organism (Olsen, Woese, & Overbeek, 1994; Stanier, 1941; Stanier & van Niel, 1962; van Niel, 1946). Although the present classification based on 16S rRNA sequences/trees, in large parts, reflects the underlying evolutionary relationships among the prokaryotic organisms, it falls short in clarifying a number of important aspects (Gupta, 1998; Gupta & Griffiths, 2002; Ludwig & Klenk, 2005; Murray, 1986; Oren, 2004; Stackebrandt, 2006; Woese, 1992). Some of these include (i) most prokaryotic taxa are presently distinguished solely on the basis of their branching in 16S rRNA trees (Ludwig & Klenk, 2005; Yarza et al., 2010). Because branching in a tree is a continuum, and it is also influenced by many variables (Ludwig & Klenk, 2005; Woese, 1991; Yarza et al., 2010) (Felsenstein, 2004; Moreira & Philippe, 2000), more reliable (stable) means are needed for demarcation of different prokaryotic taxa (Gupta & Griffiths, 2002). (ii) For most prokaryotic taxa, no genetic, biochemical or molecular properties are known that are specific for them (Ludwig & Klenk, 2005; Oren, 2004). To better understand the differences among different groups of microorganisms, it is important to discover biochemical and molecular properties that are specific for different groups (Gao & Gupta, 2012a). (iii) The phylogenetic trees based upon 16S rRNA and other genes/proteins sequences are unable to reliably resolve the branching order and interrelationships among higher prokaryotic taxa (Ludwig & Klenk, 2005; Yarza et al., 2010). These limitations are common to different phylogenetic tree-construction approaches including trees based on large datasets of concatenated proteins (Ciccarelli et al., 2006; Segata, Bornigen, Morgan, & Huttenhower, 2013; Wu et al., 2009). Hence, it is important to develop other means of phylogenetic reconstruction that can complement the traditional phylogenetic (trees based on gene or protein sequences) approaches and strengthen/advance our understanding of the evolutionary relationships among different organisms (Gao & Gupta, 2012a; Kampfer, 2012; Oren, 2010; Richter & Rossello-Mora, 2009; Woese, 1991).

2 CHARACTERISTICS THAT ARE WELL-SUITED FOR CLASSIFICATION

Our evolutionary thinking, in large part, is based on the seminal work of Charles Darwin. In his classic work 'On the Origin of Species' (Darwin, 1859), Darwin goes to a great length (Chapter XIV, Recapitulation and Conclusion provide a good summary) in describing what kind of characters/markers are ideally suited for classification of organisms, and why certain other types of characters (or markers) are unsuitable for such studies and should be rejected. According to Darwin, the characters that are most useful for classification, or for establishing evolutionary relationships among organisms, are those which have been inherited from a common parent (i.e. shared-derived characteristics or synapomorphies).

> the characters which naturalists consider as showing true affinity between any two or more species, are those which have been inherited from a common parent...

In this chapter, Darwin also stresses the unreliability for systematic studies of traits or characters, which can be readily acquired by analogical means or those which are of adaptive value. In considering molecular sequence data, the terms analogical and adaptive resemblances are equivalent to 'occurrence of similar changes independently' or by means of 'lateral gene transfers (viz. LGTs)', respectively.

> We can understand, on these views, the very important distinction between real affinities and analogical or adaptive resemblances. . . . .we can clearly understand why analogical or adaptive character, although of the utmost importance to the welfare of the being, are almost valueless to the systematist.

In the present context, the challenge before us is to identify reliable shared-derived genetic characters that can be used to clearly demarcate different groups of prokaryotic organisms and provide information regarding their evolutionary relationships. Genome sequences provide a tremendous, and likely the ultimate resource, for discovering the requisite types of molecular markers or characters (Gao & Gupta, 2012a; Kampfer, 2012; Nelson, Paulsen, & Fraser, 2001; Sutcliffe, Trujillo, & Goodfellow, 2012; Wolf, Rogozin, Grishin, & Koonin, 2002; Zhi, Zhao, Li, & Zhao, 2012). Comparative analyses of genome sequences are indeed leading to discovery of numerous shared-derived molecular markers (such as those Darwin had envisaged), which are providing powerful new means for understanding microbial phylogeny and systematics. One important category of these molecular markers is *Conserved Signature Indels* or CSIs (indel means insert or deletion) in genes/proteins sequences (Ajawatanawong & Baldauf, 2013; Baldauf & Palmer, 1993; Cutino-Jimenez et al., 2010; Gao & Gupta, 2012a; Gupta, 1998, 2010; Gupta & Griffiths, 2002; Rivera & Lake, 1992; Rokas & Holland, 2000). The significance of these markers for evolutionary and systematic studies, the methods that we use to discover them and how the results obtained from them are interpreted are discussed below.

3 CONSERVED SIGNATURE INDELS AND THEIR USEFULNESS FOR CLASSIFICATION AND EVOLUTIONARY STUDIES

CSIs in gene/proteins sequences provide an important category of molecular markers for understanding microbial phylogeny and systematics. The indels which provide useful molecular markers are generally of defined size and they are flanked on both sides by conserved regions to ensure that they constitute reliable molecular characteristics. A graphic illustrating a conserved signature indel in a protein sequence is shown in Figure 1A. In this graphic, an 18 amino acid (aa) indel in a conserved region of a protein is commonly shared by all species belonging to the group X, but this indel is absent in different species from groups A, B, C and D. The groups A, B, C, D and X in this graphic could represent different phyla or clades of other taxonomic ranks. The genetic changes that give rise to conserved indels of definite sizes, at precise locations, in gene/protein sequences are of a highly specific nature (rare genetic changes) and they are less likely to occur independently in different organisms. Hence, when a conserved indel in a specific location is found in a phylogenetically related group of organisms, its simplest and most parsimonious explanation is that the genetic change that gave rise to it occurred once in a common ancestor of the group X and then it was vertically inherited by different descendants (Figure 1A, right diagram). Additionally, based upon the absence of this CSI in all other taxa, which represent outgroup species, it can be inferred that the indel shown in Figure 1A is an insert that occurred in a common ancestor of the taxon X.

Because genetic changes leading to CSIs can occur at different evolutionary stages (see Figure 1A and B), it is possible, in principle, to identify CSIs in gene/protein sequences at different phylogenetic branch-points. Thus, the identified CSIs can be specific for different taxonomic clades ranging from phyla, orders, families

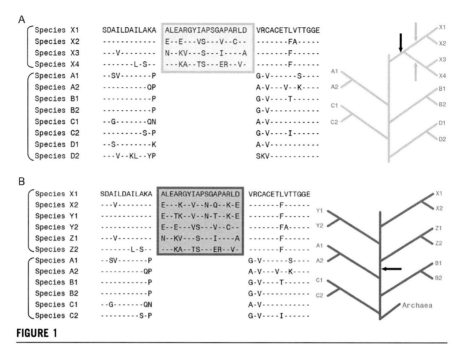

FIGURE 1

(A) A graphic showing an 18 aa conserved signature indel (boxed), which is specific for species from the taxon X, in a protein sequence alignment. The dashes in the alignments indicate identity with the amino acid on the top line. The interpretative diagram on the right shows that the genetic change responsible for this indel likely occurred in a common ancestor of the taxon X (black arrow). The grey arrows show that CSIs can be identified at multiple levels enabling distinction of different taxa in molecular terms. (B) The CSI in this graphic, which are referred to as main line signatures in our work, is commonly shared by members of the phyla X, Y and Z, but absent in the phyla A, B and C. Based upon the presence or absence of this CSI in the archaeal species, one can determine whether this CSI is an insert in X, Y and Z or a deletion in A, B and C. Based upon these types of CSIs which have occurred at important branch-points in evolution, the branching order of the main bacteria phyla can be determined (Bhandari & Gupta, 2014b; Griffiths & Gupta, 2004; Gupta, 2001, 2010). (See the colour plate.)

This figure has been adapted and modified from the Website www.bacterialphylogeny.com.

and genera to single species (or strains) and they can be used for identification and demarcation of different taxa in molecular terms (Bhandari & Gupta, 2014a; Gupta, Chander, & George, 2013; Gupta & Lali, 2013; Gupta, Mahmood, & Adeolu, 2013). Additionally, based upon the presence and absence of CSIs in different bacterial phyla, it is also possible to determine the relative branching order and the evolutionary relationships among them. A graphic illustrating a CSI shared by multiple phyla (referred to as the 'main-line' signatures in our work) is shown in Figure 1B. The 18 aa indel in this case is commonly present in members of the phyla X, Y and

Z, but it is absent in other phyla (viz. A, B and C). The genetic change responsible for this CSI likely occurred in a common ancestor of either the taxa X, Y and Z (or A, B and C) as indicated in the interpretive diagram on the right. However, based upon the distribution of this CSI in these groups, one cannot determine whether this CSI is an insert in the groups X, Y and Z or a deletion common to A, B and C. To infer this, it is necessary to obtain information regarding the ancestral state of this CSI using an outgroup taxa. If the groups shown in this figure correspond to different bacterial phyla, then the presence or absence of this CSI in the archaeal homologues can be used to determine whether the ancestral form of this protein contained or lacked this indel. If this indel is absent in the archaeal homologues, then the bacterial groups lacking this indel (viz. A, B and C) are ancestral and this indel represents an insert, which was introduced in a common ancestor of the groups X, Y and Z. On the other hand, the presence of this CSI in the archaeal homologues will indicate that this CSI in a deletion that occurred in a common ancestor of the taxa A, B and C. In earlier work, a number of CSIs of this kind in widely distributed proteins have been identified, which have been introduced at different important branch-points in bacterial evolution (Bhandari & Gupta, 2014b; Griffiths & Gupta, 2004; Gupta, 1998, 2001, 2003). Based upon these CSIs, it is possible to determine the branching order and interrelationships among many of the major phyla of bacteria, independently of the phylogenetic trees.

The interpretation of the evolutionary relationships suggested by CSIs is generally straightforward and, in most cases, it only depends upon determining the presence or absence of a given CSI in different taxa under consideration. Due to their presence in conserved regions, the presence or absence of the evolutionarily useful CSIs in gene/protein sequences is generally not affected by factors such as differences in evolutionary rates among different species, or at different sites, which can greatly influence the branching pattern of species in phylogenetic trees. However, in some cases, similar CSIs are also present in evolutionarily unrelated taxa. Such CSIs can result from either independent genetic changes or due to LGT between the indicated groups. The interpretation and significance of these CSIs for microbial phylogeny/systematics are discussed later (Section 5).

---

4 IDENTIFICATION OF CONSERVED SIGNATURE INDELS IN PROTEIN SEQUENCES

The identification of CSIs can be carried out using either gene (nucleotide) or protein (aa) sequences. However, in practise, it has been found much easier to identify evolutionarily useful CSIs working with the protein sequences. The two main considerations, which make protein sequences more useful for these studies are (i) due to the presence of 20 characters in protein sequences, sequence similarity and conservation among protein homologues is less likely to result stochastically and it can be more reliably determined than in nucleic acid homologues; (ii) in nucleic acid sequences, short indels can often result from sequencing errors; in contrast, in protein sequences, even a single aa indel is the result of an inframe 3-bp insert or deletion in the corresponding gene. Thus, such changes, particularly when they are present within a conserved region of a protein, are less likely to arise by artifactual means. The methods that we use for identification of CSIs in protein sequences involve a number of different steps and considerations, which are discussed below.

4.1 CREATION OF MULTIPLE SEQUENCE ALIGNMENTS

The creation of multiple sequence alignments (MSA) for protein homologues is the first step in the identification of CSIs. These alignments should contain representative organisms from the group of interest as well as a number of outgroup species. Although, one would think that it would be useful to include sequences for as many species as possible in these initial sequence alignments, from a practical standpoint, this will be very time consuming, and it could also lead to difficulty in identifying many useful CSIs. Sequence alignments from diverse organisms often contain more than one CSI within the same region, which are of different lengths and show different species specificities. The presence of these multiple CSIs, in the same region, can make it difficult to identify CSIs that are specific for the group of interest. Additionally, the inclusion of homologues from distantly related taxa in the alignments will reduce the overall sequence conservation, which can also adversely affect identification of some CSIs. Due to these considerations, the initial alignments for identification of CSIs that are created generally contain sequences for about 15–25 species, including those from the outgroup taxa.

The selection of taxa whose sequences are included in the initial alignments is based upon the overall objective of the project. For example, if one is interested in identifying CSIs that are specific for a group that contains only a limited number of species (e.g. phylum Thermotogae or Aquificae), the initial MSAs could include information for most or all of the sequenced species from these groups. The outgroup species in these cases should include at least two to three species each from two or more phyla. On the other hand, if one is interested in identifying CSIs that are specific for a larger taxonomic group such as Gammaproteobacteria or Actinobacteria then, due to the large number of sequences that are available from these taxa at multiple phylogenetic levels, it is difficult to include all of the sequences in the MSAs. The task of identifying CSIs for these large groups involves the creation of larger MSAs, which should include representatives from different classes and orders covering the phylogenetic diversity of these groups; in addition, these MSAs should also contain representatives from a number of phyla of bacteria. The identification of CSIs for these larger taxonomic groups was much easier, when sequence information was limited (Gao & Gupta, 2005; Gupta, 1998, 2000, 2005); however, with the large increase in the number of genome sequences that are now available for these groups, the task of identifying CSIs for these larger groups has become more difficult. For Gammaproteobacteria some CSIs have been identified at the class level, as well as many others that are specific for some of the orders (viz. *Pasteurellales*, *Xanthomonadales* and *Enterobacteriales*), and distinct subgroups within them

(Cutino-Jimenez et al., 2010; Gao, Mohan, & Gupta, 2009; Gupta, 2000; Naushad & Gupta, 2012, 2013; Naushad, Lee, & Gupta, 2014).

For creation of MSAs, the genome sequences for one or two species from the group of interest are chosen. Blastp searches are performed on most proteins (ORFs) from these genomes against the NCBI non-redundant (nr) database. Protein sequences which are <75 aa long are often omitted from Blast searches, as very few CSIs have been detected in them. For example, in our work on identification of CSIs that are specific for the phylum Aquificae, Blastp searches were performed on different proteins from the genome of *Aquifex aeolicus* (Deckert et al., 1998). Based on these searches, 10–20 high-scoring homologues (preferably with E value $<1e^{-20}$) for different proteins are retrieved from different Aquificae species as well as a limited number (6–8) of the outgroup species in the FASTA format. It is not necessary to have all the sequenced Aquificae species in every alignments and the outgroup species can also vary. For proteins, whose homologues are not found in other species, or which are present in only a limited number of species (generally <6), MSAs are not created.

The MSAs of the protein homologues are created using the Clustal_X 2.1 program (Larkin et al., 2007). However, other programs such as Mega or MUSCLE can also be used for this purpose (Chun & Hong, 2010; Edgar, 2004; Kumar, Nei, Dudley, & Tamura, 2008). In the sequence alignments created using the Clustal_X 2.1 based on FASTA files obtained from downloaded sequences (see Figure 2A), the alignment files show only the GI (Genbank identification) number and the accession numbers of different homologues (see Figure 2B). No information is displayed for the species names, which is essential for determining the taxa specificity of any CSI that may be present in an alignment. To rectify this problem, the sequences in the FASTA file should be edited so that information for the species name appears first on the information line. The edited species names also should not contain any space between the genus and the species names, otherwise only the genera names will be displayed in the created alignments. This can be problematic if multiple species from the same genus are present in an MSA. The names can also be abbreviated at this stage, if necessary. We carry out processing of sequence names prior to creation of MSA using a program 'SEQ_RENAME' that we have developed for this purpose (see Table 1; available from the Gupta Lab Evolutionary Analysis Software (Gleans) Webpage, www.gleans.net). The input and output files for this program for a representative set of sequences are shown in Figure 2A and C, respectively. If two or more sequences in the output file have the same names, they should be edited at this stage. The Clustal_X alignment for the output file generated by the 'SEQ_RENAME' program is shown in Figure 2D.

4.2 IDENTIFICATION OF POTENTIAL CONSERVED INDELS IN THE SEQUENCE ALIGNMENTS

The sequence alignments are next examined for the presence of conserved indels. This is presently carried out by visual inspection of the alignments. However, a program is currently being evaluated that may allow automated identification of indels

A
```
>gi| 15605614| ref| NP_212987.1| elongation factor Tu [Aquifex aeolicus VF5]
MAKEKFERTKEHVNVGTIGHVDHGKSTLTSAITCVLAAGLVEGGKAKCFKYEEIDKAPEEKERGITINITHVEYETAKRHYAHVDCPGHADYIKNMITGA

>gi| 288818166| ref| YP_003432514.1| elongation factor EF-Tu [Hydrogenobacter thermophilus TK-6]
MAKEKFIREKEHVNVGTIGHVDHGKSTLTSAITCVLAAGVLPGGKAKCTKYEEIDKAPEEKERGITINITHVEYETPKRHYAHVDCPGHADYIKNMITGA

>gi| 289548393| ref| YP_003473381.1| translation elongation factor Tu [Thermocrinis albus DSM 14484]
MAKEKFVREKEHVNVGTIGHVDHGKSTLTSAITCVLGAGLMPGGKAKCTKYEEIDKAPEERERGITINITHVEYETAKRHYAHVDCPGHADYIKNMITGA

>gi| 16077181| ref| NP_387994.1| elongation factor Tu [Bacillus subtilis subsp. subtilis str. 168]
MAKEKFDRSKSHANIGTIGHVDHGKTTLTAAITTVLHKKSGKGTAMAYDQIDGAPEERERGITISTAHVEYETETRHYAHVDCPGHADYVKNMITGA
```

B
```
gi| 289548393| ref| YP_003473381.    MAKEKFVREKEHVNVGTIGHVDHGKSTLTSAITCVLGAGLMPGGKAKCTK
gi| 15605614| ref| NP_212987.1|      MAKEKFERTKEHVNVGTIGHVDHGKSTLTSAITCVLAAGLVEGGKAKCFK
gi| 288818166| ref| YP_003432514.    MAKEKFIREKEHVNVGTIGHVDHGKSTLTSAITCVLAAGVLPGGKAKCTK
gi| 16077181| ref| NP_387994.1|      MAKEKFDRSKSHANIGTIGHVDHGKTTLTAAITTVLHK---KSGKGTAMA
                                     ****** * *_*_*:**********:***;*** **   .**...

gi| 289548393| ref| YP_003473381.    YEEIDKAPEERERGITINITHVEYETAKRHYAHVDCPGHADYIKNMITGA
gi| 15605614| ref| NP_212987.1|      YEEIDKAPEEKERGITINITHVEYETAKRHYAHVDCPGHADYIKNMITGA
gi| 288818166| ref| YP_003432514.    YEEIDKAPEEKERGITINITHVEYETPKRHYAHVDCPGHADYIKNMITGA
gi| 16077181| ref| NP_387994.1|      YDQIDGAPEERERGITISTAHVEYETETRHYAHVDCPGHADYVKNMITGA
                                     *:_:** ****:******. :****** .**************:*******
```

C
```
>AquifexaeolicusVF5
MAKEKFERTKEHVNVGTIGHVDHGKSTLTSAITCVLAAGLVEGGKAKCFKYEEIDKAPEEKERGITINITHVEYETAKRHYAHVDCPGHADYIKNMITGA

>Hydrogenobactertherm
MAKEKFIREKEHVNVGTIGHVDHGKSTLTSAITCVLAAGVLPGGKAKCTKYEEIDKAPEEKERGITINITHVEYETPKRHYAHVDCPGHADYIKNMITGA

>ThermocrinisalbusDSM
MAKEKFVREKEHVNVGTIGHVDHGKSTLTSAITCVLGAGLMPGGKAKCTKYEEIDKAPEERERGITINITHVEYETAKRHYAHVDCPGHADYIKNMITGA

>Bacillussubtilissubs
MAKEKFDRSKSHANIGTIGHVDHGKTTLTAAITTVLHKKSGKGTAMAYDQIDGAPEERERGITISTAHVEYETETRHYAHVDCPGHADYVKNMITGA
```

D
```
ThermocrinisalbusDSM    MAKEKFVREKEHVNVGTIGHVDHGKSTLTSAITCVLGAGLMPGGKAKCTK
AquifexaeolicusVF5      MAKEKFERTKEHVNVGTIGHVDHGKSTLTSAITCVLAAGLVEGGKAKCFK
Hydrogenobactertherm    MAKEKFIREKEHVNVGTIGHVDHGKSTLTSAITCVLAAGVLPGGKAKCTK
Bacillussubtilissubs    MAKEKFDRSKSHANIGTIGHVDHGKTTLTAAITTVLHK---KSGKGTAMA
                        ****** * *_*_*:**********:***;*** **   .**...

ThermocrinisalbusDSM    YEEIDKAPEERERGITINITHVEYETAKRHYAHVDCPGHADYIKNMITGA
AquifexaeolicusVF5      YEEIDKAPEEKERGITINITHVEYETAKRHYAHVDCPGHADYIKNMITGA
Hydrogenobactertherm    YEEIDKAPEEKERGITINITHVEYETPKRHYAHVDCPGHADYIKNMITGA
Bacillussubtilissubs    YDQIDGAPEERERGITISTAHVEYETETRHYAHVDCPGHADYVKNMITGA
                        *:_:** ****:******. :****** .**************:*******
```

FIGURE 2

(A) Protein sequences from some species retrieved from NCBI database in the FASTA format. (B) A multiple sequence alignment of the sequences in (A) created using the Clustal_X 2.1 program. No species names are depicted by this alignment file. (C) Editing of the sequence names in (A) using the SEQ_RENAME program. (D) Multiple sequence alignment created from the sequences shown in (C).

that meet specified criteria. The indels that are of interest are generally of fixed lengths and are flanked on both sides by minimally four to five conserved residues in the neighbouring 30–40 aa. The indel length can vary from 1 aa to very large indels (>20 aa). In the MSA created using the Clustal_X 2.1 program, identical and conserved positions are marked by asterisks (*) and semicolons (:) (Figure 3),

Table 1 Descriptions of the Software Programs Used for the Creation of Signature Files

Name of the Program	Program Description	Input/Output	Availability
SIG_RENAME	Renaming of the FASTA sequences so that the species names can be recognised in an alignment file	See Figure 2A and C	These programs and the instructions for their usage are available from the Gupta Lab Evolutionary Analysis Software (Gleans) Webpage, www.gleans.net
SIG_CREATE	This program extracts information from the Blast output for the species name, accession numbers and sequences of the proteins from target species	See Figure 4A and B	
SIG_STYLE	This program converts all amino acid identical to that on the top line to dashes (−) and all sequence gaps into blank spaces for easier visualisation of the sequence conservation	See Figure 4B and C and text	

respectively. In examining these MSAs, attention is paid mainly to only those sequence gaps (−) or indels, which are flanked on both sides by a number of (*) and (:) in the neighbouring 30–40 aa. The indels which are not present in conserved regions are not further investigated. As an example, complete sequence alignment of the protein EF-Tu for most of the Aquificae species and some outgroup species is shown in Figure 3. The inspection of this alignment reveals four indels or sequence gaps, which are highlighted and marked #1 to #4. Indel #1 consists of a 3–4 aa insert, which is flanked on both sides by multiple conserved residues and is limited to only certain Aquificae species. Indel #2 is also present in a conserved region, but indels of different lengths are present in members of two Aquificae families and an indel of a different length is also present in the same position in the two Thermotogae species. Indel #3 involves a 1 aa deletion, which is limited to *Staphylococcus aureus*. This indel could be useful for *Staphylococcus* or related taxa, but in the present context, where the focus is on identifying indels that are specific for the Aquificae group or those shared by Aquificae and some other bacterial phyla, this indel is not of interest. Indel #4 is 1 aa insert in a conserved region that is commonly shared by most Aquificae species as well as the two species from the phylum Thermotogae, but it is absent in the other outgroup species. The shared presence of this indel in the species from

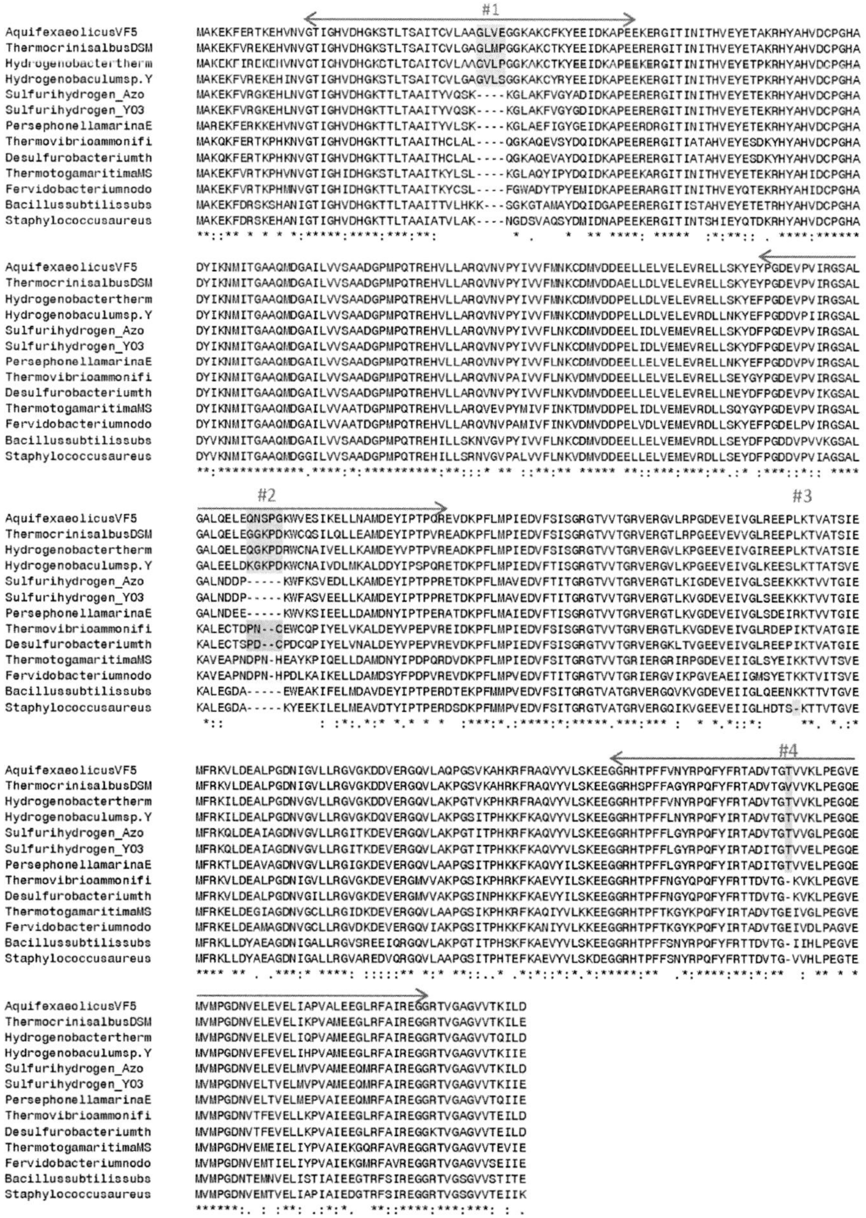

```
                        #1
                  ┌──────────────────────────────┐
AquifexaeolicusVF5    MAKEKFERTKEHVNVGTIGHVDHGKSTLTSAITCVLAAGLVEGGKAKCFKYEEIDKAPEEKERGITINITHVEYETAKRHYAHVDCPGHA
ThermocrinisalbusDSM  MAKEKFVREKEHVNVGTIGHVDHGKSTLTSAITCVLGAGLMPGGKAKCTKYEEIDKAPEERERGITINITHVEYETAKRHYAHVDCPGHA
Hydrogenobactertherm  MAKEKFIREKEHVNVGTIGHVDHGKDTLTGAITCVLAAGVLPGGKAKCTKYEEIDKAPEEKERGITINITHVEYETPKRHYAHVDCPGHA
Hydrogenobaculumsp.Y  MAKEKFVREKEHINVGTIGHVDHGKSTLTSAITCVLGAGVLSGGKAKCYRYEEIDKAPEEKERGITINITHVEYETPKRHYAHVDCPGHA
Sulfurihydrogen_Azo   MAKEKFVRGKEHLNVGTIGHVDHGKTTLTAAITYVQSK----KGLAKFVGYADIDKAPEEKERGITINITHVEYETEKRHYAHVDCPGHA
Sulfurihydrogen_YO3   MAKEKFVRGKEHLNVGTIGHVDHGKTTLTAAITYVQSK----KGLAKFVGYGDIDKAPEEKERGITINITHVEYETEKRHYAHVDCPGHA
PersephonellamarinaE  MAREKFERKKEHVNVGTIGHVDHGKTTLTAAITYVLSK----KGLAEFIGYGEIDKAPEERDRGITINITHVEYETEKRHYAHVDCPGHA
Thermovibrioammonifi  MAKQKFERTKPHKNVGTIGHVDHGKTTLTAAITHCLAL----QGKAQEVSYDQIDKAPEERERGITIATAHVEYESDKYHYAHVDCPGHA
Desulfurobacteriumth  MAKQKFERTKPHKNVGTIGHVDHGKTTLTAAITHCLAL----QGKAQEVAYDQIDKAPEEERGITIATAHVEYESDKYHYAHVDCPGHA
ThermotogamaritimaMS  MAKEKFVRTKPHVNVGTIGHIDHGKSTLTAAITKYLSL----KGLAQYIPYDQIDKAPEEKARGITINITHVGYETEKRHYAHIDCPGHA
Fervidobacteriumnodo  MAKEKFVRTKPHMNVGTIGHVDHGKTTLTAAITKYCSL----FGWADYTPYEMIDKAPEERERGITINITHVGYQTEKRHYAHIDCPGHA
Bacillussubtilissubs  MAKEKFDRSKSHANIGTIGHVDHGKTTLTAAITTVLHKK---SGKGTAMAYDQIDGAPEEERGITISTAHVEYETETRHYAHVDCPGHA
Staphylococcusaureus  MAKEKFDRSKEHANIGTIGHVDHGKTTLTAAIATVLAK----NGDSVAQSYDMIDNAPEEKERGITINTSHIEYQTDKRHYAHVDCPGHA
                      **:.:** * * *:*****.****.***:**:        *   .    *  ** ****. ***** .:*:**:. . ****.******

AquifexaeolicusVF5    DYIKNMITGAAQMDGAILVVSAADGPMPQTREHVLLARQVNVPYIVVFMNKCDMVDDEELLELVELEVRELLSKYEYPGDEVPVIRGSAL
ThermocrinisalbusDSM  DYIKNMITGAAQMDGAILVVSAADGPMPQTREHVLLARQVNVPYIVVFMNKCDMVDDAELLDLVELEVRELLSKYEYPGDEVPVIRGSAL
Hydrogenobactertherm  DYIKNMITGAAQMDGAILVVSAADGPMPQTREHVLLARQVNVPYIVVFMNKCDMVDDPELLDLVELEVRELLSKYEFPGDEVPVIRGSAL
Hydrogenobaculumsp.Y  DYIKNMITGAAQMDGAILVVSAADGPMPQTREHVLLARQVNVPYIVVFMNKCDMVDDPELLDLVELEVRDLLNKYEFPGDEVPIIRGSAL
Sulfurihydrogen_Azo   DYIKNMITGAAQMDGAILVVSAADGPMPQTREHVLLARQVNVPYIVVFLNKCDMVDDEELIDLVEMEVRELLSKYDFPGDEVPVIRGSAL
Sulfurihydrogen_YO3   DYIKNMITGAAQMDGAILVVSAADGPMPQTREHVLLARIHVPYIVVFLNKCDMVDDPELIDLVEMEVRELLKVDFPGDEVPVIRGSAL
PersephonellamarinaE  DYIKNMITGAAQMDGAILVVSAADGPMPQTREHVLLARQVNVPYIVVFLNKCDMVDDEELLELVELEVRELLNKYEFPGDDVPVIRGSAL
Thermovibrioammonifi  DYIKNMITGAAQMDGAILVVSAADGPMPQTREHVLLARQVNVPAIVVFLNKVDMVDDEELLELVELEVRELLSEYGYPGDEVPVIRGSAL
Desulfurobacteriumth  DYIKNMITGAAQMDGAILVVSAADGPMPQTREHVLLARQVNVPYIVVFLNKVDMVDDEELLELVELEVRELLNEYDFPGDEVPVIKGSAL
ThermotogamaritimaMS  DYIKNMITGAAQMDGAILVVAATDGPMPQTREHVLLARQVEVPYMIVFINKTDMVDDPELIDLVEMEVRDLLSQYGYPGDEVPVIRGSAL
Fervidobacteriumnodo  DYIKNMITGAAQMDGAILVVAATDGPMPQTREHVLLARQVNVPYMIVFINKVDMVDDPELVDLVEMEVRDLLSEYDFPGDEVPVIRGSAL
Bacillussubtilissubs  DYVKNMITGAAQMDGAILVVSAADGPMPQTREHILLSKNVGVPYIVVFLNKVDMVDDEELLELVEMEVRDLLSEYDFPGDDVPVVKGSAL
Staphylococcusaureus  DYVKNMITGAAQMDGGILVVSAADGPMPQTREHILLSRNVGVPALVVFLNKVDMVDDEELLELVEMEVRDLLSEYDFPGDDVPVIAGSAL
                      **:.***********.****.*:**********:**::.* ** .:**:** ***** **:.:***.***:**.:* :.***.:** :.****

                        #2                                                      #3
                  ┌────────────────────────────┐                          ┌
AquifexaeolicusVF5    GALQELEQNSPGKWVESIKELLNAMDEYIPTPQREVDKPFLMPIEDVFSISGRGTVVTGRVERGVLRPGDEVEIVGLREEPLKTVATSIE
ThermocrinisalbusDSM  GALQELEGGKPDKWCQSILQLLEAMDEYIPTPVREADKPFLMPIEDVFSISGRGTVVTGRVERGTLRPGEEVEVVGLREEPLKTVATSIE
Hydrogenobactertherm  GALQELEQGKPDRWCNAIVELLKAMDEYVPTPVREADKPFLMPIEDVFSISGRGTVVTGRVERGVLKPGEEVEIVGIREEPLKTVATSIE
Hydrogenobaculumsp.Y  GALEELDKGKPDKWCNAIVDLMKALDDYIPSPQRETDKPFLMPIEDVFTISGRGTVVTGRVERGVLKPGEEVEIVGLKEESLKTTATSVE
Sulfurihydrogen_Azo   GALNDDP-----KWFKSVEDLLKAMDEYIPTPPRETDKPFLMAVEDVFTITGRGTVVTGRVERGTLKIGDEVEIVGLSEEKKKTVVTGIE
Sulfurihydrogen_YO3   GALNDDP-----KWFASVEELLKAMDEYIPTPPRETDKPFLMAVEDVFTITGRGTVVTGRVERGTLKVGDEVEIVGLSEEKKKTVVTGIE
PersephonellamarinaE  GALNDEE-----KWVKSIEELLDAMDNYIPTPERATDKPFLMAIEDVFTISGRGTVVTGRVERGVLSDEIRKTVVTGIE
Thermovibrioammonifi  KALECTDPN--CEWCQPIYELVKALDEYVPEPVREIDKPFLMPIEDVFSISGRGTVVTGRVERGTLKVGDEVEIVGLRDEPIKTVATGIE
Desulfurobacteriumth  KALECTSPD--CPDCQPIYELVNALDEYVPEPVREVDKPFLMPIEDVFSISGRGTVVTGRVERGVLKVGDEVEIVGLREEPIKTVATGIE
ThermotogamaritimaMS  KAVEAPNDPN-HEAYKPIQELLDAMDNYIPDPQRDVDKPFLMPIEDVFSITGRGTVVTGRIERGRIRPGDEVEIIGLSYEIKKTVVTSVE
Fervidobacteriumnodo  KAVEAPNDPN-HPDLKAIKELLDAMDSYFPDPVREVDKPFLMPIEDVFTITGRGTVVTGRIERGVIKPGVEAEIIGMSYETKKTVITSVE
Bacillussubtilissubs  KALEGDA-----EWEAKIFELMDAVDEYIPTPERDTEKPFMMPVEDVFSITGRGTVATGRVERGQVKVGDEVEIIGLQEENKKTTVTGVE
Staphylococcusaureus  KALEGDA-----KYEEKILELMEAVDTYIPTPERDSDKPFMMPVEDVFSITGRGTVATGRVERGQIKVGEEVEIIGLHDTS-KTTVTGVE
                      *::         : :*:.*:* *.* *.* :***:*:.:****:*:****.****:*** * * *.*:*:  *** .**..*.:*

                        #4
                          ┌─────────────────────────────
AquifexaeolicusVF5    MFRKVLDEALPGDNIGVLLRGVGKDDVERGQVLAQPGSVKAHKRFRAQVYVLSKEEGGRHTPFFVNYRPQFYFRTADTVGTVVKLPEGVE
ThermocrinisalbusDSM  MFRKVLDEALPGDNIGVLLRGVGKDDVERGQVLAKPGSVKAHRKFRAQVYVLSKEEGGRHSPFFAGYRPQFYFRTADVTGVVVKLPEGVE
Hydrogenobactertherm  MFRKILDEALPGDNVGVLLRGVGKDDVERGQVLAKPGTVKPHKRFRAQVYVLSKEEGGRHTPFFVNYRPQFYFRTADVTGTVVKLPEGQE
Hydrogenobaculumsp.Y  MFRKILDEALPGDNVGVLLRGVGKDQVERGQVLAKPGSITPHKKFKAQVYVLSKEEGGRHTPFFLNYRPQFYIRTADVTGTVVKLPEGQE
Sulfurihydrogen_Azo   MFRKQLDEAIAGDNVGVLLRGITKDEVERGQVLAKPGTITPHKRFKAQVYVLSKEEGGRHTPFFLGYRPQFYIRTADVTGTVVGLPEGQE
Sulfurihydrogen_YO3   MFRKQLDEAIAGDNVGVLLRGITKDEVERGQVLAKPGTITPHKKFKAQVYVLSKEEGGRHTPFFLGYRPQFYIRTADITGTVVELPEGQE
PersephonellamarinaE  MFRKTLDEAVAGDNVGVLLRGIGKDEVERGQVLAAPGSITPHKKFKAQVYILSKEEGGRHTPFFLGYRPQFYIRTADITGTVVKLPEGQE
Thermovibrioammonifi  MFRKVLDEALPGDNIGVLLRGVGKDEVERGMVVAKPGSIKPHRKFKAEVYILSKEEGGRHTPFFNGYQPQFYFRTTDVTG-KVKLPEGVE
Desulfurobacteriumth  MFRKVLDEALPGDNVGILLRGVGKDEVERGMVVAKPGSIKPHRKFKAEVYILSKEEGGRHTPFFNGYQPQFYFRTTDVTG-KVKLPEGVE
ThermotogamaritimaMS  MFRKELDEGIAGDNVGCLLRGIDKDEVERGQVLAAPGSIKPHKRFKAQIYVLKKEEGGRHTPFTKGYKPQFYIRTADVTGEIVGLPEGVE
Fervidobacteriumnodo  MFRKELDEAMAGDNVGCLLRGVDKDEVERGQVIAKPGSITPHKKFKANIYVLKKEEGGRHTPFTKGYKPQFYIRTADVTGEIVDLPEGVE
Bacillussubtilissubs  MFRKLLDYAEAGDNIGALLRGVSREEIQRGQVLAKPGTITPHSKFKAEVYVLSKEEGGRHTPFFSNYRPQFYFRTTDVTG-IIHLPEGVE
Staphylococcusaureus  MFRKLLDYAEAGDNIGALLRGVAREDVQRGQVLAKPGSITPHTEFKAEVYVLSKDEGGRHTPFFSNYRPQFYFRTTDVTG-VVHLPEGTE
                      **** ** . .***:* ****: :::::**.*:* **::..* *:*:::*:*.*:******:** .*:*****:**:*:** : ** * *

AquifexaeolicusVF5    MVMPGDNVELEVELIAPVALEEGLRFAIREGGRTVGAGVVTKILD
ThermocrinisalbusDSM  MVMPGDNVELEVELIKPVAMEEGLRFAIREGGRTVGAGVVTKILE
Hydrogenobactertherm  MVMPGDNVELEVELIGPVAMEEGLRFAIREGGRTVGAGVVTQILD
Hydrogenobaculumsp.Y  MVMPGDNVEFEVELIHPVAMEEGLRFAIREGGRTVGAGVVTKIIE
Sulfurihydrogen_Azo   MVMPGDNVELEVELMVPVAMEEQMRFAIREGGRTVGAGVVTKILD
Sulfurihydrogen_YO3   MVMPGDNVELTVELMVPVAMEEQMRFAIREGGRTVGAGVVTKIIE
PersephonellamarinaE  MVMPGDNVELTVELMEPVAIEEQMRFAIREGGRTVGAGVVTQIIE
Thermovibrioammonifi  MVMPGDNVTFEVELLKPVAIEEGLRFAIREGGRTVGAGVVTEILD
Desulfurobacteriumth  MVMPGDNVTFEVELLKPVAIEEGLRFAIREGGRTVGAGVVTEILD
ThermotogamaritimaMS  MVMPGDHVEMEIELIYPVAIEKGQRFAVREGGRTVGAGVVTEVIE
Fervidobacteriumnodo  MVMPGDNVEMTIELIYPVAIEKGMRFAVREGGRTVGAGVVSEIIE
Bacillussubtilissubs  MVMPGDNTEMNVELISTIAIEEGTRFSIREGGRTVGSGVVSTITE
Staphylococcusaureus  MVMPGDNVEMTVELIAPIAIEDGTRFSIREGGRTVGSGVVTEIIK
                      ******:. : :**: .:*:*. **:.****:***:***: : .
```

FIGURE 3

Multiple sequence alignment of the protein (EF-Tu) from several Aquificae and some outgroup species. Four indels that are present in conserved regions in this sequence alignment are marked #1 to #4. The suggested lengths of the query sequences for further Blast searches for indels #1, #2 and #4 are marked on top of the sequence alignment by arrows. (See the colour plate.)

these two phyla could be due to LGT. Thus, based upon the visual inspection and preliminary assessment of different indels in the EF-Tu alignment, the indels #1, #2 and #4 are considered potentially useful for further investigation.

The potential indels identified above are further evaluated by detailed Blast searches to determine their species distribution and specificity. The query sequences for these Blast searches are taken from one of the species that is of interest (in the present example from one of the Aquificae species) and they should contain the indels as well as their flanking conserved regions. The lengths of the query sequence for further Blast searches are generally between 40 and 100 aa long and they are based on the following considerations: (i) the query sequence should retain sufficient sequence similarity on both sides to show the reliability of any identified CSI; (ii) for query sequences >100 aa, the signature files are not conveniently displayed in a preferred portrait format; (iii) query sequences <40 aa often do not retrieve many homologous sequences; and (iv) additionally, the query sequences for further Blast searches, where possible, should begin and end on both sides by one or more completely conserved residues. Based on these considerations, the suggested lengths of query sequences for further Blast searches for the three potentially useful indels (#1, #2 and #4) in the EF-Tu sequence alignment are marked by arrows on top of the sequence in Figure 3.

4.3 BLAST SEARCHES ON POTENTIAL CONSERVED INDELS TO IDENTIFY USEFUL CONSERVED INDELS

Blastp searches on the query sequences are performed using the NCBI nr database and default parameters, except that, in most cases, the maximum number of target sequence (found under algorithm parameters) is changed from 100 to 250, which returns 250 matches for the Blast search. Generally, if the number of target sequences examined is about 5–10 times larger than the total number of entries in the clade, for which the indel is specific, this is adequate to determine specificity. The rationale for this is that, if an indel is specific for a certain group then other related species containing the indel, or those which have acquired the gene containing this indel by LGTs, should all exhibit highest similarity to the query sequence. Hence, it is expected that the sequences from all taxa containing the indel, due to either shared evolutionary history or LGTs, should be recovered in the Blast hits that are 5–10 times larger than the group of interest.

After Blast searches on the query sequences, the next step involves evaluation of the results to determine which of the indels provide useful molecular signatures. For evaluation as well as further formatting of the indels, the results of Blast searches are initially saved as a .txt file. This file should contain information for only the pairwise sequence alignments with the query sequence. This is done by going to the 'Formatting options' on the Blast results screen and then changing the following options: Show alignment as → plain text, Alignment view → pairwise (default), Description → change to 0, Graphical overview → change to 0, Alignment → (default is 100), if the number of target analysed is 250 or more, this should be changed accordingly. In addition, the box marked NCBI-gi should also be checked. After choosing these options, the results are reformatted and saved as a .txt file. Excerpts from a .txt file based on Blast search with the query #1 are shown in Figure 4A. The results from this file are further processed using the 'SIG_CREATE' program (see Table 1; available from the Gupta Lab Evolutionary Analysis Software (Gleans) Webpage, www.gleans.net). This program extracts information from the Blast results file for the species name, accession numbers and sequences of the proteins from target species. The output of the 'SIG_CREATE' program from the file shown in Figure 4A is shown in Figure 4B.

The results from different query sequences are assessed individually. Only some of these files contain useful indels (or useful signatures). These indels are present within conserved regions and they are primarily found in the species from the group of interest, or those which in addition are also shared by species from some other taxa. However, in the results obtained from a vast majority of the query sequences, no conserved indel of any kind can be discerned, or indels of different lengths are present in assorted taxa, showing no apparent specificity or pattern. The results of these kinds are deemed to contain no useful signature(s). The lack of any useful signatures in these cases is explained by the fact that the original alignment, based on which the indel queries were identified, contained sequences for only a limited number of species. If the sequence region where the identified query is located is not conserved in other species or if other indels are present in this region in different taxa, the indel will not show any apparent specificity in more detailed Blast searches. Further work is carried out only on those files which contain useful indels/signatures.

4.4 FORMATTING OF THE CONSERVED INDELS

The formatting of the files containing useful signatures involves the following steps. (i) All of the species containing the 'indel' in the group of interest are grouped together at the top of the alignment. (ii) If the indel is a deletion in the group of interest, then gaps (−) are inserted in these sequences at the position where this deletion is present (based on the results of Blast searches and ensuring maximal sequence conservation on both sides). (iii) If the indel is an insert in the group of interest, then sequence gaps in all other sequences are aligned again ensuring maximal conservation on both sides. The alignment of the gaps is based on the premise that the observed insert present in a conserved region is the result of a single genetic change. (iv) If the sequences for some species are missing a few amino acids at either end, which will happen if the amino acids in target species are dissimilar (not conserved) from the query sequence at the ends, then gaps (−) are inserted at these positions to properly align the sequences and keeping the total number of characters in each row the same (this step is not necessary, except for using the SEQ_STYLE program, described later). These changes should be saved as edited versions of the original file. (v) At this stage, if the signature file is to be formatted for publication purpose, the total number of taxa in it may also be reduced. In doing so, it is recommended to keep all of the species from the group of interest as well as those

A

```
BLASTP 2.2.29+

RID: F27DYSU7014

Database: All non-redundant GenBank CDS
translations+PDB+SwissProt+PIR+PRF excluding environmental samples
from WGS projects
           36,381,662 sequences; 12,841,114,591 total letters
Query=
Length=45

ALIGNMENTS
>gi| 15605614| ref| NP_212987.1| elongation factor Tu [Aquifex aeolicus VF5]
Length=405

 Score = 94.4 bits (233),  Expect = 7e-22, Method: Compositional matrix adjust.
 Identities = 45/45 (100%), Positives = 45/45 (100%), Gaps = 0/45 (0%)

Query  1   GTIGHVDHGKSTLTSAITCVLAAGLVEGGKAKCFKYEEIDKAPEE  45
           GTIGHVDHGKSTLTSAITCVLAAGLVEGGKAKCFKYEEIDKAPEE
Sbjct  16  GTIGHVDHGKSTLTSAITCVLAAGLVEGGKAKCFKYEEIDKAPEE  60

>gi| 3913574| sp| O50293.1| EFTU_AQUPY RecName: Full=Elongation factor Tu; Short=EF-Tu
 gi| 2687828| emb| CAA75781.1| elongation factor Tu [Aquifex pyrophilus]
Length=405

 Score = 94.4 bits (233),  Expect = 7e-22, Method: Compositional matrix adjust.
 Identities = 45/45 (100%), Positives = 45/45 (100%), Gaps = 0/45 (0%)

Query  1   GTIGHVDHGKSTLTSAITCVLAAGLVEGGKAKCFKYEEIDKAPEE  45
           GTIGHVDHGKSTLTSAITCVLAAGLVEGGKAKCFKYEEIDKAPEE
Sbjct  16  GTIGHVDHGKSTLTSAITCVLAAGLVEGGKAKCFKYEEIDKAPEE  60
```

B

```
Aquifex aeolicus VF5            15605614     GTIGHVDHGKSTLTSAITCVLAAGLVEGGKAKCFKYEEIDKAPEE
Aquifex pyrophilus              3913574      GTIGHVDHGKSTLTSAITCVLAAGLVEGGKAKCFKYEEIDKAPEE
Hydrogenivirga sp. 128-5-R1-1   495562947     GTIGHVDHGKSTLTSAITCTLAAGLVEGGKAECYKYEEIDKAPEE
Hydrogenobaculum sp. HO         452943463    GTIGHVDHGKSTLTSAITCVLAAGVLPGGKAKCYRYEEIDKAPEE
Thermocrinis ruber DSM 12173    570725183    GTIGHVDHGKSTLTSAITCVLAAGVLPGGKAKCMKYEEIDKAPEE
Hydrogenobacter hydrogenophilu  62240386      TIGHVDHGKSTLTSAITCVLAAGIMPGGKARCTRYEEIDKAPEE
Hydrogenobacter thermophilus T  288818166    GTIGHVDHGKSTLTSAITCVLAAGVLPGGKAKCTKYEEIDKAPEE
Thermocrinis albus DSM 14484    289548406    GTIGHVDHGKSTLTSAITCVLGAGLMPGGKAKCTKYEEIDKAPEE
Clostridium perfringens         489574790    GTIGHVDHGKTTLTAAITTVLA----QAGGAEAFKYDEIDKAPEE
Alicyclobacillus acidocaldariu  495617083    GTIGHVDHGKTTLTAAITTVLAA----KGKAKAQRYEDIDKAPEE
Clostridium carboxidivorans     494125187    GTIGHVDHGKTTLTAAITMVLA----KEGKAEAFKYDEIDKAPEE
Candidatus Arthromitus sp. SFB  496359979    GTIGHVDHGKTTLTAAITTVLA----NKGYAEAFKYDEIDKAPEE
Thermobrachium celere           517493183    GTIGHVDHGKTTLTAAITTVLA----QYGRAQATKYDEIDKAPEE
Clostridium cellulovorans 743B  302876536    GTIGHVDHGKTTLTAAITSVLA---LKGG-AKAFKYDEIDKAPEE
Alicyclobacillus acidocaldariu  384136733    GTIGHVDHGKTTLTAAITTVLAA----KGKAKAQRYEDIDKAPEE
```

C

```
Aquifex aeolicus VF5            15605614     GTIGHVDHGKSTLTSAITCVLAAGLVEGGKAKCFKYEEIDKAPEE
Aquifex pyrophilus              3913574      ---------------------------------------------
Hydrogenivirga sp. 128-5-R1-1   495562947    -------------------T-----------E-Y-----------
Hydrogenobaculum sp. HO         452943463    ------------------------VLP------YR----------
Thermocrinis ruber DSM 12173    570725183    ------------------------VLP------M-----------
Hydrogenobacter hydrogenophilu  62240386      -----------------------IMP----R-TR----------
Hydrogenobacter thermophilus T  288818166    ------------------------VLP------T-----------
Thermocrinis albus DSM 14484    289548406    ---------------------G---MP------T-----------
Clostridium perfringens         489574790    ----------T---A---T---    QA-G-EA---D--------
Alicyclobacillus acidocaldariu  495617083    ----------T---A---T---    AK----AQR--D-------
Clostridium carboxidivorans     494125187    ----------T---A---M---    KE---EA---D--------
Candidatus Arthromitus sp. SFB  496359979    ----------T---A---T---    NK-Y-EA---D--------
Thermobrachium celere           517493183    ----------T---A---T---    QY-R-QAT--D--------
Clostridium cellulovorans 743B  302876536    ----------T---A---S---    LK-G--A---D--------
Alicyclobacillus acidocaldariu  384136733    ----------T---A---T---    AK----AQR--D-------
```

FIGURE 4

(A) Partial results of the Blast searches with the query sequence for indel #1 in Figure 3 saved in the suggested .txt format. (B) Creation of a draft signature file from the results of Blast searches such as that shown in (A) using the SIG_CREATE program. (C) The signature file shown in (B) was reformatted using the SIG_STYLE program after alignments of the gaps.

containing CSIs of similar lengths. If the CSI is specific for a group with large numbers of sequenced members, then the sequences for some species, whose omission will not affect the significance or interpretation of the observed CSI, may also be deleted. All signature files should also contain information for sufficient number of outgroup species to interpret whether the observed CSI is an insert or a deletion in the group(s) of interest. In cases, where a given CSI is specific for a lower level taxonomic group (e.g. genus, family or order), sequences from other groups within that phylum can also serve as suitable outgroups. (vi) If the sequences for some species of interest are not retrieved by the initial Blast search, or if only draft genomes are available for some species, than local or specific Blast searches with the query sequences are carried out against these genomes, and sequence information for the new species are added to the signature files.

Lastly, the signature file, formatted as above, is converted into a format where the CSI and the sequence conservation can be readily visualised. This is carried out using the 'SIG_STYLE' program (see Table 1; available from www.gleans.net), which converts all amino acid residues that are identical to that present on the top line to dashes (−) and all sequence gaps into blank spaces. In cases, where a deletion is present in the group of interest (i.e. sequences shown on the top), the gaps in the top line are initially substituted with xx, which are removed after the styling step.

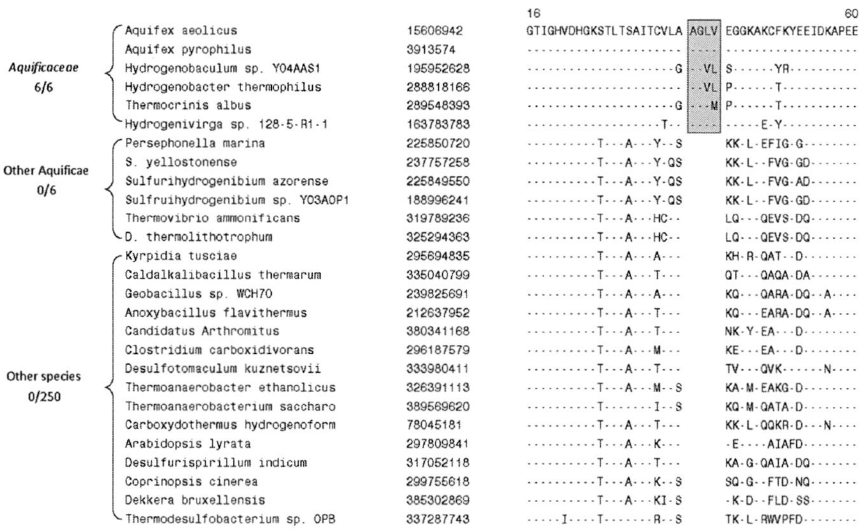

FIGURE 5

A signature file formatted for publication (for indel #1) showing the group and species specificity of the identified indel. If the indels of other lengths are present in this region in other groups, they are not considered relevant here due to their different lengths and species specificities (i.e. resulting from independent genetic events).

The output from the SIG_STYLE program for the signature file shown in Figure 4B, after alignment of the gaps, is presented in Figure 4C. If this signature is to be displayed and further edited by word-processing software, then a mono-space font such as Courier or Swiss Monospace should be used. After styling, a column of space is generally inserted on both two sides of the indel and the indel region is boxed. The position of the sequence in the alignment is marked on the top line and different groups of bacteria containing the CSI or those lacking the CSI are appropriately marked (see Figure 5).

5 INTERPRETING THE SIGNIFICANCE OF CONSERVED INDELS

Although the evolutionary significance of a vast majority of the identified CSIs, particularly those which are limited to members of specific taxa, can be evaluated independently of phylogenetic trees, it is recommended that the interpretation of the evolutionary significances of different CSIs should be done in conjunction with the phylogenetic trees. For this purpose, we generally construct, rooted, bootstrapped, phylogenetic trees, based upon concatenated sequences for 15–30 highly conserved proteins for different genome-sequenced members from the group of interest, using both neighbour-joining and maximum-likelihood algorithms. The trees based on 16S rRNA gene sequences for different species are available from the all-species living tree of life project site (www.arb-silva.de/projects/living-tree/). These trees provide a phylogenetic framework for interpreting the evolutionary significance of different identified CSIs.

The CSIs, which are useful for understanding evolutionary relationships, are generally of three main kinds. These include (i) the CSIs which are completely specific (as judged by the results of Blast searches) for a phylogenetically well-defined group of species. If some CSIs are mainly specific for a group, except that they contain a few isolated exceptions from some unrelated or poorly characterized taxon, they are also included in this category. Among other possibilities, such exceptions can result from incorrect taxonomic assignment of the species in question, as well as errors at the sequencing/annotation level; (ii) the CSIs which are limited to a single species/strain; (iii) the CSIs which are commonly shared by members from two or more distinct taxa, but absent in all other organisms. Of these CSIs, the CSIs which are uniquely present in different strongly supported clades (>90% bootstrap scores) corresponding to known taxa, their interpretation is relatively straightforward. The genetic changes responsible for them likely occurred in the common ancestors of these clades and then vertically acquired by all descendants. The CSIs of this kind are present at multiple phylogenetic levels and, due to their specificities for members of different clades; they provide molecular synapomorphies for the identification/demarcation of these clades in molecular terms. Of the three potential queries identified in the EF-Tu sequence alignment (Figure 3), further analysis with the indel #1 showed that this represents a 4 aa insert that is specifically found in members of the family *Aquificaceae* (Figure 5). Further studies with the indel #2 indicate that in this position inserts of different lengths are present in two different families from the phylum Aquificae. All of the members from the family *Aquificaceae* contain a 5 aa insert, whereas those from the family *Desulfurobacteriaceae* have a 3 aa insert in this position (see Supplementary figure 7 in Gupta & Lali, 2013). In addition, a 4 aa insert is also present in the same position in some Thermotogae species. Due to the different lengths and group specificities of these inserts, genetic changes leading to them have occurred independently in the common ancestors of these families, and these CSIs provide molecular markers for the two families. The second classes of CSIs, which are limited to a single sequenced species or strain can provide useful molecular markers for identification and diagnostics. However, more detailed information for related species/strains is needed in these cases to fully understand the evolutionary taxonomic/significance of such CSIs.

The presence of similar CSIs in members of different taxa can result from at least three distinct possibilities: (i) The groups containing these CSIs are evolutionarily related and the shared presence of the CSIs by them reflect their unique shared ancestry. A number of CSIs have been identified previously that are commonly shared by members of the phyla Fibrobacteres, Chlorobi and Bacteroidetes (FCB group) (Gupta, 2004; Gupta & Lorenzini, 2007), or by members of the phyla Planctomycetes, Verrucomicrobia and Chlamydiae (PVC group) (Griffiths & Gupta, 2007; Gupta, Bhandari, & Naushad, 2012). A close relationship between the members of these phyla is also strongly supported by phylogenetic studies thereby providing evidence that the shared presence of the CSIs in members of these phyla is due to their unique shared ancestry (Ciccarelli et al., 2006; Gupta et al., 2012; Jenkins & Fuerst, 2001; Segata et al., 2013; Teeling, Lombardot, Bauer, Ludwig, & Glockner, 2004; Wagner & Horn, 2006; Ward et al., 2000). Similarly, earlier studies have identified a number of main-line signatures (CSIs) in widely distributed proteins that are commonly shared by members of different bacteria phyla at successive branch-points in evolution, providing information regarding their relative branching order (Griffiths & Gupta, 2004; Gupta, 1997, 2001). (ii) The observed CSIs have originated independently in these different taxa. This could occur if similar genetic changes in a given gene/protein are required in different lineages for similar functional needs. (iii) The presence of similar CSIs in different taxa can also result from lateral transfer of a gene containing CSI from one taxon to the other (Beiko & Ragan, 2008; Gogarten, Doolittle, & Lawrence, 2002; Griffiths & Gupta, 2002, 2006a; Lima, Paquola, Varani, Van Sluys, & Menck, 2008; Nesbo et al., 2009; Olendzenski et al., 2000). These possibilities can be distinguished from each other by means of phylogenetic analysis based upon the protein sequences containing the shared CSIs as well as other widely distributed genes/proteins (Cutino-Jimenez et al., 2010; Huang & Gogarten, 2009; Naushad & Gupta, 2013). If the shared presence of the CSI in these groups is due to independent genetic lesions, then the members of these groups should branch separately from each other in different phylogenetic trees including that constructed based on the protein containing the shared CSI. On the other hand, if the gene containing the CSI has been laterally transferred between the two taxa, then these two taxa should branch together in the tree based upon this specific protein, and should branch separately in phylogenetic trees based upon other proteins.

Several examples have been identified showing that the shared presence of similar CSIs in different taxa could result via independent genetic changes as well as by LGTs. A recent study on members of the order *Xanthomonadales* identified 28 CSIs that were uniquely present in either all or most genome-sequenced members of this order (Naushad & Gupta, 2013). However, seven CSIs identified in this work in assorted proteins (viz. ValRS, CarB, PyrE, GlyS, RnhB, MinD and X001065) were commonly shared by different *Xanthomonadales* as well as by specific subclades of *Alpha-* and *Betaproteobacteria*. In phylogenetic trees constructed based upon these protein sequences, the members of the order *Xanthomonadales* branched distinctly from the *Alpha-* and/or *Betaproteobacteria* groups that contained similar CSIs (Naushad & Gupta, 2013). The latter groups of species branched with the other *Alpha* and *Betaproteobacteria*, as may be expected. The observed results showed that the shared presence of similar CSIs in these groups of Proteobacteria was not due to LGTs, but due to independent occurrence of similar genetic lesions in the common ancestors of these groups (Naushad & Gupta, 2013). Similar results have been obtained in large number of other cases, where similar CSIs were present in unrelated groups of bacteria (Bhandari & Gupta, 2012, 2014a; Bhandari, Naushad, & Gupta, 2012). A number of other CSIs provide examples, where the shared presence of similar CSIs in different groups is due to LGTs. In the GlyA and MurA proteins, a number of CSIs are commonly shared by members of the phylum Chlamydiae and a subgroup of Actinobacteria. In phylogenetic trees constructed based upon these protein sequences, chlamydiae species branch within the Actinobacteria and they group with the insert containing Actinobacteria (Griffiths & Gupta, 2006a). Similarly, the shared presence of several CSIs in the bacterio-chlorophyll biosynthesis proteins in different phyla of photosynthetic prokaryotes has also been shown to be due to LGTs (Gupta, 2012; Raymond, Zhaxybayeva, Gogarten, Gerdes, & Blankenship, 2002). For Aquificae species, further analysis of the 1 aa insert in the EF-Tu protein (indel #4) shows that it is commonly shared by all genome-sequenced members of the phylum Thermotogae as well as by members of the families *Aquificaceae* and *Hydrogenothermaceae* (Bhandari & Gupta, 2014b; Gupta & Bhandari, 2011). As discussed elsewhere (Bhandari & Gupta, 2014b), the shared presence of this CSI in these two phyla of bacteria is again very likely due to lateral transfer of the gene for the EF-Tu protein from a member of the phylum Thermotogae to a common ancestor of the families *Aquificaceae* and *Hydrogenothermaceae*. Further evidence supporting this inference and the significance of LGT between these two phyla of thermophilic–hyperthermophilic bacteria is discussed in detail in our recent work (Bhandari & Gupta, 2014b). Thus, the analysis of CSI which are commonly shared by unrelated taxa provides a potential means to identify interesting cases of LGTs.

6 CORRESPONDENCE OF THE RESULTS OBTAINED FROM CSIs WITH rRNA AND OTHER PHYLOGENETIC APPROACHES

A reliable understanding of microbial phylogeny or systematics should, in most cases, be supported by different lines of evidences (Kampfer, 2012; Ludwig & Klenk, 2005; Oren & Stackebrandt, 2002; Woese, 1992). Hence, it is important to consider how the inferences based upon the CSIs correlate with other commonly used approaches such as the phylogenetic trees based upon 16S rRNA gene sequences or different datasets of concatenated proteins (Ciccarelli et al., 2006; Segata et al., 2013; Wu et al., 2009; Yarza et al., 2010). Detailed studies have been carried out, in the past few years, on the identification of CSIs for a number of bacterial phyla (viz. Thermotogae, Aquificae, Spirochaetes, Fusobacteria, Chloroflexi, Actinobacteria, etc.). These studies have identified numerous CSIs at multiple phylogenetic levels that are either specific for members of these phyla, or those which were commonly shared with some other taxa (Bhandari & Gupta, 2014a; Gao & Gupta, 2012b; Gupta, Chander, et al., 2013; Gupta, Chen, Adeolu, & Chai, 2013; Gupta & Lali, 2013; Gupta, Mahmood, et al., 2013). In parallel, phylogenetic trees were also constructed for members of these phyla based upon 16S rRNA gene sequences and concatenated protein sequences. The results of these studies for the phylum Thermotogae are briefly described below.

The phylum Thermotogae contains 10 genera harbouring 41 species all of which, until recently, were part of a single family *Thermotogaceae* (within the order *Thermotogales*) (Frock, Notey, & Kelly, 2010; Gupta & Bhandari, 2011; Huber & Hannig, 2006; L'Haridon et al., 2006; LPSN Web Resource, 2013; Reysenbach et al., 2013). Recent work on Thermotogae genomes has identified 85 CSIs that are specific for members of this phylum at multiple phylogenetic levels (Bhandari & Gupta, 2014a, 2014b; Gupta & Bhandari, 2011). A summary of the species specificities of the discovered CSIs is presented in Figure 6. Eleven of these CSIs were uniquely present in all genome-sequenced Thermotogae species, whereas the remainders were specific for multiple clades encompassing different members of this phylum. Different subclades of Thermotogae identified by these CSIs are also supported by phylogenetic trees based upon 16S rRNA gene and concatenated protein sequences (Bhandari & Gupta, 2014a, 2014b). The latter trees provide greater resolving power and they are now considered to more accurately depict the species relationships than those based on single genes and proteins (Ciccarelli et al., 2006; Rokas, Williams, King, & Carroll, 2003; Segata et al., 2013). The existence of the same Thermotogae clades is also independently strongly supported by the 'character compatibility approach', a method in which the molecular sequence data are analysed to identify the largest clique(s) of compatible characters (Estabrook, Johnson, & McMorris, 1976; Gupta & Sneath, 2007; Pisani, 2004). This method identified a single largest clique of compatible characters, where all of the observed subclades were supported by multiple specific amino acid substitutions (Bhandari & Gupta, 2014a). Thus, the evolutionary relationships among the Thermotogae species inferred based upon the discovered CSIs are independently strongly supported by

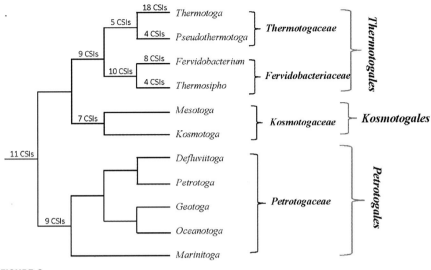

FIGURE 6

Summary of the clade specificities of different CSIs that are specific for members of the phylum Thermotogae. Based upon consistent evidence obtained from different lines of evidence (viz. branching in the 16S rRNA and concatenated protein trees, character compatibility analysis, species distribution of different CSIs), the taxonomy of this phylum was recently revised (Bhandari & Gupta, 2014a) and the newly proposed hierarchical arrangement of the groups within this phylum is shown here.

different phylogenetic/evolutionary methods and they are deemed to be reliable. Thus, based upon the discovered CSIs, the phylum Thermotogae as well as a number of its subclades at multiple phylogenetic depths can now be reliably distinguished in more definitive molecular terms (Figure 6). Based upon the consistent evidence obtained from different lines of evidence, the phylum (class) Thermotogae has been now divided into three orders (viz. *Thermotogales*, *Kosmotogales* and *Petrotogales*) containing four families (viz. *Thermotogaceae*, *Fervidobacteriaceae*, *Kosmotogaceae* and *Petrotogaceae*). Additionally, the members of the genus *Thermotoga* have also been divided into two distinct genera viz. *Thermotoga* and *Pseudothermotoga*, differing from each other in molecular, phylogenetic and physiological characteristics (Figure 6) (Bhandari & Gupta, 2014a; Frock et al., 2010).

The results shown in Figure 6 for Thermotogae are typical of a number of other bacterial taxa ranging from phylum (viz. Spirochaetes, Aquificae, Chloroflexi, Synergistetes, Actinobacteria, Fusobacteria) (Bhandari & Gupta, 2014a; Gao & Gupta, 2012b; Gupta, Chander, et al., 2013; Gupta & Lali, 2013; Gupta, Mahmood, et al., 2013), class (viz. *Coriobacteriia*, *Dehalococcoidia*) (Gupta, Chen, et al., 2013; Ravinesan & Gupta, 2014), order (viz. *Neisseriales*,

Pasteurellales, *Xanthomonadales*, *Enterobacteriales*) (Adeolu & Gupta, 2013; Cutino-Jimenez et al., 2010; Naushad & Gupta, 2012, 2013; Naushad et al., 2014) and genus level (*Bacillus*, *Clostridium* and *Borrelia*) (Adeolu & Gupta, 2014; Bhandari & Gupta, 2014b; Gupta & Gao, 2009) that we have studied using this approach. In all of these cases, multiple highly specific CSIs have been identified for members of these groups at different phylogenetic levels, and the relationships inferred from them are strongly supported by different phylogenetic approaches.

While the CSI-based approach has provided important insights into the evolutionary relationships amongst different groups of bacteria that we have studied, it is difficult to make accurate predictions as to how many CSIs will be discovered for any clade of a particular taxonomic level. Nevertheless, a common observation based upon these studies is that for those clades which are strongly supported by phylogenetic analyses (particularly in trees based on concatenated sequences for multiple conserved proteins) and which are separated from others by a long branch, the number of CSIs that are discovered are more numerous in comparison to the clades that are poorly resolved in the phylogenetic trees. Additionally, the total number of CSIs identified for a given group will also depend upon its phylogenetic depth, the amount of sequence information that is available for different members from that group and the level of detail at which the analysis for identification of CSIs is carried out.

It is important now to consider how the CSIs that are commonly shared by members of unrelated taxa, either due to independent genetic changes, or due to LGTs, affect the overall understanding of microbial phylogeny and systematics. The impact of LGTs on understanding prokaryotic phylogeny is a hotly debated and controversial issue (Raoult & Koonin, 2012). While some authors hold the view that LGTs among the prokaryotes is so profuse that it has abolished all vertical signal emanating from the Darwinian mode of evolution (Bapteste, Boucher, Leigh, & Doolittle, 2004; Doolittle & Bapteste, 2007; Gogarten et al., 2002; Koonin, 2009; Merhej & Raoult, 2012; Zhaxybayeva et al., 2009), others have inferred that the overall incidences of LGTs and their impact on prokaryotic phylogeny is very minimal (Beiko & Ragan, 2008; Bhandari et al., 2012; Gao & Gupta, 2012a; Kurland, 2005; Kurland, Canback, & Berg, 2003; Puigbo, Wolf, & Koonin, 2010). Extensive work on CSIs for different bacterial groups strongly supports the latter view (Bhandari et al., 2012). In all of the bacterial phyla (or groups) studied by this means (Adeolu & Gupta, 2013; Bhandari & Gupta, 2014a; Dutilh, Huynen, Bruno, & Snel, 2004; Gao & Gupta, 2012b; Gupta, Chander, et al., 2013; Gupta, Chen, et al., 2013; Gupta & Lali, 2013; Gupta, Mahmood, et al., 2013; Lerat, Daubin, Ochman, & Moran, 2005; Naushad & Gupta, 2012, 2013; Naushad et al., 2014), a vast majority (>80%) of the discovered evolutionarily useful CSIs are specific for different monophyletic clades of the studied phyla at multiple phylogenetic levels. The clades for which these CSIs show specificities are also independently strongly supported by different forms of phylogenetic analyses. The remaining 15–20% of the identified CSIs are present in assorted groups of bacteria (Bhandari et al., 2012; Naushad & Gupta, 2013). Most importantly, in all of these cases, based upon these CSIs, no specific relationships can be inferred between the group that is being studied and any other group(s) of bacteria. The question may be asked how should these latter CSIs be interpreted and do they pose significant problem in the understanding of microbial phylogeny and systematics? Although during Darwin's life time there was no knowledge of genes, genomes or even microorganisms, the answer to this question can be found in his work. The characteristics (in our case CSIs), which are commonly shared by unrelated group of organisms are referred to by Darwin as analogical or adaptive traits, and his views regarding their impact on classification are stated in the following quote.

> *the very same characters are analogical when one class or order is compared with another, but give true affinities when the members of the same class or order are compared one with another: ... "in comparing one group with a distinct group, we summarily reject analogical or adaptive characters, and yet use these same characters within the limit of the same group".*

In the present context, if a CSI is commonly shared by two unrelated groups of microbes, then it is a useful diagnostic characteristic when the two groups are individually considered, but it is not a useful property when these two groups considered together, or to infer that they are related to each other. Thus, the CSIs that are commonly shared by more than one group of microbes do not pose any problem in understanding microbial phylogeny and classification that is supported by large numbers of CSIs that are unique properties of different taxa.

7 IMPORTANCE OF THE DISCOVERED CSIs FOR UNDERSTANDING MICROBIAL CLASSIFICATION AND PHYLOGENY

The inferences based upon different CSIs, in most parts, show excellent correlation with the current understanding of microbial phylogeny and systematics based upon 16S rRNA gene sequences, or those inferred from concatenated protein trees (Quast et al., 2013; Segata et al., 2013; Wu et al., 2009; Yarza et al., 2010). The high degree of concordance in the relationships observed using multiple independent approaches gives confidence that the understanding of the microbial phylogeny or systematics that is emerging from these studies is reliable. The question may be asked how the discovery of CSIs is advancing the understanding of microbiology in comparison to its current state based on the 16S rRNA trees. As noted earlier, the understanding of microbial phylogeny and systematics, based upon phylogenetic trees has some limitations, which are now being supplemented by the discovered CSIs. Thus, some of the advances that are resulting from the discoveries of the CSIs are as follows: (i) Unlike the phylogenetic trees, where the members of different prokaryotic taxa are distinguished primarily on the basis of their branching in the trees, the discovered CSIs are enabling distinction and demarcation of members of different clades or prokaryotes based upon multiple highly specific molecular markers. Extensive work on CSIs for a number of major groups/phyla of bacteria (e.g. Cyanobacteria, Actinobacteria, Chlamydiae, Alphaproteobacteria, Gammaproteobacteria) has shown that they exhibit high degree of predictive ability for the indicated groups; many CSIs for these groups were identified when sequence information was available for only a small number of genomes (Gao & Gupta, 2005; Gupta, 1998, 2000, 2001, 2005; Gupta, Pereira, Chandrasekera, & Johari, 2003). However, despite >10–50-fold increase in the number of genomes sequenced for these groups and other prokaryotes, most of the originally identified CSIs have retained their specificity for the indicated groups (Bhandari et al., 2012; Gao & Gupta, 2012b; Gao et al., 2009; Gupta, 2009; Gupta & Mathews, 2010; Gupta & Mok, 2007). Additionally, in a number of cases where genomes sequences were not available for the group of interest, PCR amplification and sequencing studies confirmed the presence of the group-specific CSIs in the examined species from these groups (Gao & Gupta, 2005; Griffiths & Gupta, 2006b; Griffiths, Petrich, & Gupta, 2005). (ii) For most prokaryotic taxa at different taxonomic levels, no genetic, biochemical or physiological properties are known which are specific for them and phylogenetic trees provide no useful insights for discovering such characteristics. In contrast, the work on a number of CSIs has shown that these rare genetic changes present in conserved regions are essential for the groups of species where they are found and deletions, or most changes in them, lead to failure of cell growth (Akiva, Itzhaki, & Margalit, 2008; Schoeffler, May, & Berger, 2010; Singh & Gupta, 2009). Thus, it is expected that most of the discovered CSIs, which are specific for different bacterial groups, will also be involved in functions that are specific and very likely essential for these groups of bacteria. Hence, further studies on understanding the cellular functions of these highly specific genetic changes should to lead to identification of novel functional characteristics that are specific for different groups of prokaryotes. (iii) The species distribution of CSIs (or main-line signatures) in several important proteins described previously provides a mean to determine the relative branching order and interrelationships among different bacterial phyla, which are not resolved by the 16S rRNA trees (Bhandari & Gupta, 2014b; Griffiths & Gupta, 2001, 2004; Gupta, 2003; Ludwig & Klenk, 2005; Yarza et al., 2010). (iv) The identified CSIs, due to specificity for different bacterial groups, also provide novel diagnostic tools (Ahmod, Gupta, & Shah, 2011). The primary sequences of the genes/proteins containing many of these CSIs are highly conserved. Based upon the conserved regions found in these genes/proteins, suitable PCR primers can be designed that should lead to rapid and specific amplification of the members of these groups of bacteria enabling their identification in clinical as well as other settings including in metagenomes (Gao, Parmanathan, & Gupta, 2006; Griffiths et al., 2005). (v) Additionally, due to the predicted essential roles of these CSIs in the bacterial groups where they are found, they provide potential targets for development of novel antibacterial agents that can specifically target different groups of bacteria (Gupta & Griffiths, 2006; Nandan et al., 2007; Naushad & Gupta, 2013).

ACKNOWLEDGEMENTS

I thank Mobolaji Adeolu and Hafiz Sohail Naushad for critical reading of the chapter and providing useful comments.

REFERENCES

Adeolu, M., & Gupta, R. S. (2013). Phylogenomics and molecular signatures for the order *Neisseriales*: Proposal for division of the order *Neisseriales* into the emended family *Neisseriaceae* and *Chromobacteriaceae* fam. nov. *Antonie van Leeuwenhoek, 104*, 1–24.

Adeolu, M., & Gupta, R. S. (2014). A phylogenomic and molecular marker based proposal for the division of the genus Borrelia into two genera: The emended genus Borrelia containing only the members of the relapsing fever Borrelia, and the genus Borreliella gen. nov. containing the members of the Lyme disease Borrelia (Borrelia burgdorferi sensu lato complex). *Antonie van Leeuwenhoek, 105*, 1049–1072.

Ahmod, N. Z., Gupta, R. S., & Shah, H. N. (2011). Identification of a *Bacillus anthracis* specific indel in the yeaC gene and development of a rapid pyrosequencing assay for distinguishing *B. anthracis* from the *B. cereus* group. *Journal of Microbiological Methods, 87*, 278–285.

Ajawatanawong, P., & Baldauf, S. L. (2013). Evolution of protein indels in plants, animals and fungi. *BMC Evolutionary Biology, 13*, 140.

Akiva, E., Itzhaki, Z., & Margalit, H. (2008). Built-in loops allow versatility in domain-domain interactions: Lessons from self-interacting domains. *Proceedings of the National Academy of Sciences of the United States of America, 105*, 13292–13297.

Baldauf, S. L., & Palmer, J. D. (1993). Animals and fungi are each other's closest relatives: Congruent evidence from multiple proteins. *Proceedings of the National Academy of Sciences of the United States of America, 90*, 11558–11562.

Bapteste, E., Boucher, Y., Leigh, J., & Doolittle, W. F. (2004). Phylogenetic reconstruction and lateral gene transfer. *Trends in Microbiology, 12*, 406–411.

Beiko, R. G., & Ragan, M. A. (2008). Detecting lateral genetic transfer: A phylogenetic approach. *Methods in Molecular Biology, 452*, 457–469.

Bhandari, V., & Gupta, R. S. (2012). Molecular signatures for the phylum Synergistetes and some of its subclades. *Antonie van Leeuwenhoek, 102*, 517–540.

Bhandari, V., & Gupta, R. S. (2014a). Molecular signatures for the phylum (class) Thermotogae and a proposal for its division into three orders (Thermotogales, Kosmotogales ord. nov. and Petrotogales ord. nov.) containing four families (Thermotogaceae, Fervidobacteriaceae fam. nov., Kosmotogaceae fam. nov. and Petrotogaceae fam. nov.) and a new genus Pseudothermotoga gen. nov. with five new combinations. *Antonie van Leeuwenhoek, 105*, 143–168.

Bhandari, V., & Gupta, R. S. (2014b). Phylum thermotogae. In E. Rosenberg, E. F. Delong, E. Stackebrandt, & F. Thompson (Eds.), *The prokaryotes: Vol. 11* (4th ed.). Heidelberg: Springer, in press.

Bhandari, V., Naushad, H. S., & Gupta, R. S. (2012). Protein based molecular markers provide reliable means to understand prokaryotic phylogeny and support Darwinian mode of evolution. *Frontiers in Cellular and Infection Microbiology, 2*, 98.

Brenner, D. J., Staley, J. T., & Krieg, N. R. (2005). Classification of procaryotic organisms and the concept of bacterial speciation. In D. J. Brenner, N. R. Krieg, & J. T. Staley (Eds.), *Part*

C, Introductory essays: Vol. 2. Bergey's Manual of Systematic Bacteriology (pp. 27–32). New York: Springer.

Chun, J., & Hong, S. G. (2010). Methods and programs for calculation of phylogenetic relationships from molecular sequences. In A. Oren & R. T. Papke (Eds.), *Molecular phylogeny of organisms* (pp. 23–40). Norfolk: Caister Academic Press.

Ciccarelli, F. D., Doerks, T., von Mering, C., Creevey, C. J., Snel, B., & Bork, P. (2006). Toward automatic reconstruction of a highly resolved tree of life. *Science, 311*, 1283–1287.

Cole, J. R., Wang, Q., Cardenas, E., Fish, J., Chai, B., Farris, R. J., et al. (2009). The Ribosomal Database Project: Improved alignments and new tools for rRNA analysis. *Nucleic Acids Research, 37*, D141–D145.

Cutino-Jimenez, A. M., Martins-Pinheiro, M., Lima, W. C., Martin-Tornet, A., Morales, O. G., & Menck, C. F. (2010). Evolutionary placement of Xanthomonadales based on conserved protein signature sequences. *Molecular Phylogenetics and Evolution, 54*, 524–534.

Darwin, C. (1859). *The origin of species by means of natural selection, or the preservation of favoured races in the struggle for life.* London: John Murray.

Deckert, G., Warren, P. V., Gaasterland, T., Young, W. G., Lenox, A. L., Graham, D. E., et al. (1998). The complete genome of the hyperthermophilic bacterium *Aquifex aeolicus. Nature, 392*, 353–358.

Doolittle, W. F., & Bapteste, E. (2007). Pattern pluralism and the Tree of Life hypothesis. *Proceedings of the National Academy of Sciences of the United States of America, 104*, 2043–2049.

Dutilh, B. E., Huynen, M. A., Bruno, W. J., & Snel, B. (2004). The consistent phylogenetic signal in genome trees revealed by reducing the impact of noise. *Journal of Molecular Evolution, 58*, 527–539.

Edgar, R. C. (2004). MUSCLE: Multiple sequence alignment with high accuracy and high throughput. *Nucleic Acids Research, 32*, 1792–1797.

Estabrook, G. F., Johnson, C. S., Jr., & McMorris, F. R. (1976). A mathematical foundation for the analysis of cladistic character compatibility. *Mathematical Biosciences, 29*, 181–187.

Felsenstein, J. (2004). *Inferring phylogenies.* Sunderland, Massachusetts: Sinauer Associates, Inc.

Frock, A. D., Notey, J. S., & Kelly, R. M. (2010). The genus Thermotoga: Recent developments. *Environmental Technology, 31*, 1169–1181.

Gao, B., & Gupta, R. S. (2005). Conserved indels in protein sequences that are characteristic of the phylum *Actinobacteria. International Journal of Systematic and Evolutionary Microbiology, 55*, 2401–2412.

Gao, B., & Gupta, R. S. (2012a). Microbial systematics in the post-genomics era. *Antonie van Leeuwenhoek, 101*, 45–54.

Gao, B., & Gupta, R. S. (2012b). Phylogenetic framework and molecular signatures for the main clades of the phylum Actinobacteria. *Microbiology and Molecular Biology Reviews, 76*, 66–112.

Gao, B., Mohan, R., & Gupta, R. S. (2009). Phylogenomics and protein signatures elucidating the evolutionary relationships among the *Gammaproteobacteria. International Journal of Systematic and Evolutionary Microbiology, 59*, 234–247.

Gao, B., Parmanathan, R., & Gupta, R. S. (2006). Signature proteins that are distinctive characteristics of *Actinobacteria* and their subgroups. *Antonie van Leeuwenhoek, 90*, 69–91.

Garrity, G. M., Bell, J. A., & Lilburn, T. G. (2005). The revised road map to the manual. In D. J. Brenner, N. R. Krieg, & J. T. Staley (Eds.), *Bergey's manual of systematic bacteriology* (pp. 159–220). New York: Springer.

Gogarten, J. P., Doolittle, W. F., & Lawrence, J. G. (2002). Prokaryotic evolution in light of gene transfer. *Molecular Biology and Evolution, 19*, 2226–2238.

Griffiths, E., & Gupta, R. S. (2001). The use of signature sequences in different proteins to determine the relative branching order of bacterial divisions: Evidence that *Fibrobacter* diverged at a similar time to *Chlamydia* and the *Cytophaga- Flavobacterium-Bacteroides* division. *Microbiology, 147*, 2611–2622.

Griffiths, E., & Gupta, R. S. (2002). Protein signatures distinctive of chlamydial species: Horizontal transfer of cell wall biosynthesis genes *glmU* from *Archaebacteria* to *Chlamydiae*, and *murA* between *Chlamydiae* and *Streptomyces*. *Microbiology, 148*, 2541–2549.

Griffiths, E., & Gupta, R. S. (2004). Signature sequences in diverse proteins provide evidence for the late divergence of the order *Aquificales*. *International Microbiology, 7*, 41–52.

Griffiths, E., & Gupta, R. S. (2006a). Lateral transfers of serine hydroxymethyl transferase (*glyA*) and UDP-N-acetylglucosamine enolpyruvyl transferase (*murA*) genes from free-living *Actinobacteria* to the parasitic chlamydiae. *Journal of Molecular Evolution, 63*, 283–296.

Griffiths, E., & Gupta, R. S. (2006b). Molecular signatures in protein sequences that are characteristics of the Phylum Aquificales. *International Journal of Systematic and Evolutionary Microbiology, 56*, 99–107.

Griffiths, E., & Gupta, R. S. (2007). Phylogeny and shared conserved inserts in proteins provide evidence that *Verrucomicrobia* are the closest known free-living relatives of chlamydiae. *Microbiology, 153*, 2648–2654.

Griffiths, E., Petrich, A., & Gupta, R. S. (2005). Conserved indels in essential proteins that are distinctive characteristics of *Chlamydiales* and provide novel means for their identification. *Microbiology, 151*, 2647–2657.

Gupta, R. S. (1997). Protein phylogenies and signature sequences: Evolutionary relationships within prokaryotes and between prokaryotes and eukaryotes. *Antonie van Leeuwenhoek, 72*, 49–61.

Gupta, R. S. (1998). Protein phylogenies and signature sequences: A reappraisal of evolutionary relationships among archaebacteria, eubacteria, and eukaryotes. *Microbiology and Molecular Biology Reviews, 62*, 1435–1491.

Gupta, R. S. (2000). The phylogeny of *Proteobacteria*: Relationships to other eubacterial phyla and eukaryotes. *FEMS Microbiology Review, 24*, 367–402.

Gupta, R. S. (2001). The branching order and phylogenetic placement of species from completed bacterial genomes, based on conserved indels found in various proteins. *International Microbiology, 4*, 187–202.

Gupta, R. S. (2003). Evolutionary relationships among photosynthetic bacteria. *Photosynthesis Research, 76*, 173–183.

Gupta, R. S. (2004). The phylogeny and signature sequences characteristics of *Fibrobacters, Chlorobi* and *Bacteroidetes*. *Critical Reviews in Microbiology, 30*, 123–143.

Gupta, R. S. (2005). Protein signatures distinctive of Alpha proteobacteria and its subgroups and a model for Alpha proteobacterial evolution. *Critical Reviews in Microbiology, 31*, 135.

Gupta, R. S. (2009). Protein signatures (molecular synapomorphies) that are distinctive characteristics of the major cyanobacterial clades. *International Journal of Systematic and Evolutionary Microbiology, 59*, 2510–2526.

Gupta, R. S. (2010). Applications of conserved indels for understanding microbial phylogeny. In A. Oren & R. T. Papke (Eds.), *Molecular phylogeny of microorganisms* (pp. 135–150). Norfolk, UK: Caister Academic Press.

Gupta, R. S. (2012). Origin and spread of photosynthesis based upon conserved sequence features in key bacteriochlorophyll biosynthesis proteins. *Molecular Biology and Evolution, 29*, 3397–3412.

Gupta, R. S., & Bhandari, V. (2011). Phylogeny and molecular signatures for the phylum thermotogae and its subgroups. *Antonie van Leeuwenhoek, 100*, 1–34.

Gupta, R. S., Bhandari, V., & Naushad, H. S. (2012). Molecular signatures for the PVC clade (Planctomycetes, Verrucomicrobia, Chlamydiae, and Lentisphaerae) of bacteria provide insights into their evolutionary relationships. *Frontiers in Microbiology, 3*, 327.

Gupta, R. S., Chander, P., & George, S. (2013). Phylogenetic framework and molecular signatures for the class *Chloroflexi* and its different clades; proposal for division of the class Chloroflexi class. nov. into the suborder *Chloroflexineae* subord. nov., consisting of the emended family *Oscillochloridaceae* and the family *Chloroflexaceae* fam. nov., and the suborder *Roseiflexineae* subord. nov., containing the family *Roseiflexaceae* fam. nov. *Antonie van Leeuwenhoek, 103*, 99–119.

Gupta, R. S., Chen, W. J., Adeolu, M., & Chai, Y. (2013). Molecular signatures for the class *Coriobacteriia* and its different clades; proposal for division of the class *Coriobacteriia* into the emended order *Coriobacteriales*, containing the emended family *Coriobacteriaceae* and *Atopobiaceae* fam. nov., and *Eggerthellales* ord. nov., containing the family *Eggerthellaceae* fam. nov. *International Journal of Systematic and Evolutionary Microbiology, 63*, 3379–3397.

Gupta, R. S., & Gao, B. (2009). Phylogenomic analyses of clostridia and identification of novel protein signatures that are specific to the genus *Clostridium* sensu stricto (cluster I). *International Journal of Systematic and Evolutionary Microbiology, 59*, 285–294.

Gupta, R. S., & Griffiths, E. (2002). Critical issues in bacterial phylogeny. *Theoretical Population Biology, 61*, 423–434.

Gupta, R. S., & Griffiths, E. (2006). *Chlamydiae*-specific proteins and indels: Novel tools for studies. *Trends in Microbiology, 14*, 527–535.

Gupta, R. S., & Lali, R. (2013). Molecular signatures for the phylum Aquificae and its different clades: Proposal for division of the phylum Aquificae into the emended order *Aquificales*, containing the families *Aquificaceae* and *Hydrogenothermaceae*, and a new order *Desulfurobacteriales* ord. nov., containing the family *Desulfurobacteriaceae*. *Antonie van Leeuwenhoek, 104*, 349–368.

Gupta, R. S., & Lorenzini, E. (2007). Phylogeny and molecular signatures (conserved proteins and indels) that are specific for the *Bacteroidetes* and *Chlorobi* species. *BMC Evolutionary Biology, 7*, 71.

Gupta, R. S., Mahmood, S., & Adeolu, M. (2013). A phylogenomic and molecular signature based approach for characterization of the phylum Spirochaetes and its major clades: Proposal for a taxonomic revision of the phylum. *Frontiers in Microbiology, 4*, 217.

Gupta, R. S., & Mathews, D. W. (2010). Signature proteins for the major clades of *Cyanobacteria*. *BMC Evolutionary Biology, 10*, 24.

Gupta, R. S., & Mok, A. (2007). Phylogenomics and signature proteins for the alpha proteobacteria and its main groups. *BMC Microbiology, 7*, 106.

Gupta, R. S., Pereira, M., Chandrasekera, C., & Johari, V. (2003). Molecular signatures in protein sequences that are characteristic of Cyanobacteria and plastid homologues. *International Journal of Systematic and Evolutionary Microbiology, 53*, 1833–1842.

Gupta, R. S., & Sneath, P. H. A. (2007). Application of the character compatibility approach to generalized molecular sequence data: Branching order of the proteobacterial subdivisions. *Journal of Molecular Evolution, 64*, 90–100.

Huang, J., & Gogarten, J. P. (2009). Ancient gene transfer as a tool in phylogenetic reconstruction. *Methods in Molecular Biology, 532,* 127–139.

Huber, R., & Hannig, M. (2006). Thermotogales. *The Prokaryotes, 7,* 899–922.

Jenkins, C., & Fuerst, J. A. (2001). Phylogenetic analysis of evolutionary relationships of the planctomycete division of the domain bacteria based on amino acid sequences of elongation factor Tu. *Journal of Molecular Evolution, 52,* 405–418.

Kampfer, P. (2012). Systematics of prokaryotes: The state of the art. *Antonie van Leeuwenhoek, 101,* 3–11.

Koonin, E. V. (2009). Darwinian evolution in the light of genomics. *Nucleic Acids Research, 37,* 1011–1034.

Kumar, S., Nei, M., Dudley, J., & Tamura, K. (2008). MEGA: A biologist-centric software for evolutionary analysis of DNA and protein sequences. *Briefings in Bioinformatics, 9,* 299–306.

Kurland, C. G. (2005). What tangled web: Barriers to rampant horizontal gene transfer. *Bioessays, 27,* 741–747.

Kurland, C. G., Canback, B., & Berg, O. G. (2003). Horizontal gene transfer: A critical view. *Proceedings of the National Academy of Sciences of the United States of America, 100,* 9658–9662.

Larkin, M. A., Blackshields, G., Brown, N. P., Chenna, R., McGettigan, P. A., McWilliam, H., et al. (2007). Clustal W and Clustal X version 2.0. *Bioinformatics, 23,* 2947–2948.

Lerat, E., Daubin, V., Ochman, H., & Moran, N. A. (2005). Evolutionary origins of genomic repertoires in bacteria. *PLoS Biology, 3,* e130.

L'Haridon, S., Reysenbach, A. L., Tindall, B. J., Schonheit, P., Banta, A., Johnsen, U., et al. (2006). Desulfurobacterium atlanticum sp. nov., Desulfurobacterium pacificum sp. nov. and Thermovibrio guaymasensis sp. nov., three thermophilic members of the Desulfurobacteriaceae fam. nov., a deep branching lineage within the Bacteria. *International Journal of Systematic and Evolutionary Microbiology, 56,* 2843–2852.

Lima, W. C., Paquola, A. C., Varani, A. M., Van Sluys, M. A., & Menck, C. F. (2008). Laterally transferred genomic islands in Xanthomonadales related to pathogenicity and primary metabolism. *FEMS Microbiology Letters, 281,* 87–97.

LPSN Web Resource. (2013). List of prokaryotic names with standing in nomenclature. http://www.bacterio.net/.

Ludwig, W., & Klenk, H.-P. (2005). Overview: A phylogenetic backbone and taxonomic framework for prokaryotic systematics. In D. J. Brenner, N. R. Krieg, J. T. Staley, & G. M. Garrity (Eds.), *Bergey's manual of systematic bacteriology* (pp. 49–65). Berlin: Springer-Verlag.

Merhej, V., & Raoult, D. (2012). Rhizome of life, catastrophes, sequence exchanges, gene creations, and giant viruses: How microbial genomics challenges Darwin. *Frontiers in Cellular and Infection Microbiology, 2,* 113.

Moreira, D., & Philippe, H. (2000). Molecular phylogeny: Pitfalls and progress. *International Microbiology, 3,* 9–16.

Murray, R. G. E. (1986). The higher taxa, or, a place for everything...? In P. H. A. Sneath, N. S. Mair, M. E. Sharpe, & J. G. Holt (Eds.), *Bergey's manual of systematic bacteriology* (pp. 31–34). Baltimore: The Williams and Wilkins.

Nandan, D., Lopez, M., Ban, F., Huang, M., Li, Y., Reiner, N. E., et al. (2007). Indel-based targeting of essential proteins in human pathogens that have close host orthologue(s): Discovery of selective inhibitors for Leishmania donovani elongation factor-1alpha. *Proteins, 67,* 53–64.

Naushad, H. S., & Gupta, R. S. (2012). Molecular signatures (conserved indels) in protein sequences that are specific for the order Pasteurellales and distinguish two of its main clades. *Antonie van Leeuwenhoek, 101*, 105–124.

Naushad, H. S., & Gupta, R. S. (2013). Phylogenomics and molecular signatures for species from the plant pathogen-containing order xanthomonadales. *PLoS One, 8*, e55216.

Naushad, H. S., Lee, B., & Gupta, R. S. (2014). Conserved signature indels and signature proteins as novel tools for understanding microbial phylogeny and systematics: Identification of molecular signatures that are specific for the phytopathogenic genera Dickeya, Pectobacterium and Brenneria. *International Journal of Systematic and Evolutionary Microbiology, 64*, 366–383.

Nelson, K. E., Paulsen, I. T., & Fraser, C. M. (2001). Microbial genome sequencing: A window into evolution and physiology. *ASM News, 67*, 310–317.

Nesbo, C. L., Bapteste, E., Curtis, B., Dahle, H., Lopez, P., MacLeod, D., et al. (2009). The genome of Thermosipho africanus TCF52B: Lateral genetic connections to the Firmicutes and Archaea. *Journal of Bacteriology, 191*, 1974–1978.

Olendzenski, L., Liu, L., Zhaxybayeva, O., Murphey, R., Shin, D. G., & Gogarten, J. P. (2000). Horizontal transfer of archaeal genes into the deinococcaceae: Detection by molecular and computer-based approaches. *Journal of Molecular Evolution, 51*, 587–599.

Olsen, G. J., Woese, C. R., & Overbeek, R. (1994). The winds of (evolutionary) change: Breathing new life into microbiology. *Journal of Bacteriology, 176*, 1–6.

Oren, A. (2004). Prokaryote diversity and taxonomy: Current status and future challenges. *Philosophical Transactions of the Royal Society of London Series B: Biological Sciences, 359*, 623–638.

Oren, A. (2010). Microbial systematics. In L. K. Wang, V. Ivanov, J.-H. Tay, & Y. T. Hung (Eds.), *Environmental biotechnology: Vol. 10. Handbook of environmental engineering* (pp. 81–120). New York: Springer Science+Business Media.

Oren, A., & Stackebrandt, E. (2002). Prokaryote taxonomy online: Challenges ahead. *Nature, 419*, 15.

Pace, N. R. (2009). Mapping the tree of life: Progress and prospects. *Microbiology and Molecular Biology Reviews, 73*, 565–576.

Pisani, D. (2004). Identifying and removing fast-evolving sites using compatibility analysis: An example from the Arthropoda. *Systematic Biology, 53*, 978–989.

Puigbo, P., Wolf, Y. I., & Koonin, E. V. (2010). The tree and net components of prokaryote evolution. *Genome Biology and Evolution, 2*, 745–756.

Quast, C., Pruesse, E., Yilmaz, P., Gerken, J., Schweer, T., Yarza, P., et al. (2013). The SILVA ribosomal RNA gene database project: Improved data processing and web-based tools. *Nucleic Acids Research, 41*, D590–D596.

Raoult, D., & Koonin, E. V. (2012). Microbial genomics challenge Darwin. *Frontiers in Cellular and Infection Microbiology, 2*, 127.

Ravinesan, D. A., & Gupta, R. S. (2014). Molecular signatures for members of the genus Dehalococcoides and the class Dehalococcoidia. *International Journal of Systematic and Evolutionary Microbiology*, Epub March 27, 2014, PMID: 24676731.

Raymond, J., Zhaxybayeva, O., Gogarten, J. P., Gerdes, S. Y., & Blankenship, R. E. (2002). Whole-genome analysis of photosynthetic prokaryotes. *Science, 298*, 1616–1620.

Reysenbach, A. L., Liu, Y., Lindgren, A. R., Wagner, I. D., Sislak, C. D., Mets, A., et al. (2013). *Mesoaciditoga lauensis* gen. nov., sp. nov., a moderate thermoacidophilic Thermotogales from a deep-sea hydrothermal vent. *International Journal of Systematic and Evolutionary Microbiology, 63*, 4724–4729.

Richter, M., & Rossello-Mora, R. (2009). Shifting the genomic gold standard for the prokaryotic species definition. *Proceedings of the National Academy of Sciences of the United States of America, 106*, 19126–19131.

Rivera, M. C., & Lake, J. A. (1992). Evidence that eukaryotes and eocyte prokaryotes are immediate relatives. *Science, 257*, 74–76.

Rokas, A., & Holland, P. W. (2000). Rare genomic changes as a tool for phylogenetics. *Trends in Ecology & Evolution, 15*, 454–459.

Rokas, A., Williams, B. L., King, N., & Carroll, S. B. (2003). Genome-scale approaches to resolving incongruence in molecular phylogenies. *Nature, 425*, 798–804.

Schoeffler, A. J., May, A. P., & Berger, J. M. (2010). A domain insertion in *Escherichia coli* GyrB adopts a novel fold that plays a critical role in gyrase function. *Nucleic Acids Research, 38*, 7830–7844.

Segata, N., Bornigen, D., Morgan, X. C., & Huttenhower, C. (2013). PhyloPhlAn is a new method for improved phylogenetic and taxonomic placement of microbes. *Nature Communications, 4*, 2304.

Singh, B., & Gupta, R. S. (2009). Conserved inserts in the Hsp60 (GroEL) and Hsp70 (DnaK) proteins are essential for cellular growth. *Molecular Genetics and Genomics, 281*, 361–373.

Stackebrandt, E. (2006). Defining taxonomic ranks. In M. Dworkin, S. Falkow, E. Rosenberg, K.-H. Schleifer, & E. Stackebrandt (Eds.), *The prokaryotes* (pp. 29–57): Heidelberg: Springer.

Stanier, R. Y. (1941). The main outlines of bacterial classification. *Journal of Bacteriology, 42*, 437–466.

Stanier, R. Y., & van Niel, C. B. (1962). The concept of a bacterium. *Archiv für Mikrobiologie, 42*, 17–35.

Sutcliffe, I. C., Trujillo, M. E., & Goodfellow, M. (2012). A call to arms for systematists: Revitalising the purpose and practises underpinning the description of novel microbial taxa. *Antonie van Leeuwenhoek, 101*, 13–20.

Teeling, H., Lombardot, T., Bauer, M., Ludwig, W., & Glockner, F. O. (2004). Evaluation of the phylogenetic position of the planctomycete 'Rhodopirellula baltica' SH 1 by means of concatenated ribosomal protein sequences, DNA-directed RNA polymerase subunit sequences and whole genome trees. *International Journal of Systematic and Evolutionary Microbiology, 54*, 791–801.

van Niel, C. B. (1946). The classification and natural relationships of bacteria. *Cold Spring Harbor Symposium of Quantitative Biology, 11*, 285–301.

Wagner, M., & Horn, M. (2006). The Planctomycetes, Verrucomicrobia, Chlamydiae and sister phyla comprise a superphylum with biotechnological and medical relevance. *Current Opinion in Biotechnology, 17*, 241–249.

Ward, N. L., Rainey, F. A., Hedlund, B. P., Staley, J. T., Ludwig, W., & Stackebrandt, E. (2000). Comparative phylogenetic analyses of members of the order *Planctomycetales* and the division *Verrucomicrobia*: 23S rRNA gene sequence analysis supports the 16S rRNA gene sequence-derived phylogeny. *International Journal of Systematic Bacteriology, 50*, 1965–1972.

Woese, C. R. (1987). Bacterial evolution. *Microbiological Reviews, 51*, 221–271.

Woese, C. R. (1991). The use of ribosomal RNA in reconstructing evolutionary relationships among bacteria. In R. K. Selander, A. G. Clark, & T. S. Whittmay (Eds.), *Evolution at molecular level* (pp. 1–24). Sunderland, MA: Sinauer Associates Inc.

Woese, C. R. (1992). Prokaryote systematics: The evolution of a science. In A. Balows, H. G. Trüper, M. Dworkin, W. Harder, & K. H. Schleifer (Eds.), *The prokaryotes* (pp. 3–18). New York: Springer-Verlag.

Wolf, Y. I., Rogozin, I. B., Grishin, N. V., & Koonin, E. V. (2002). Genome trees and the Tree of Life. *Trends in Genetics, 18*, 472–479.

Wu, D., Hugenholtz, P., Mavromatis, K., Pukall, R., Dalin, E., Ivanova, N. N., et al. (2009). A phylogeny-driven genomic encyclopaedia of *Bacteria* and *Archaea. Nature, 462*, 1056–1060.

Yarza, P., Ludwig, W., Euzeby, J., Amann, R., Schleifer, K. H., Glockner, F. O., et al. (2010). Update of the all-species living tree project based on 16S and 23S rRNA sequence analyses. *Systematic and Applied Microbiology, 33*, 291–299.

Zhaxybayeva, O., Swithers, K. S., Lapierre, P., Fournier, G. P., Bickhart, D. M., DeBoy, R. T., et al. (2009). On the chimeric nature, thermophilic origin, and phylogenetic placement of the Thermotogales. *Proceedings of the National Academy of Sciences of the United States of America, 106*, 5865–5870.

Zhi, X. Y., Zhao, W., Li, W. J., & Zhao, G. P. (2012). Prokaryotic systematics in the genomics era. *Antonie van Leeuwenhoek, 101*, 21–34.

Reconciliation Approaches to Determining HGT, Duplications, and Losses in Gene Trees

Olga K. Kamneva*,[1], Naomi L. Ward[†]

*Department of Biology, Stanford University, Stanford, California, USA
[†]Department of Molecular Biology, University of Wyoming, Laramie, Wyoming, USA
[1]Corresponding author: e-mail address: okamneva@stanford.edu

1 INTRODUCTION

Dramatic gene loss and gain, corresponding to physiology and lifestyle shifts, has been documented for a number of bacterial lineages (Lefébure & Stanhope, 2007). Such changes are of interest to a variety of disciplines, including microbiology, molecular and cellular biology, medicine, and biotechnology. Changes in genomic content occur due to evolutionary processes such as gene duplication, loss, and horizontal gene transfer (HGT). These events lead to differences between the histories of different genes, as well as discrepancies between the evolutionary history of individual genes and the genomes overall (Pamilo & Nei, 1988). Analyses of different gene family histories, or *genealogies*, can also be used to infer the kinds of events that have taken place in the evolutionary history of a genome, providing insight into the ways in which genomes have changed over time, including functional changes (Kamneva, Knight, Liberles, & Ward, 2012).

Explaining inconsistencies between the evolutionary histories of genes and the species in which they evolve is called *reconciliation*. Reconciliation has a number of important practical applications. For instance, reconciliation is the most comprehensive way to describe the dynamics of gene family evolution in terms of gene copy number (David & Alm, 2011; Kamneva et al., 2012). It is also the most reliable way to identify truly orthologous genes between different genomes (Åkerborg, Sennblad, Arvestad, & Lagergren, 2009), which is important when using information about a gene in one organism to understand the function of related genes in other organisms, an important part of genome annotation and sequence analysis.

Therefore, comparisons between the family histories of genes and the history of the species in which they evolve are becoming a common practice in microbiology research. The aims of these comparisons are to (1) predict the functions and

Methods in Microbiology, Volume 41, ISSN 0580-9517, http://dx.doi.org/10.1016/bs.mim.2014.08.004

properties of newly characterized genes and genomes, (2) characterize the evolutionary history of individual genes, (3) characterize genome evolution in terms of gene family content, and (4) predict ancestral gene family composition.

In this chapter, we describe how to apply some of the methods developed for gene and genome history reconciliation. We include several real data analysis examples to illustrate the different reconciliation techniques, and we include explicit analysis protocols with every example.

2 BACTERIAL SPECIES TREE

A *biological species* is a group of genetically similar organisms that are capable of interbreeding and producing fertile offspring. The evolutionary relationships among different species are represented by a tree-like graph called a "species tree," in which every branching point represents the divergence of one species into two or more new species. It is generally assumed that each divergence, or *speciation event*, occurs at a fixed point in time (Figure 1A). However, in bacteria, the classical definition of a species is challenged because gene flow can occur between closely related strains via recombination during the time of species separation, and between distantly related species via HGT (for a review, see Ochman, Lawrence, & Groisman, 2000; Retchless & Lawrence, 2010). Processes of hybridization and HGT lead to different evolutionary histories in different parts of a genome. Therefore, in any particular set of species, the classical model of species evolution (a bifurcating tree) is somewhat invalidated (McInerney, Pisani, Bapteste, & O'Connell, 2011). The accumulating evidence for non-tree-like patterns in the evolutionary histories of different organisms has led to the view that species histories are best represented by networks, rather than trees (Figure 1B). However, the implementation of software tools based on this new view of species evolution is still under development. Many approaches for genomic analysis relying on the assumption that horizontal processes (recombination, HGT, hybridization) occur in the context of a species history that is fundamentally tree-like often lead to satisfactory results in molecular evolutionary studies (David & Alm, 2011; Kamneva et al., 2012).

The evolutionary history of a set of species, or the *species tree*, is one of the necessary components of the analyses described in this chapter. Several methods are currently used to infer species trees from DNA and protein sequences. Classical approaches in molecular taxonomy utilize sequences of universally distributed genes

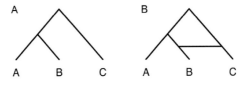

FIGURE 1

Example species tree (A) and network (B).

encoding the rRNA of the small subunit of the ribosome (16S rRNA in bacteria and archaea and 18S rRNA in eukaryotes). Since the 1970s, when 16S rRNA genes were first used in molecular phylogenetics (Woese & Fox, 1977), comprehensive protocols for 16S rRNA sequencing and further data processing have been developed (see Ludwig, Oliver Glöckner, & Yilmaz, 2011 for further details). However, the limited phylogenetic information available within one molecular marker often leads to unresolved or poorly supported phylogenies. This difficulty has been addressed by using multiple universally distributed protein-coding genes for phylogeny reconstruction (Rokas, Williams, King, & Carroll, 2003). The development of whole-genome sequencing and re-sequencing projects has provided data that are well suited for multi-locus species tree reconstruction.

Several software tools for gene and genome history reconciliation also require dated species trees, in which every node in the tree has a date assigned to it and the branch lengths of the tree are measured in units of time, rather than in the numbers of nucleotide or amino acid substitutions occurring along each branch. The protocol to produce such a tree is described and utilized in David and Alm (2011), Kamneva et al. (2012), and Parfrey, Lahr, Knoll, and Katz (2011) and relies on the assumption of a molecular clock, whereby the amount of mutational change is a linear function of time, which is calibrated using dated archaeological fossils. This type of analysis is complicated and time consuming; therefore, the use of an existing dated species trees from previous analyses is often accepted (Ciccarelli et al., 2006).

3 GENE FAMILY

Another central component of gene–genome history reconciliation is the gene family, a group of sequences in different genomes that share a common history and which presumably carry out the same functions. Homologous gene families are constructed in different ways, described in detail in Pertea et al. (2003) and Tatusov et al. (2003). The process mostly relies on sequence comparison between genes from the same or different genomes, following the clustering of genes into different gene families (Li, Stoeckert, & Roos, 2003).

Two canonical representations of gene family are the gene tree (Figure 2A) and the phyletic pattern or profile (Figure 2B). A phyletic profile is a very simple

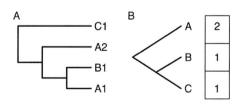

FIGURE 2

Phylogenetic tree representing evolution of a gene family (A) and its phylogenetic profile representation (B).

representation of gene family. It conveys information only about the presence or absence of a gene in contemporary genomes (Figure 2B). On the other hand, a phylogenetic gene tree reflects how sequences of different genes sampled from extant organisms are related to each other through various evolutionary events, such as gene duplication, HGT, speciation, and divergence (Figure 2A). The phylogenetic gene tree provides more information about the gene family. The reconstruction of the gene tree phylogeny is a challenging task on its own, especially when very closely or very distantly related sequences are considered. An example protocol for reconstructing gene phylogenies using either protein or DNA sequences is described here (Rokas, 2001) and include the following logical steps: (1) collecting sequences belonging to the gene family of interest; (2) aligning sequences in the data set to each other to generate a multiple sequence alignment; (3) testing for the evolutionary model of best fit; and (4) building a gene phylogeny.

It is important to identify gene families accurately, as including distantly related paralogous genes as members of the same family will lead to a number of problems. In the case where a phylogenetic profile is used to represent a gene family, additional duplication events might be predicted and erroneously placed onto lineages of the species tree. Including distantly related genes will also cause problems with gene phylogeny inference due to, for instance, long-branch attraction artefacts (Bergsten, 2005). Over-prediction of duplication and loss events or HGT events, and erroneous placement of the events over species tree lineages are also possible in this case. On the other hand, excluding some sequences from the gene family is also problematic, as then the gene family might be predicted to originate on the wrong lineage of the species tree, or additional gene loss or HGT events might be erroneously inferred.

4 EVOLUTION OF GENES IN BACTERIAL GENOMES

Multiple evolutionary processes affect genes in bacterial genomes over time. The set of events and processes that affect a gene family constitute its evolutionary history. Ultimately, the history of each gene should be considered in light of the history of the genome. As we discussed previously, speciation is a significant event in genome evolution (Figure 3A). Genes that originate from a common ancestor and that have been separated by speciation events are called orthologous genes (Figure 3B).

Other evolutionary processes include gene duplication, recombination, HGT, and gene loss. Gene duplication occurs when a second copy of a gene is created in a genome. Genes separated by an ancestral duplication event are called paralogs (Figure 3C). In the simplest case, duplication can be inferred from the gene history when two clades in the gene tree closely resemble the complete species tree or a part of it. This resemblance is the result of the fact that, after a duplication event, both copies of the gene follow similar evolutionary paths along the species tree. Different orders in the events of duplication and speciation give rise to in- and out-paralogs (Figure 3C). Divergence after gene duplication might lead to the emergence of more

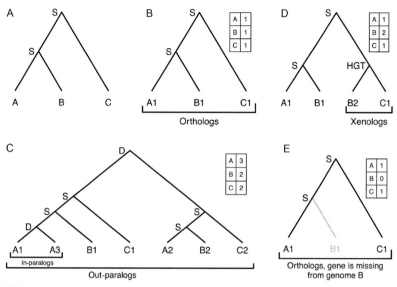

FIGURE 3

Species tree (A), gene tree relating orthologs and corresponding phyletic pattern (B); gene tree relating in-, out-paralogs and orthologs and corresponding phyletic pattern (C); gene tree relating xenologs and orthologs and corresponding phyletic pattern (D); gene tree relating orthologous genes and corresponding phyletic pattern; gene loss has removed the gene from genome B (E). Events are denoted as follows: S, speciation; D, duplication; HGT, horizontal gene transfer.

specialized or slightly different biological functions within the paralogous gene pairs. A set of orthologs and in-paralogs represents a gene family in a classic orthology analysis used in comparative genomics-based gene function prediction. The converse of gene duplication is gene loss. Gene loss can be detected in the gene history due to missing lineages observed in a gene tree but present in the species tree (Figure 3E).

When characterizing gene evolution in bacteria, it is important to account for recombination and HGT as well as other events. Recombination is a process that affects an ensemble of recombining bacterial strains that are often associated with distinct ecological niches, but which nevertheless maintain a sufficient amount of genetic similarity to recombine with one another.

With time, bacterial lineages diverge and genetic isolation will be established for the major parts of the genome, although this does not completely prohibit the further exchange of genetic information. HGT is often observed between distantly related bacterial genomes. It can be facilitated by mobile genetic elements and bacteriophages after transformation, conjugation, or transduction. Different kinds of mobile elements are also major forces driving genomic rearrangements and allowing for high levels of genomic plasticity. Genes arising from HGT events in the gene history are called xenologs (Figure 3D). Xenologs can be detected in the gene phylogeny, as they will cluster with the genes to which they are more closely related from organisms that are often distantly related.

In reality, the number of events occurring on different genetic lineages at different times might confound patterns of duplication, loss, and HGT. Thus, rigorous methods are needed in order to characterize gene family evolution in light of these large-scale evolutionary processes. A variety of such methods currently exist. They reconcile detailed gene trees or gene family-specific presence/absence profiles with a species tree through the inference of gene gain, loss, duplication, and HGT events. Methods that can perform inference of HGT events are particularly relevant for the analysis of bacterial genome evolution.

It is important to point out that alternative scenarios of gene family evolution exist. Ideally, one should evaluate which scenario is the most likely one; however, the majority of currently used, parsimony-based, methods do not allow such evaluation and require the user to rely on very arbitrary measures. Therefore, it is sometimes challenging to evaluate how realistic any given inferred gene history is.

5 GENE TREE/SPECIES TREE RECONCILIATION

In essence, gene tree/species tree reconciliation procedures aim to explain inconsistencies between gene and genome history using various evolutionary events. Figure 4 illustrates this process. It is important to note the existence of a number of alternative scenarios consistent with the given species and gene history. While running the analysis it is important to be able to choose the most plausible scenario. For that a number of software tools to analyze gene evolution in the context of species trees have been developed (Table 1). Various factors should be considered when performing such analyses. One important component of such an analysis is an accurate estimate of the species tree. The majority of existing methods for gene tree/species tree reconciliation require fully resolved, bifurcating, species trees which are sometimes hard to generate. However, several well-resolved and sufficiently supported species trees for a wide range of organisms have been reported and are sometimes used (Ciccarelli et al., 2006).

Another important thing to consider is the statistical framework that is used. Generally, Bayesian and likelihood-based methods allow reliable estimation of model parameters; however, they generally take a long time to run, which can be problematic in the case of large data sets or genome-level studies. In addition, many existing tools that use Bayesian or likelihood approaches do not consider lateral transfer events, which can be an especially important point to consider in the case of bacteria. Some parsimony-based methods do address transfer events, which makes them very useful for bacterial comparative genomics. They use some kind of event penalties to evaluate prospective evolutionary scenarios. The more events occur, the less parsimonious and less favourable the scenario is taken to be.

Representation of the gene family in use is another factor to take into account. A gene tree provides a more detailed description of a gene family than a phyletic

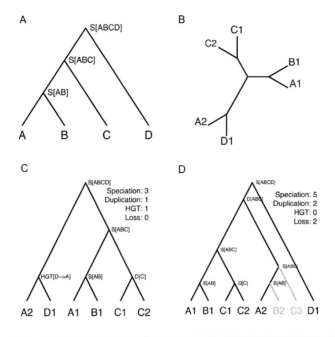

FIGURE 4

Species tree, with speciation events denoted by the letter S and the name of the emerging lineage given in squared brackets (A); unrooted gene tree (B); one plausible scenario of gene family evolution consistent with the species relationships (C); alternative scenario (D). Speciation, duplication, and HGT events are marked by letters S, D, and HGT sign, ancestral lineages where events are predicted to occur are indicated in squared brackets. Gene loss is indicated by a grey colour of the lineage and corresponding leaves.

profile. The use of phyletic patterns can be problematic, especially when analyzing genes from distantly related organisms because duplications, losses, and transfers that occur over long time periods could cancel each other out and they might not be reflected in the phyletic profile. On the other hand, gene tree reconciliation methods that use detailed gene histories should be robust to errors in gene tree inference. This is especially relevant for analyses involving distantly, or very closely, related sequences.

Another issue to consider is the taxonomy of the organisms included in the analysis. It is appealing to consider large number of genomes to obtain the most comprehensive view of gene evolution. However, including many species in analyses often results in long computation time. This is especially relevant in cases of whole-genome analyses. It is also important to make sure that taxonomic sampling is not biased towards one or several taxonomic groups or species with certain lifestyles. It is considered the best practice to include evolutionarily and taxonomically divergent species associated with various ecological habitats in the analysis.

Table 1 Some Programs for Gene Tree Species/Tree Reconciliation and Species Tree-Guided Gene Tree Reconstruction

Program	Events Considered	Framework	Time Consistent Transfers	Input
AnGST (David & Alm, 2011)	Duplication Loss HGT	Parsimony	Yes (if dated species tree is provided)/ No	Bifurcating species tree One or several gene trees Event penalties
MPR (Doyon, Hamel, & Chauve, 2010)	Duplication Loss HGT	Parsimony	Yes	Dated, bifurcating species tree One or several gene trees Event penalties
Ranger-dtl (Bansal, Alm, & Kellis, 2012)	Duplication Loss HGT	Parsimony	No	Bifurcating species tree One or several gene trees Event penalties
DLCpa (Wu, Rasmussen, Bansal, & Kellis, 2014)	Duplication Loss ILS	Parsimony	No	
Prime dltrs (Åkerborg et al., 2009)	Duplication Loss HGT	Bayesian	Yes	Dated, bifurcating species tree Gene sequence alignment Model of evolution

Here, we describe protocols for running AnGST, a parsimony-based program for gene tree–species tree reconciliation which requires python as well as basic familiarity with a command-line interface. AnGST was introduced by David and Alm (2011), and it can be obtained from http://almlab.mit.edu/angst/files/angst.tar.gz or run *via* the web application at http://almlab.mit.edu/angst/. AnGST implements an algorithm that uses gene birth, duplication, losses, and HGT to explain discrepancies between gene trees and species trees, assuming that evolutionary events that lead to discordance between a gene tree and the species tree carry fixed costs. Here, we examine the evolution of DUF70 domain proteins from several archaeal genomes using a locally installed version of AnGST. The protocol for running the same analysis on the server does not differ significantly.

The costs of evolutionary events can potentially be set to arbitrary values, which have significantly different effects on the inference. Here, we will examine the effect that penalty values have on the results by conducting the inference using various event costs.

AnGST also allows the user to incorporate information about gene tree uncertainty into the analysis. This is done via a bootstrap amalgamation procedure when the gene tree with the lowest reconciliation cost is chosen from a collection of trees that are consistent with the set of bipartitions present in all of the input gene trees.

5.1 PROTOCOL FOR RUNNING AnGST

1. Download the online supplementary files from http://dx.doi.org/10.1016/bs.mim. 2014.08.004. File AnGST_ST contains a dated, bifurcating species tree for 100 organisms. It can be viewed using a number of tree viewers, for instance FigTree. File AnGST_GTs contains a set of gene trees from several bootstrap runs (this file might also contain a single gene tree). Individual gene trees can also be viewed using any phylogeny visualization tool. AnGST_penalties is a penalties file where penalties for birth, duplication, loss, and HGT events are defined. AnGST_run1.input and AnGST_run2.input are control files for AnGST. Familiarize yourself with these files. Note that gene naming follows a certain convention where gene names include species names separated by the symbol ".". However, other symbols can be used as well (see the AnGST manual for details).

2. Create a directory (folder) where you want your results to go, and place all of the files (AnGST_ST, AnGST_GTs, AnGST_costs) within it. Now open a terminal and move to the directory that contains those files. Assuming that AnGST is in the directory located at /path_to_AnGST/, run the program using following command in the command-line prompt:

   `/path_to_AnGST/python angst_lib/AnGST.py AnGST_rin1.input`

3. Familiarize yourself with the results (they are also available at AnGST_run1.zip). If you have not edited the control file, the results will be written to a directory called AnGST_run1. The AnGST.nexus file contains a single resulting gene tree with every node annotated with the genome name in which it is predicted to be present. Additionally, the AnGST.events file contains a list of all the events asserted to have happened in the history of the gene under consideration.

4. Now use a text editor to change the event costs within the AnGST_costs file and re-run the program. The objective here is to compute the reconciliation for the example gene family, given different sets of event costs, and observe how predicted evolutionary scenario changes with the change of event penalties.
   a. Use HGT cost equal to 1, leaving all the other penalties unchanged
   b. Run AnGST again using AnGST_run2.input control file:

   `/path_to_AnGST/python angst_lib/AnGST.py AnGST_rin2.input`

5. Examine results of two additional runs, pay special attention to AnGST.events file, and note varying number of different events inferred for the data set.

5.2 INTERPRETING THE RESULTS OF AnGST ANALYSES

Results of these three AnGST runs are also visualized in Figure 5. It is clear that they differ quite a bit. When the penalty for HGT decreases relatively to the cost of duplication, more transfer events are inferred. This on its own indicates that every possible reconciliation can potentially be recovered with varying event penalties under parsimonious inference. This calls for additional justification of event cost values. In the original article introducing AnGST, the authors used genome flux (average change in genome size over every lineage of species tree) analysis to justify duplication, loss, and transfer costs (David & Alm, 2011). They obtained values included in the original AnGST_costs file. Very similar values for event penalties were recovered using genome flux analysis in a different study, on the data set including only bacterial genomes (Kamneva et al., 2012). Therefore, since optimization of event penalties using genome flux is computationally intensive, use of previously reported event penalties might be acceptable.

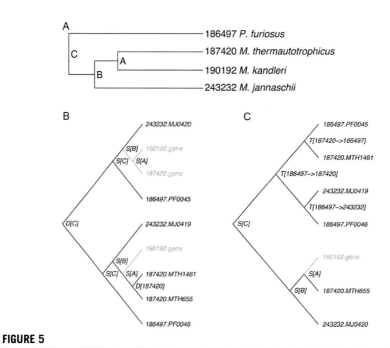

FIGURE 5

Species tree, names of ancestral genomes are shown as node labels and NCBI taxonomy IDs are given as tip labels along with the abbreviated species name (A); one plausible scenario of gene family evolution consistent with the species relationships as inferred in the first run of AnGST (B); alternative scenario, as inferred in the second AnGST run (C). Speciation, duplication, and HGT events are marked by letters S, D, and T, ancestral lineages where events are predicted to occur are indicated in squared brackets. Gene loss is indicated by a grey colour of the lineage and corresponding tip labels.

6 ANALYSIS AT THE GENOME SCALE

Whole-genome-level analysis has become very popular as a result of the accumulation of a large number of sequenced bacterial genomes. Here, we illustrate how this type of analysis is performed using the COUNT program, as it was designed to work on genome-scale data sets. However, genome-level analyses can be performed on a family-by-family basis using one of the protocols described in the previous sections.

The COUNT software tool was introduced by Csurös and Miklós (2009) and later by Csűös (2010). It is available from the authors' website: http://www.iro.umontreal.ca/~csuros/gene_content/count.html. It implements a variety of methods for evolutionary analysis of phyletic patterns or other types of integer-valued evolutionary characters, including parsimony, and probabilistic methods based on a phylogenetic birth-and-death model. The latter functionality is the most interesting within the framework of this chapter.

Ignoring explicit gene phylogenies within COUNT allows one to fit more sophisticated models of gene family evolution and estimate model parameters using genome-wide data. The implemented birth-death model assumes gene loss, duplication, and gain occurs along each branch of the species tree according to a stochastic process. Gene losses, gains, and duplications are assigned gene family- and edge-specific rates (k_f, k_e, lambda$_f$, lambda$_e$, mu$_f$, mu$_e$), and the processes run along the species tree edges for the time $t_e * t_f$. Edge-specific parameters are either held constant or vary along the branches of the species tree. Gene family-specific parameters are also assumed to be either constant or have a discretized Gamma distribution.

6.1 PROTOCOL FOR RUNNING COUNT

Here, we provide a protocol for running probabilistic inference of genome content using COUNT:

1. Obtain the online supplementary files from http://dx.doi.org/10.1016/bs.mim. 2014.08.004. The file COUNT_ST contains a species tree for five archaeal organisms. It can be viewed using any phylogeny visualization program. The page COUNT_table contains a list of gene families in the form of phyletic patterns and can be viewed in any text editor or Microsoft Excel. Familiarize yourself with these files. The file COUNT_commands contains command-prompt lines to be used to execute COUNT. The file content can be viewed using text editor.

2. Create a directory where you want your results to go, and place both data files (COUNT_ST, COUNT_table) within it. Now open a terminal and move to the directory that contains those files. Assuming that count.jar is in the directory located at /PATH_TO_COUNT/, run the analysis using commands included in the COUNT_commands file. This will allow you to fit a general model of genome evolution in a hierarchical fashion from the simplest to more and more complex.

3. Examine the resulting rate files, which are also enclosed within the COUNT_results.zip. Identify the line within the results file that gives the likelihood score for the data set. It is denoted as final likelihood but it appears as the negative logarithm of the likelihood score.

4. In the next step, we would like to compare models to each other to find the least complex model that fits the data reasonably well. We will use AIC for this purpose. A general review of this topic is provided by Burnham and Anderson (2002). In order to employ AIC, knowledge of the number of parameters estimated by the model in every COUNT run is required (here we denote the model with the name in the generated rate file):

    run1.rates: 3 edge-specific parameters + 1 parameter for gene family size distribution at the root of the species tree $= 4 = k$

    run2.rates: $3 * 8$ edge-specific parameters + 1 parameter for gene family size distribution at the root of the species tree $= 25 = k$

    run3.rates: $3 * 8$ edge-specific parameters + 1 parameter for gene family size distribution at the root of the species tree + 1 parameter for Gamma distribution for family-specific branch length adjustment factor $= 26 = k$

    run4.rates: $3 * 8$ edge-specific parameters + 1 parameter for gene family size distribution at the root of the species tree + 1 parameter for Gamma distribution for family-specific branch length adjustment factor + 1 parameter for Gamma distribution for family-specific duplication rate $= 27 = k$

    run5.rates: $3 * 8$ edge-specific parameters + 1 parameter for gene family size distribution at the root of the species tree + 1 parameter for Gamma distribution for family-specific branch length adjustment factor + 1 parameter for Gamma distribution for family-specific duplication rate + 1 parameter for Gamma distribution for family-specific loss rate $= 28 = k$

    run6.rates: $3 * 8$ edge-specific parameters + 1 parameter for gene family size distribution at the root of the species tree + 1 parameter for Gamma distribution for family-specific branch length adjustment factor + 1 parameter for Gamma distribution for family-specific duplication rate + 1 parameter for Gamma distribution for family-specific loss rate + 1 parameter for Gamma distribution for family-specific gain rate $= 29 = k$

    Use Microsoft Excel or other tool to calculate AIC as $2k - 2\ln(L)$. Model number 6 (run6.rates) with the lowest AIC value (42,283.77) should be recognized as the best fitting one.

5. Now in order to infer the ancestral composition of gene families for every node of the species tree and characterize the evolutionary dynamics of every gene family, we will use the COUNT application which has a graphical interface and allows both posterior analysis and data visualization. Start the COUNT application obtained with the COUNT package. Choose *new session* option under *Session* menu, load the species tree from COUNT_ST file; choose *open table* option under *Data* menu, load the gene family table from COUNT_table file; and lastly, choose *load rates* option under *Rates* menu from run6.rates file. To run the analysis, select *family history by posterior probability* option within the *Analysis*

menu. Alternatively, to obtain data for further manipulation and visualization using custom tools, the Posterior application can be run with the following syntax:

```
java -Xmx2048M -cp /PATH_TO_COUNT/Count.jar
ca.umontreal.iro.evolution.genecontent.Posteriors -max_paralogs m
OUNT_ST COUNT_table run6.rates > run6.posteriors
```

The run6.posteriors file will contain information on the probability of gene family (zero, one, or multiple genes are predicted to be present at every node of gene tree) and dynamics (gene family expansion, contraction, loss, or gain is predicted) at every node of the species tree for every gene family in the data set. For simplicity, here we proceed with the GUI version of the workflow.

After the analysis is finished, gene family dynamics can be viewed under the *Data* tab within the application. Family names will be shown along with the information on family size, number of extant lineages containing at least one member of the gene, and phyletic pattern, and information on gene family composition and dynamics at every node of the species tree will also be shown for every gene family as well (Figure 6A). Information about every gene family can be viewed individually for every gene family in tree display (Figure 6B) or in a tabular format (Figure 6C). The ArCOG00175 family is predicted to have been present in the ancestor of all the class *Methanomicrobia* as a multigene family and then to have been contracted and lost on the lineage leading to *Methanococcoides burtonii* DSM 6242.

Additionally, a summary across the species tree for a set of gene families can be obtained by selecting desired gene families from the data table. Figure 7 shows the genome-wide summary obtained by selecting all the gene families, visualized over the tree (Figure 7A) and in the tabular format (Figure 7B). It appears that on the lineage leading to the ancestor of all species in the order Methanomicrobiales, some gene families were gained while a lot more lost, and twice as many gene families were predicted to shrink in size compared to expanded families. Both trends are consistent with genome contraction happening on the lineage.

CONCLUDING REMARKS

The field of comparative and evolutionary genomics is developing fast, driven by the availability of genome sequence data. However, new techniques and analysis protocols should be used with caution. With complicated multi-stage analyses, errors and biases can be introduced at any step, leading to systematic biases in the results and conclusions. Additionally, benchmarking studies are rarely published alongside the methods for gene tree/species tree reconciliation, which constitutes a significant knowledge gap and calls for those benchmarking studies.

In this chapter, we have discussed issues that should be considered when performing gene tree/species tree reconciliation to identify duplications, losses, and transfers. They include (1) the nature of the taxonomic and ecological sampling of

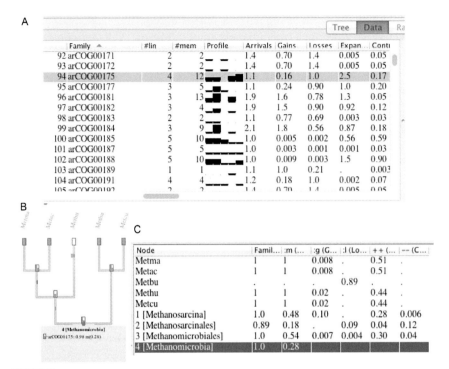

FIGURE 6

Gene family evolution as inferred by COUNT, general information across the families (A); and gene family history for arCOG00175 over the species tree (B) and in tabular format (C). Species are indicated as follows: Methma (*Methanosarcina mazei*), Methac (*Methanosarcina acetivorans*), Methbu (*Methanococcoides burtonii* DSM 6242), Methu (*Methanospirillum hungatei* JF-1), and methma (*Methanoculleus marisnigri* JR1). (See the colour plate.)

genomes included in the analysis; for instance, including taxonomically divergent but ecologically similar species in the data set might lead to underestimation of the number of HGT events within a gene family of niche-specific genes; (2) errors in gene family identification; for instance, inclusion of distant paralogs not only complicates reconciliation but complicates inference of gene tree topology; (3) uncertainties in gene and species tree estimation; and (4) choice of computer program or algorithms with realistic underlying assumptions and justified parameter values. The last point is, in our opinion, the most influential one.

Some additional points that have not been discussed here include the inference of duplication, loss, and transfer events using gene order/synteny. This practice seems to lead to good results on empirical data, but protocols for such studies are not established.

One additional phenomenon that is widely acknowledged in other fields of evolutionary biology, but rarely discussed by empiricists in the gene tree/species tree

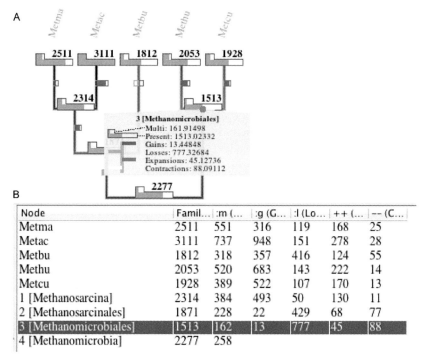

FIGURE 7

Genome content evolution as inferred by COUNT, genome-wide summary is shown over the species tree and in tabular format. Species are indicated as follows: Methma (*Methanosarcina mazei*), Methac (*Methanosarcina acetivorans*), Methbu (*Methanococcoides burtonii* DSM 6242), Methu (*Methanospirillum hungatei* JF-1), and methma (*Methanoculleus marisnigri* JR1). (See the colour plate.)

reconciliation community, is incomplete lineage sorting (ILS). ILS is the discordance between gene trees and species trees due to stochasticity in the coalescence times of ancestral genetic lineages. In a number of studies, a large amount of HGT has been identified between closely related organisms; however, much of the signal for HGT could in fact be due to ILS, due to the large effective population sizes in bacteria. Although some methods exist for inferring horizontal transmission of genetic information while accounting for ILS (Yu, Barnett, & Nakhleh, 2013), they do not treat duplication and loss events and the methodology in this area is still very much under development.

ACKNOWLEDGEMENTS

The authors thank Ethan Jewett and Miklós Csűrös for helpful discussions. This work was supported by The Stanford Center for Computational, Evolutionary and Human Genomics

Postdoctoral Fellowship and National Science Foundation (MCB-0920667 to N. L. W. and DBI-1146722 to O. K. K.). The content of this chapter is solely the responsibility of the authors and does not necessarily represent the official views of the National Science Foundation.

REFERENCES

Åkerborg, Ö., Sennblad, B., Arvestad, L., & Lagergren, J. (2009). Simultaneous Bayesian gene tree reconstruction and reconciliation analysis. *Proceedings of the National Academy of Sciences of the United States of America*, *106*(14), 5714–5719. http://dx.doi.org/10.1073/pnas.0806251106.

Bansal, M. S., Alm, E. J., & Kellis, M. (2012). Efficient algorithms for the reconciliation problem with gene duplication, horizontal transfer and loss. *Bioinformatics*, *28*(12), i283–i291. http://dx.doi.org/10.1093/bioinformatics/bts225.

Bergsten, J. (2005). A review of long-branch attraction. *Cladistics*, *21*(2), 163–193. http://dx.doi.org/10.1111/j.1096-0031.2005.00059.x.

Burnham, K., & Anderson, D. R. (2002). *Model selection and multimodel inference—A practical information-theoretic approach*. Retrieved from, http://www.springer.com/statistics/statistical+theory+and+methods/book/978-0-387-95364-9.

Ciccarelli, F. D., Doerks, T., von Mering, C., Creevey, C. J., Snel, B., & Bork, P. (2006). Toward automatic reconstruction of a highly resolved tree of life. *Science (New York, N.Y.)*, *311*(5765), 1283–1287. http://dx.doi.org/10.1126/science.1123061.

Csűös, M. (2010). Count: Evolutionary analysis of phylogenetic profiles with parsimony and likelihood. *Bioinformatics*, *26*(15), 1910–1912. http://dx.doi.org/10.1093/bioinformatics/btq315.

Csurös, M., & Miklós, I. (2009). Streamlining and large ancestral genomes in archaea inferred with a phylogenetic birth-and-death model. *Molecular Biology and Evolution*, *26*(9), 2087–2095. http://dx.doi.org/10.1093/molbev/msp123.

David, L. A., & Alm, E. J. (2011). Rapid evolutionary innovation during an Archaean genetic expansion. *Nature*, *469*(7328), 93–96. http://dx.doi.org/10.1038/nature09649.

Doyon, J.-P., Hamel, S., & Chauve, C. (2010). *An efficient method for exploring the space of gene tree/species tree reconciliations in a probabilistic framework*. Retrieved from, http://hal-lirmm.ccsd.cnrs.fr/lirmm-00448486.

Kamneva, O. K., Knight, S. J., Liberles, D. A., & Ward, N. L. (2012). Analysis of genome content evolution in PVC bacterial super-phylum: Assessment of candidate genes associated with cellular organization and lifestyle. *Genome Biology and Evolution*, *4*(12), 1375–1390. http://dx.doi.org/10.1093/gbe/evs113.

Lefébure, T., & Stanhope, M. J. (2007). Evolution of the core and pan-genome of *Streptococcus*: Positive selection, recombination, and genome composition. *Genome Biology*, *8*(5), R71. http://dx.doi.org/10.1186/gb-2007-8-5-r71.

Li, L., Stoeckert, C. J., Jr., & Roos, D. S. (2003). OrthoMCL: Identification of ortholog groups for eukaryotic genomes. *Genome Research*, *13*(9), 2178–2189. http://dx.doi.org/10.1101/gr.1224503.

Ludwig, W., Oliver Glöckner, F., & Yilmaz, P. (2011). *16—The use of rRNA gene sequence data in the classification and identification of prokaryotes*. In F. Rainey & A. Oren (Eds.), *Methods in microbiology: Vol. 38* (pp. 349–384). Waltham, MA, USA: Academic Press. Retrieved from, http://www.sciencedirect.com/science/article/pii/B9780123877307000164.

McInerney, J. O., Pisani, D., Bapteste, E., & O'Connell, M. J. (2011). The Public Goods Hypothesis for the evolution of life on earth. *Biology Direct*, *6*, 41. http://dx.doi.org/10.1186/1745-6150-6-41.

Ochman, H., Lawrence, J. G., & Groisman, E. A. (2000). Lateral gene transfer and the nature of bacterial innovation. *Nature*, *405*(6784), 299–304. http://dx.doi.org/10.1038/35012500.

Pamilo, P., & Nei, M. (1988). Relationships between gene trees and species trees. *Molecular Biology and Evolution*, *5*(5), 568–583.

Parfrey, L. W., Lahr, D. J. G., Knoll, A. H., & Katz, L. A. (2011). Estimating the timing of early eukaryotic diversification with multigene molecular clocks. *Proceedings of the National Academy of Sciences of the United States of America*, *108*(33), 13624–13629. http://dx.doi.org/10.1073/pnas.1110633108.

Pertea, G., Huang, X., Liang, F., Antonescu, V., Sultana, R., Karamycheva, S., et al. (2003). TIGR Gene Indices clustering tools (TGICL): A software system for fast clustering of large EST datasets. *Bioinformatics*, *19*(5), 651–652. http://dx.doi.org/10.1093/bioinformatics/btg034.

Retchless, A. C., & Lawrence, J. G. (2010). Phylogenetic incongruence arising from fragmented speciation in enteric bacteria. *Proceedings of the National Academy of Sciences of the United States of America*, *107*(25), 11453–11458. http://dx.doi.org/10.1073/pnas.1001291107.

Rokas, A. (2001). Phylogenetic analysis of protein sequence data using the Randomized Axelerated Maximum Likelihood (RAXML) program. In F. M. Ausubel & R. E. Kingston, et al. (Eds.), *Current protocols in molecular biology: 96:19.11.1-14.* Hoboken, NJ, USA: Wiley. Retrieved from, http://onlinelibrary.wiley.com/doi/10.1002/0471142727.mb1911s96/abstract.

Rokas, A., Williams, B. L., King, N., & Carroll, S. B. (2003). Genome-scale approaches to resolving incongruence in molecular phylogenies. *Nature*, *425*(6960), 798–804. http://dx.doi.org/10.1038/nature02053.

Tatusov, R. L., Fedorova, N. D., Jackson, J. D., Jacobs, A. R., Kiryutin, B., Koonin, E. V., et al. (2003). The COG database: An updated version includes eukaryotes. *BMC Bioinformatics*, *4*(1), 41. http://dx.doi.org/10.1186/1471-2105-4-41.

Woese, C. R., & Fox, G. E. (1977). Phylogenetic structure of the prokaryotic domain: The primary kingdoms. *Proceedings of the National Academy of Sciences of the United States of America*, *74*(11), 5088–5090. http://dx.doi.org/10.1073/pnas.74.11.5088.

Wu, Y.-C., Rasmussen, M. D., Bansal, M. S., & Kellis, M. (2014). Most parsimonious reconciliation in the presence of gene duplication, loss, and deep coalescence using labeled coalescent trees. *Genome Research*, *24*(3), 475–486. http, //dx.doi.org/10.1101/gr.161968.113.

Yu, Y., Barnett, R. M., & Nakhleh, L. (2013). Parsimonious inference of hybridization in the presence of incomplete lineage sorting. *Systematic Biology*, *62*(5), 738–751. http://dx.doi.org/10.1093/sysbio/syt037.

Multi-Locus Sequence Typing and the Gene-by-Gene Approach to Bacterial Classification and Analysis of Population Variation

10

Alison J. Cody, Julia S. Bennett, Martin C.J. Maiden[1]

Department of Zoology, University of Oxford, Oxford, United Kingdom

[1]Corresponding author: e-mail address: martin.maiden@zoo.ox.ac.uk

1 INTRODUCTION

1.1 HISTORICAL PERSPECTIVE

Microbiology is dependent on the reproducible and systematic classification of bacteria into types, but the great diversity of the bacterial and archeal domains (Krieg, Brenner, & Staley, 2005) has hindered the development of a single characterisation scheme applicable to all bacteria at all levels of relatedness. Genotypic methods have become increasingly important in bacterial characterisation and for nearly 30 years the 16S ribosomal RNA gene (Woese, 1987) has provided a general framework for molecular studies of taxonomic relationships and the identification and cataloguing of bacterial diversity (Clarridge, 2004; Vamosi, Heard, Vamosi, & Webb, 2009; Woese, Kandler, & Wheelis, 1990). Considering its small size, this gene has provided an enormous amount of information and has had a major impact on prokaryotic systematics. Despite its success in defining higher taxa, however, the 16S ribosomal RNA classification system lacks the resolution required to distinguish among closely related bacteria, in particular, failing to resolve species belonging to some medically important genera (Bennett et al., 2012; Harmsen et al., 2001; Walcher, Skvoretz, Montgomery-Fullerton, Jonas, & Brentano, 2013).

1.2 MULTI-LOCUS POPULATION ANALYSES

The development of high-throughput sequencing approaches that Frederick Sanger pioneered (Sanger, Nicklen, & Coulson, 1977) has permitted the implementation of other methods including, for example, the estimation of average nucleotide identity (ANI) and multi-locus sequence typing (MLST). As with previous DNA-based approaches, such as DNA–DNA hybridisation, ANI indicates the overall level of similarity between the genomes of two bacterial isolates (Konstantinidis & Tiedje, 2005). Typically, genomes of the same species have ANI values greater than 95% but, whilst this technique is useful for the investigation of closely related bacteria, it is limited to isolates which share a large pool of genes. MLST, based on the concept of multi-locus enzyme electrophoresis (Selander et al., 1986), indexes the nucleotide sequence variation of approximately 500 bp of internal fragments of seven housekeeping genes (Maiden et al., 1998), has a wide range of applications and is used extensively in global epidemiological studies.

2 MULTI-LOCUS SEQUENCE TYPING

MLST indexes the nucleotide sequence variation of approximately 500 bp internal fragments of seven housekeeping genes. For each of the loci, any novel nucleotide sequence is assigned an arbitrary and unique allele number. Thus, in each isolate studied the genetic variation present at the MLST loci is represented by a seven digit 'allelic profile', also referred to as a sequence type (ST), which summarises several thousand base pairs of sequence data. These data are stored along with isolate phenotype, provenance and epidemiological data (metadata) for at least one example of each ST recorded. Isolates with STs that are similar (usually sharing four or more alleles) to a heuristically identified central genotype are referred to as 'clonal complexes' and reflect lineages that share a recent common ancestor (Maiden, 2006).

One feature of MLST is that allele changes are regarded as single genetic events, providing a simple adjustment for the observation that, in many bacteria, horizontal genetic transfer (HGT) may account for more genetic variation than point mutations (Didelot & Maiden, 2010). Analyses using allelic profiles are informative when examining relationships within a recombining species, such as *Campylobacter jejuni* (Dingle et al., 2002) and *Neisseria meningitidis* (Holmes, Urwin, & Maiden, 1999) where allele fragments have been shown to spread independently (Jolley et al., 2002). However, analyses using allelic profiles rather than nucleotide sequences are inappropriate when comparing relationships among species, as alleles are shared only infrequently among different species within a genus (Bennett et al., 2007). As MLST databases also record the nucleotide sequences of allelic variants, these data can be used for sequenced-based analyses (Bennett et al., 2007; Mo et al., 2013; Sawabe et al., 2013) and this latter approach has been termed 'multilocus sequence analysis' (Thompson et al., 2013).

As housekeeping genes across different bacteria are genetically diverse, it has been necessary to develop separate MLST schemes even for quite closely related bacteria (Maiden, 2006). For each species or group of species, these definitions are stored in distinct curated Web-accessible databases, e.g., those found in the PubMLST database collection (http://pubmlst.org/databases.shtml). At the time of writing (January, 2014), there were in excess of 50 MLST schemes hosted at http://pubmlst.org, widely used for evolutionary, population and epidemiological studies of the organisms. Among these, schemes developed for *Neisseria* and the *C. jejuni/coli* databases contain among the largest number of isolates and will be used as exemplar organisms for this chapter.

MLST does not, however, provide sufficient discrimination to define all variants of all bacteria; for example, it cannot resolve differences among recently emergent, low-diversity single-clone organisms, such as bacteria within the *Bacillus cereus* group (Priest, Barker, Baillie, Holmes, & Maiden, 2004) and *Yersinia pestis* (Achtman et al., 2004). It is also unable to discriminate among all species of *Neisseria* (Bennett et al., 2012) and provides insufficient resolution among *Campylobacter* isolates for the detection of outbreaks (Dingle, McCarthy, Cody, Peto, & Maiden, 2008). Under these circumstances, data have been supplemented by the addition of additional housekeeping loci (Didelot, Urwin, Maiden, & Falush, 2009), or by examining and cataloguing variation at one or more diverse loci, such as the variable regions of antigen genes known to be under diversifying selection (Dingle et al., 2008; Jolley, Brehony, & Maiden, 2007).

3 WHOLE-GENOME DATA ANALYSES

Despite these extended MLST schemes, there has remained a requirement for more extensive multi-locus approaches that are able to address all isolate identification and typing needs. The advent of low cost, highly parallel and redundant 'next-generation' sequencing technologies have facilitated whole-genome sequencing (WGS) of a wide variety of bacterial species (Medini et al., 2008) for which rapidly increasing amounts of data are now available (Chain et al., 2009). These rapid methods have transformed bacterial identification and typing as they have reduced the practical constraints that have prevented the development of large-scale MLST schemes.

Several approaches have been successfully used to detect variation among bacterial samples using WGS, but most methodologies have limitations, which must be clearly understood for correct interpretation of their outputs. For example, the analysis of single nucleotide polymorphisms has been used in evolutionary and epidemiological studies of closely related and single-clone pathogens with great effect (Bryant et al., 2013; Cui et al., 2013; Grad et al., 2012; He et al., 2010; Köser et al., 2012; Mutreja et al., 2011; Walker et al., 2013) but difficulties may develop upon its application to more distantly related isolates where there is extensive HGT. Furthermore, this technique requires the development of a standard methodology and needs a reference genome against which short reads or *de novo*-assembled sequences can be mapped (Croucher, Harris, Grad, & Hanage, 2013). Software packages are also available that do not require a reference genome, for example, simultaneous *de novo* assembly of multiple samples compared by colored de Bruijn graphs (Iqbal, Turner, & McVean, 2013), but, neither of these types of approach produce open source, Web-accessible information that is available to the wider, collaborative community (Maiden et al., 2013).

3.1 GENE-BY-GENE ANALYSIS OF WGS DATA

The analysis of *de novo*-assembled short-read data using a gene-by-gene approach has no requirement for mapping to a reference genome and is flexible such that the level of resolution required in any study is dependent on the question asked and can extend from (i) one or a few genes (Enright et al., 2002), for example, to identify the species of an organism (Didelot, Bowden, Wilson, Peto, & Crook, 2012; Harmsen, Rothganger, Frosch, & Albert, 2002; Koser et al., 2012) or its membership of a lineage (Enright et al., 2002); up to (ii) all the genes in a genome, enabling high-level discrimination for investigation of disease outbreaks or analysis of within-patient variation (Cody et al., 2013; Harris et al., 2012) (Figure 1). The gene-by-gene approach therefore permits a wide range of scalable and hierarchical studies and permits comparison not only of functional protein encoding genes but also paralogous loci, pseudogenes and intergenic regions, among sample populations. It is facilitated by a Web-accessible database, the Bacterial Isolate Genome Sequence Database (BIGSdb), which has been developed as an extension of the mlstdbNet software (Jolley, Chan, & Maiden, 2004), originally designed for the storage and distribution of MLST data.

3.2 THE BACTERIAL ISOLATE GENOME SEQUENCE DATABASE

BIGSdb databases can be established for any bacterial species and comprise isolate provenance and phenotype data, sequence data from any source for each isolate and a reference database which details each allele at each locus and their known combinations, for that organism (Figure 2). These databases have 'users', who can access the data, and 'curators' who can access and edit data and who have the responsibility of ensuring data integrity and accuracy. As with users, curators access the databases through a Web interface, such as PubMLST.org, and can be sited in any location with internet access.

3.3 ISOLATE AND SEQUENCE DATABASES

Isolate databases can contain any number of isolate records, being limited only by computing capacity, and contain phenotype, provenance and epidemiological data (metadata) for the isolates under study. Nucleotide sequence data of any size or scale from one or more sources or sequencing technologies are stored in a 'sequence-bin' linked to the relevant isolate identifier. As well as sequences obtained using Sanger sequencing, data can include short reads obtained directly from isolates or from the sequence read archive, downloaded from databases such as the European Nucleotide Archive at the European Molecular Biology Laboratory-European Bioinformatics Institute Website, and assembled using algorithms such as VELVET (Zerbino, 2010; Zerbino & Birney, 2008). Assembled WGS data for complete genomes, or genes and gene fragments can also be downloaded from public databases such as the Integrated Microbial Genomes (IMG database) (Markowitz et al., 2010), GenBank, EMBL and the DNA Data Bank of Japan. Data acquired from external repositories

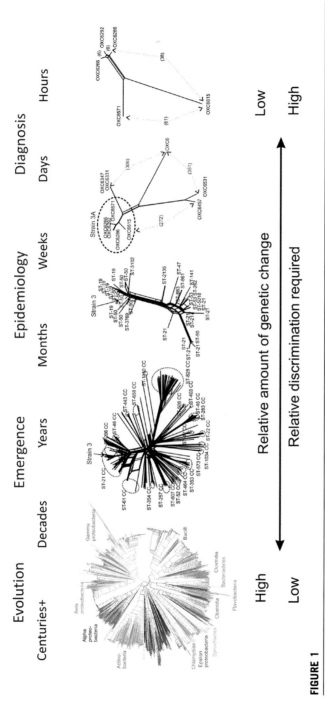

FIGURE 1

Hierarchical gene-by-gene analysis from domain to strain; analysis of genetically divergent populations requires information from fewer loci than highly similar isolates. (See the colour plate.)

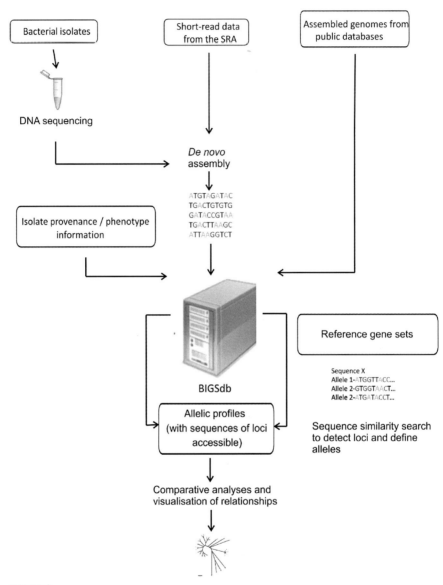

FIGURE 2

Schematic representation of the Bacterial Isolate Genome Sequence Database platform and the gene-by-gene approach to nucleotide sequence analysis. (See the colour plate.)

remain linked via the original accession number to the BIGSdb isolate identifier. In this way, data extracted from sequences obtained from a range of sources can be analysed in an integrated manner. Isolates from single or multiple studies can be identified collectively by their inclusion in a named project, to expedite population analyses.

3.4 **REFERENCE SEQUENCE AND PROFILE DEFINITIONS DATABASE**

The 'Sequence Definitions and Profiles Database' consists of a repository for all known nucleotide sequences of each defined locus and the corresponding integer by which they are described. The addition of new variants provides a comprehensive catalogue of the diversity of alleles for both known loci and newly identified genes. Although loci are usually genes (particularly protein-coding genes), any sequence string, nucleotide or peptide-coding region can be defined as a locus, so that intergenic regions, for example, can be included for analysis.

The initially identified nucleotide sequence for each gene can be obtained by manually retrieving annotated data from, for example, IMG or GenBank, and the sequence assigned an allele number, usually allele one. When subsequent sequences are added to the isolate database of BIGSdb, they are automatically scanned against allele sequences contained in the definitions databases using search tools, such as BLAST (Altschul et al., 1997). Identical matches with existing alleles are automatically recorded in the isolate record with the designation for that locus and the position of the sequence in the contig is marked (tagged). Novel sequences, including those identified from next-generation sequence data, are also automatically assigned new allele numbers, provided that the identified sequence fulfils predetermined criteria relating to start and stop codons, length and percentage identity compared to the closest pre-existing allele. For sequences that do not fit these criteria, a database curator must determine the accuracy and plausibility of the data by comparison with other alleles, and where appropriate the gene function, to determine whether it can be assigned an allele number. This is of particular importance when identifying, for example, paralogous and duplicated gene sequences and requires expertise in particular areas of the organism's biology. As a consequence, allele definitions can be viewed by all users, but only assigned curators can add and modify data within defined and controlled limits to support these community maintained databases. Users may query unidentified sequences by pasting the unknown sequence into the sequence definitions database for comparison against a specified locus or against all loci without the need to acquire curator status or deposit isolate details.

3.5 **DATABASE INTEGRITY**

The integrity of both the isolate and sequence definitions databases is maintained by curators, who assess sequence quality prior to its incorporation into the database. Conventional dideoxy-chain terminating Sanger sequencing (Sanger et al., 1977), whether done by autoradiography or with colored dyes detected optically, presents data that are derived from many thousands of individual extension reactions. As a consequence minor base errors are rarely observed, although more evenly mixed samples are apparent by ambiguous chromatogram peaks at some sites (Junemann et al., 2013) (Figure 3). Organism-specific curators therefore appraise the entire length of novel allele sequences for unambiguous polymorphic sites on

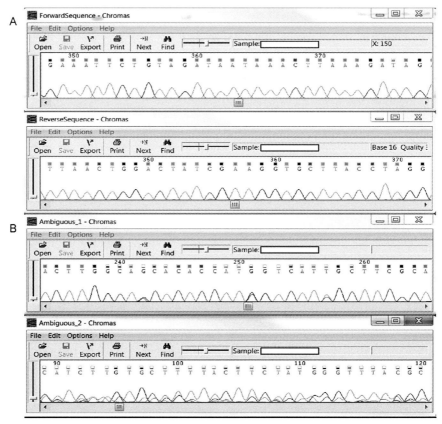

FIGURE 3

Example chromatogram files submitted for assignment of new alleles; A) clean aligned forward and reverse sequences, and B) ambiguous sequences indicating mixed DNA samples. (See the colour plate.)

both forward and reverse trace files before assigning them an allele number. Next-generation sequencing data are also a consensus of many reads and can give rise to technology-specific systematic errors, for example, regions containing conformational ordering of amino acids may be problematic for one methodology, whereas sequencing of long homopolymeric tracts may defeat another. Additionally, different assembly algorithms may inherently miscalculate particular regions. Whole-genome sequence data are therefore assessed by comparison of the number of contiguous sequences (contigs), the N50 statistic and depth of coverage obtained, for any given species. Curators appraising regions of anticipated or known complexity therefore require knowledge of these potential sources of error and the ability to evaluate their probable causes (Chain et al., 2009).

3.6 GENE NOMENCLATURE

As loci from different annotations of an organism may vary in their order and nomenclature, all loci in BIGSdb are assigned an arbitrary identifier by the addition of a genus specific prefix and number, in addition to common gene names and any other aliases. The prefix is based on the name and order of previous coding sequence identifiers, for example, in the *Neisseria* database, the nomenclature is based on the annotation for the *N. meningitidis* FAM18 genome (Bentley et al., 2007), with the prefix NEIS replacing NMC. The initial arbitrary numbering sequence reflects that of the cited annotation, but this system facilitates the subsequent addition of loci from other sources. Similarly, in the *C. jejuni/coli* database the CAMP prefix indicates loci defined in the re-annotation of *C. jejuni* NCTC 11168 (Gundogdu et al., 2007), with the addition of genes identified and annotated from other genomes. However, the loci encoding the 53 ribosomal proteins, which are common to all species within the bacterial domain, have the prefix BACT rather than a species-related identifier as part of the ribosomal MLST scheme which is described in more detail below. Thus, the use of prefixes simplifies cataloguing of genes regardless of their position in any genome, facilitating comparisons between isolates or species within a genus regardless of annotation. The link between this novel nomenclature and common gene names, or other aliases, ensures that they all are backwards (and indeed, forwards) compatible and transparent to all users.

3.7 TYPING AND ANALYSIS SCHEMES

The ability to develop flexible schemes of loci is essential for all levels of population analyses. Whereas a single locus can often provide sufficient information to distinguish between groups of organisms from the phylum to family or genus level, higher levels of resolution are required to determine the species or sub-species (Figure 4). This is achieved by increasing the number of loci included in the scheme up to the level of the whole-genome MLST (wgMLST) (Maiden et al., 2013), which is suitable for investigation of single-clone pathogens or closely related variants of diverse bacteria. Here, we describe several practical applications of the gene-by-gene approach to bacterial whole-genome sequence analysis.

4 EXAMPLES OF GENE-BY-GENE ANALYSIS: *NEISSERIA*

As the data stored in BIGSdb are backwards compatible, it is possible to investigate the 16S rRNA gene fragment, for example, in a population of *Neisseria spp.*, by using the locus identified on PubMLST.org/neisseria as 16S_rRNA (alias SSU_rRNA). Concatenated nucleotide sequences exported for any collection of isolates can be exported in FASTA format suitable for analysis within the MEGA program (Tamura et al., 2011) to reconstruct a neighbour-joining phylogeny (Saitou & Nei, 1987).

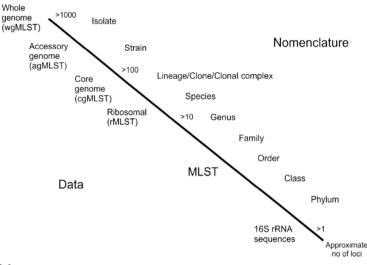

FIGURE 4

The hierarchical nature of the gene-by-gene approach to bacterial classification and population variation allows the use of nucleotide sequence data from a single locus for bacterial characterisation to analysis of whole genome sequence data to distinguish among closely related isolates.

4.1 RIBOSOMAL MULTI-LOCUS SEQUENCE TYPING

More robust species identification can be obtained by analysis of the ribosomal protein gene loci, which are ideal targets for universal, systematic, bacterial classification as they are (i) present in all bacteria; (ii) distributed around the chromosome; and (iii) encode proteins which are under stabilising selection for functional conservation (Jolley et al., 2012). Collectively, the ribosomal protein gene loci exhibit variation that resolves bacteria into groups at all taxonomic and most typing levels providing significantly more resolution than 16S rRNA gene phylogenies. The scheme enables rapid species and strain identification directly from WGS data and data far in excess of 1000 species are publicly available at PubMLST.org/rmlst, against which unknown samples can be compared.

Phylogenies reconstructed from 53 concatenated ribosomal protein genes indexed in rMLST produce clusters that are consistent with named *Neisseria* species, enabling rapid and accurate species identification. This has resolved anomalies relating to previous species assignments and misidentifications of isolates (Bennett, Jolley, & Maiden, 2013). With the online tools available from BIGSdb in PubMLST. org/neisseria, a curator is able to upload a genome sequence from an unidentified *Neisseria* species, run a database scan to identify and tag the 53 genes at the ribosomal protein gene loci used in the rMLST scheme. Users can then download the concatenated ribosomal protein gene sequences from the unidentified isolate and from the genomes of reference strains such as those obtained from type strains within the database.

Nucleotide sequence data can be downloaded in FASTA format suitable for analysis within the MEGA program (Tamura et al., 2011), to reconstruct a neighbour-joining phylogeny (Saitou & Nei, 1987), for example, or to export in NEXUS format for analyses such as split decomposition (Huson & Bryant, 2006) (Figure 5). These methods of phylogenetic reconstruction will cluster the isolate within a species-specific group if it is a known species, or separately if it is a new, undefined species.

Genes that are present in all the members of a defined group of isolates are referred to as the 'core genome' for that group and species assignments can be confirmed by reconstructing phylogenies from these core genes; for example, around 246 genes have been identified as belonging to the core genome of diverse *Neisseria* species and a phylogeny reconstructed from these concatenated whole-gene sequences has confirmed the species identity of isolates characterised by rMLST (Bennett et al., 2012). The rMLST approach to speciation requires nucleotide sequence variation data from 53 loci and, although these are readily extracted from the WGS data in BIGSdb, obtaining such data is not always economical or practical for all specimens. Subsequent investigation of these loci has led to the development of a single-locus analysis capable of accurate species identification, and this approach has the potential for application in other bacteria.

4.2 *NEISSERIA rplF* ASSAY

The phylogeny produced from the 53 concatenated genes used in rMLST microbial species differentiation was compared for congruence with any of the 53 individual phylogenies reconstructed from nucleotide sequences of the ribosomal protein gene loci of 44 diverse *Neisseria* species. This analysis identified a 413 bp fragment of the 50S ribosomal protein L6 (*rplF*) gene, that could be used in an assay to identify *Neisseria* species, in much the same way as the 16S rRNA gene has been used to differentiate organisms (Woese et al., 1990). It provides sufficient discrimination to identify most species within the genus accurately, rapidly and inexpensively. Most importantly, this assay is able to discriminate among the pathogenic and commensal *Neisseria*, which is not always possible when a fragment of the 16S rRNA gene is used to distinguish between species (Bennett, Watkins, Jolley, Harrison, & Maiden, 2014). This assay was successfully used in the MenAfiCar study to investigate carried meningococci in the African meningitis belt during vaccine implementation by sequencing gene fragments from heat-killed cell suspensions sent to Oxford from Chad (Doumagoum Moto et al., 2013).

To identify *Neisseria* species using an *rplF* fragment, the PubMLST *Neisseria* sequence definitions database can be queried using the sequence query interface. Users can select '*rplF* species' from the locus/scheme drop-down menu and then paste in their nucleotide sequence. If there is an exact match, an *rplF* genospecies designation is returned. If there are polymorphisms present, the closest match is shown and any nucleotide differences are identified and shown in an alignment, which can then be translated. All known *rplF* fragment alleles can be downloaded from the *Neisseria* locus/sequence definitions database in PubMLST, as can the *rplF*

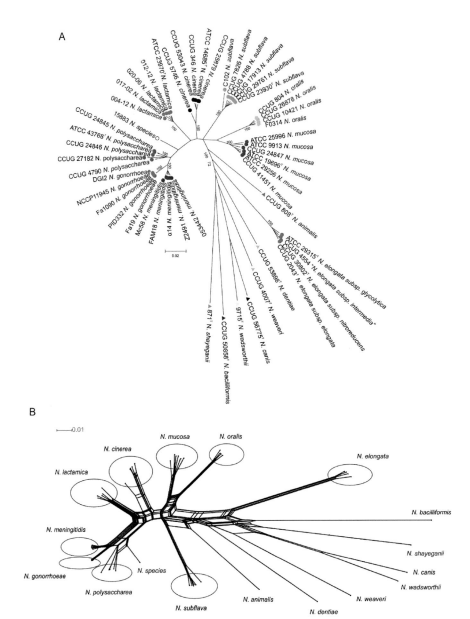

FIGURE 5

Speciation of 53 *Neisseria* spp. using concatenated nucleotide sequence data from rMLST loci; (A) by reconstruction of a neighbour-joining phylogeny and (B) by split decomposition analysis. (See the colour plate.)

profiles. The isolate database can also be searched for any related provenance data. In order to assign a new allele, novel *rplF* sequences can be submitted via PubMLST to a curator who will then assign a provisional species identity by comparing the percentage identity to known species-specific alleles within the database and reconstructing a phylogeny using all known *rplF* fragment alleles and the novel allele.

5 EXAMPLES OF GENE-BY-GENE ANALYSIS: *CAMPYLOBACTER*

5.1 CORE GENOME MULTI-LOCUS SEQUENCE TYPING

Most comparative analyses are dependent on the presence of all loci within a particular scheme, and it is therefore important to be able to identify the common subset of genes, or core genome, for the set of isolates under investigation. As outlined above, the definition of a core genome defines those loci present in all the members of a population, whereas the accessory genes are those absent from one or more isolates. The greater the genetic diversity within a population the smaller the size of the core genome, which will tend to decrease further as the population size increases. Accessory genes may be present in most or only a few members of the population and the ability to identify the proportion of isolates in which they are present is paramount to some investigations. As the stringency with which the definition of core and accessory genes can vary it may be necessary to optimise analyses by varying the cut-off values, depending on the analysis in hand.

The ability to identify the core and accessory genomes of an isolate population is facilitated by the GENOME COMPARATOR module of BIGSdb. For example, using the publicly available data at PubMLST.org/campylobacter, a study population of 933 clinical *C. jejuni* and *C. coli* isolated from patients between June 2011 and June 2012 can be identified from the project, 'Oxfordshire Human Surveillance YR1'. By exporting this dataset as a tab-delimited text file, it can be pasted and sorted in a spreadsheet to list 839 *C. jejuni*. Re-submission of these isolate identifiers to a 'List query' for this database enables the identification of a core genome via the 'Genome Comparator' analysis function. By this method 1471 core genes are identified using a 90% threshold, by comparison with the 'scheme' defined by 1643 coding sequences identified in the re-annotation of the NCTC 11168 genome (Gundogdu et al., 2007). Increasing the core threshold incrementally, to a value of 100%, reduces the core to 1348 loci.

5.2 WHOLE-GENOME MULTI-LOCUS SEQUENCE TYPING

Increasing the number of loci included in gene-by-gene analyses of bacterial populations enables higher resolution investigations, which can have practical applications in the study of potential disease outbreaks. Highly similar organisms can be identified from a genetically diverse population by comparing routine clinical isolates to each other and to the known diversity of the organism, in a step-wise process. For illustrative purposes, this methodology can be applied to WGS data from 376 clinical *Campylobacter* isolates, obtained from patients in Oxfordshire, UK, between June 27 and October 26, 2013 (Cody et al., 2013), data for which are available via (http://pubmlst.org/perl/bigsdb/bigsdb.pl?db=pubmlst_campylobacter_isolates& page=query).

An initial comparison of these isolates at the rMLST level identifies sequences for 52 ribosomal protein subunits and identifies 212 ribosomal sequence types (rSTs), a 53rd locus (*rpmD*: BACT000059) is absent from all (Cody et al., 2013). The ST-21 complex is represented 89 times and resolves into three groups when re-analysed, to identify 34 rSTs. If one of these (strain 3) is further analysed by wgMLST, genome comparator identifies 1595 of the 1643 loci identified in NCTC 11168 (Gundogdu et al., 2007), that are common to these 10 isolates and defines four groups which differ from each other at between 272 and 351 loci (Figure 1). The largest of the four groups contains five isolates which when re-analysed vary at only nine of their shared 1605 (0.56%) loci. Examination of the isolate metadata has shown that two of the three most closely related isolates (OXC6266 and OXC6292) were obtained from the same patient on the morning and the evening of the same day, whereas the third (OXC6286) was taken from a different patient and received by the microbiology laboratory 3 minutes after the morning sample from the first patient (Cody et al., 2013). No other geographical or epidemiological features were shared between the two patients, indicating a likely laboratory cross-contamination of samples, from a population of isolates whose genetic diversity is not exceeded by a collection of *C. jejuni* and *C. coli* samples obtained from a variety of hosts from several countries over a time period in excess of 10 years (Figure 6) (Cody et al., 2013; Lefebure, Pavinski Bitar, Suzuki, & Stanhope, 2010).

CONCLUSIONS

Advances in sequencing technologies have greatly increased the volume of nucleotide sequence data available and collection of these data is no longer a limiting step in the study of bacterial populations. Although it is now possible to undertake whole-genome studies on multiple isolates, the study of bacterial populations now faces the challenge of exploiting these data, potentially from whole genomes of thousands or tens of thousands of isolates. To do this, it is necessary to link provenance and phenotype information with the available sequence data. The success of multi-locus approaches, such as MLST in bacterial isolate characterisation was greatly facilitated by the accessibility of data via the Internet, enabling community participation in data collection and analysis (Maiden, 2006). The original MLST software provided a hierarchical and structured approach to population analysis, linking the ST to phenotype and provenance information (Jolley et al., 2004). BIGSdb has extended this approach, enabling whole genomes, or fragments of them, to be stored and the data organised and interpreted by any number of schemes, comprising any number of loci (Jolley & Maiden, 2010).

A

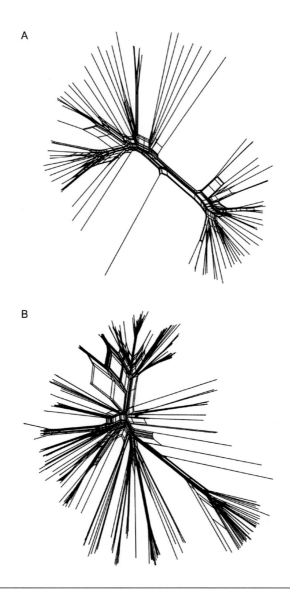

B

FIGURE 6

Comparison of neighbour-net phylogenies generated for alleles of rMLST for (A)
83 *Campylobacter* isolates from various host species from across Europe, Canada and the
United States, over more than a decade and (B) 376 clinical *Campylobacter* isolates from
humans in Oxfordshire, UK, between June and October 2011.

An advantage of the gene-by-gene analysis approach is that it is both backwards and forwards compatible: 16S rRNA and seven-locus MLST information can be extracted from WGS data and compared with that from existing Sanger derived data (Cody et al., 2013), and will not become redundant regardless of the method by which it has been obtained. Using BIGSdb, genomic data can be used to characterise isolates in any number of different ways, which can be exploited for evolutionary or functional studies. As this platform permits indexing of loci on a functional basis, by treating loci or groups of loci as independent units of analysis, it facilitates a community-based genome annotation process. As exemplified by BIGSdb, as implemented on the PubMLST.org Website this approach represents an open-access and freely available resource that will assist the microbiology research community in the elucidation of bacterial structure and function by means of the population genomics approach (Maiden et al., 2013).

REFERENCES

Achtman, M., Morelli, G., Zhu, P., Wirth, T., Diehl, I., Kusecek, B., et al. (2004). Microevolution and history of the plague bacillus, *Yersinia pestis*. *Proceedings of the National Academy of Sciences of the United States of America*, *101*(51), 17837–17842.

Altschul, S. F., Madden, T. L., Schaffer, A. A., Zhang, J., Zhang, Z., Miller, W., et al. (1997). Gapped BLAST and PSI-BLAST: A new generation of protein database search programs. *Nucleic Acids Research*, *25*(17), 3389–3402.

Bennett, J. S., Jolley, K. A., Earle, S. G., Corton, C., Bentley, S. D., Parkhill, J., et al. (2012). A genomic approach to bacterial taxonomy: An examination and proposed reclassification of species within the genus *Neisseria*. *Microbiology*, *158*(Pt. 6), 1570–1580.

Bennett, J. S., Jolley, K. A., & Maiden, M. C. (2013). Genome sequence analyses show that *Neisseria oralis* is the same species as '*Neisseria mucosa var. heidelbergensis*'. *International Journal of Systematic and Evolutionary Microbiology*, *63*(10), 3920–3926.

Bennett, J. S., Jolley, K. A., Sparling, P. F., Saunders, N. J., Hart, C. A., Feavers, I. M., et al. (2007). Species status of *Neisseria gonorrhoeae*: Evolutionary and epidemiological inferences from MLST. *BMC Biology*, *5*(1), 35.

Bennett, J. S., Watkins, E. R., Jolley, K. A., Harrison, O. B., & Maiden, M. C. (2014). Identifying *Neisseria* species using the 50S ribosomal protein L6 (*rplF*) gene. *Journal of Clinical Microbiology*, *52*(5), 1375–1381.

Bentley, S. D., Vernikos, G. S., Snyder, L. A., Churcher, C., Arrowsmith, C., Chillingworth, T., et al. (2007). Meningococcal genetic variation mechanisms viewed through comparative analysis of serogroup C strain FAM18. *PLoS Genetics*, *3*(2), e23.

Bryant, J. M., Grogono, D. M., Greaves, D., Foweraker, J., Roddick, I., Inns, T., et al. (2013). Whole-genome sequencing to identify transmission of *Mycobacterium abscessus* between patients with cystic fibrosis: A retrospective cohort study. *Lancet*, *381*(9877), 1551–1560.

Chain, P. S., Grafham, D. V., Fulton, R. S., Fitzgerald, M. G., Hostetler, J., Muzny, D., et al. (2009). Genome project standards in a new era of sequencing. *Science*, *326*(5950), 236–237.

Clarridge, J. E. (2004). Impact of 16S rRNA gene sequence analysis for identification of bacteria on clinical microbiology and infectious diseases. *Clinical Microbiology Reviews*, *17*(4), 840–862.

Cody, A. J., McCarthy, N. D., Jansen van Rensburg, M., Isinkaye, T., Bentley, S., Parkhill, J., et al. (2013). Real-time genomic epidemiology of human *Campylobacter* isolates using whole genome multilocus sequence typing. *Journal of Clinical Microbiology*, *51*(8), 2526–2534.

Croucher, N. J., Harris, S. R., Grad, Y. H., & Hanage, W. P. (2013). Bacterial genomes in epidemiology-present and future. *Philosophical Transactions of the Royal Society of London. Series B, Biological Sciences*, *368*(1614), 1–7.

Cui, Y. J., Yu, C., Yan, Y. F., Li, D. F., Li, Y. J., Jombart, T., et al. (2013). Historical variations in mutation rate in an epidemic pathogen, *Yersinia pestis*. *Proceedings of the National Academy of Sciences of the United States of America*, *110*(2), 577–582.

Didelot, X., Bowden, R., Wilson, D. J., Peto, T. E. A., & Crook, D. W. (2012). Transforming clinical microbiology with bacterial genome sequencing. *Nature Reviews. Genetics*, *13*(9), 601–612.

Didelot, X., & Maiden, M. C. (2010). Impact of recombination on bacterial evolution. *Trends in Microbiology*, *18*(7), 315–322.

Didelot, X., Urwin, R., Maiden, M. C., & Falush, D. (2009). Genealogical typing of *Neisseria meningitidis*. *Microbiology*, *155*(10), 3176–3186.

Dingle, K. E., Colles, F. M., Ure, R., Wagenaar, J., Duim, B., Bolton, F. J., et al. (2002). Molecular characterisation of *Campylobacter jejuni* clones: A rational basis for epidemiological investigations. *Emerging Infectious Diseases*, *8*(9), 949–955.

Dingle, K. E., McCarthy, N. D., Cody, A. J., Peto, T. E., & Maiden, M. C. (2008). Extended sequence typing of *Campylobacter* spp., United Kingdom. *Emerging Infectious Diseases*, *14*(10), 1620–1622.

Doumagoum Moto, D., Gami, J. P., Gamougam, K., Naibei, N., Mbainadji, L., Narbé, M., et al. (2013). The impact of a serogroup A meningococcal conjugate vaccine (PsA-TT) on serogroup A meningococcal meningitis and carriage in Chad. *Lancet*, *383*(9911), 40–47.

Enright, M. C., Robinson, D. A., Randle, G., Feil, E. J., Grundmann, H., & Spratt, B. G. (2002). The evolutionary history of methicillin-resistant *Staphylococcus aureus* (MRSA). *Proceedings of the National Academy of Sciences of the United States of America*, *99*(11), 7687–7692.

Grad, Y. H., Lipsitch, M., Feldgarden, M., Arachchi, H. M., Cerqueira, G. C., Fitzgerald, M., et al. (2012). Genomic epidemiology of the *Escherichia coli* O104:H4 outbreaks in Europe, 2011. *Proceedings of the National Academy of Sciences of the United States of America*, *109*(8), 3065–3070.

Gundogdu, O., Bentley, S. D., Holden, M. T., Parkhill, J., Dorrell, N., & Wren, B. W. (2007). Re-annotation and re-analysis of the *Campylobacter jejuni* NCTC11168 genome sequence. *BMC Genomics*, *8*, 162.

Harmsen, D., Rothganger, J., Frosch, M., & Albert, J. (2002). RIDOM: Ribosomal differentiation of medical micro-organisms database. *Nucleic Acids Research*, *30*(1), 416–417.

Harmsen, D., Singer, C., Rothganger, J., Tonjum, T., de Hoog, G. S., Shah, H., et al. (2001). Diagnostics of *Neisseriaceae* and *Muraxsellaceae* by ribosomal DNA sequencing: Ribosomal differentiation of medical microorganisms. *Journal of Clinical Microbiology*, *39*(3), 936–942.

Harris, S. R., Cartwright, E. J. P., Török, M. E., Holden, M. T. G., Brown, N. M., Ogilvy-Stuart, A. L., et al. (2012). Whole-genome sequencing for analysis of an outbreak of methicillin-resistant *Staphylococcus aureus*: A descriptive study. *Lancet Infectious Diseases*, *31*(2), 130–136.

He, M., Sebaihia, M., Lawley, T. D., Stabler, R. A., Dawson, L. F., Martin, M. J., et al. (2010). Evolutionary dynamics of *Clostridium difficile* over short and long time scales. *Proceedings of the National Academy of Sciences of the United States of America*, *107*(16), 7527–7532.

Holmes, E. C., Urwin, R., & Maiden, M. C. J. (1999). The influence of recombination on the population structure and evolution of the human pathogen *Neisseria meningitidis*. *Molecular Biology and Evolution*, *16*(6), 741–749.

Huson, D. H., & Bryant, D. (2006). Application of phylogenetic networks in evolutionary studies. *Molecular Biology and Evolution*, *23*(2), 254–267.

Iqbal, Z., Turner, I., & McVean, G. (2013). High-throughput microbial population genomics using the Cortex variation assembler. *Bioinformatics*, *29*(2), 275–276.

Jolley, K. A., Bliss, C. M., Bennett, J. S., Bratcher, H. B., Brehony, C. M., Colles, F. M., et al. (2012). Ribosomal multi-locus sequence typing: Universal characterization of bacteria from domain to strain. *Microbiology*, *158*, 1005–1015.

Jolley, K. A., Brehony, C., & Maiden, M. C. (2007). Molecular typing of meningococci: Recommendations for target choice and nomenclature. *FEMS Microbiology Reviews*, *31*(1), 89–96.

Jolley, K. A., Chan, M. S., & Maiden, M. C. (2004). mlstdbNet distributed multi-locus sequence typing (MLST) databases. *BMC Bioinformatics*, *5*(1), 86.

Jolley, K. A., Kalmusova, J., Feil, E. J., Gupta, S., Musilek, M., Kriz, P., et al. (2002). Carried meningococci in the Czech Republic: A diverse recombining population. *Journal of Clinical Microbiology*, *40*(9), 3549–3550.

Jolley, K. A., & Maiden, M. C. (2010). BIGSdb: Scalable analysis of bacterial genome variation at the population level. *BMC Bioinformatics*, *11*(1), 595.

Junemann, S., Sedlazeck, F. J., Prior, K., Albersmeier, A., John, U., Kalinowski, J., et al. (2013). Updating benchtop sequencing performance comparison. *Nature Biotechnology*, *31*(4), 294–296.

Konstantinidis, K. T., & Tiedje, J. M. (2005). Genomic insights that advance the species definition for prokaryotes. *Proceedings of the National Academy of Sciences of the United States of America*, *102*(7), 2567–2572.

Koser, C. U., Ellington, M. J., Cartwright, E. J., Gillespie, S. H., Brown, N. M., Farrington, M., et al. (2012). Routine use of microbial whole genome sequencing in diagnostic and public health microbiology. *PLoS Pathogens*, *8*(8), e1002824.

Köser, C. U., Holden, M. T., Ellington, M. J., Cartwright, E. J., Brown, N. M., Ogilvy-Stuart,-A. L., et al. (2012). Rapid whole-genome sequencing for investigation of a neonatal MRSA outbreak. *New England Journal of Medicine*, *366*(24), 2267–2275.

Krieg, N. R., Brenner, D. J., & Staley, J. R. (2005). *Bergey's manual of systematic bacteriology* (2nd revised ed.). New York: Springer-Verlag New York Inc..

Lefebure, T., Pavinski Bitar, P. D., Suzuki, H., & Stanhope, M. J. (2010). Evolutionary dynamics of complete *Campylobacter* pan-genomes and the bacterial species concept. *Genome Biology and Evolution*, *2*, 646–655.

Maiden, M. C. (2006). Multilocus sequence typing of bacteria. *Annual Review of Microbiology*, *60*, 561–588.

Maiden, M. C. J., Bygraves, J. A., Feil, E., Morelli, G., Russell, J. E., Urwin, R., et al. (1998). Multilocus sequence typing: A portable approach to the identification of clones within populations of pathogenic microorganisms. *Proceedings of the National Academy of Sciences of the United States of America*, *95*(6), 3140–3145.

Maiden, M. C., van Rensburg, M. J., Bray, J. E., Earle, S. G., Ford, S. A., Jolley, K. A., et al. (2013). MLST revisited: The gene-by-gene approach to bacterial genomics. *Nature Reviews. Microbiology*, *11*(10), 728–736.

Markowitz, V. M., Chen, I. M., Palaniappan, K., Chu, K., Szeto, E., Grechkin, Y., et al. (2010). The integrated microbial genomes system: An expanding comparative analysis resource. *Nucleic Acids Research*, *38*(Database issue), D382–D390.

Medini, D., Serruto, D., Parkhill, J., Relman, D. A., Donati, C., Moxon, R., et al. (2008). Microbiology in the post-genomic era. *Nature Reviews. Microbiology, 6*(6), 419–430.

Mo, S., You, M., Su, Y. C., Lacap-Bugler, D. C., Huo, Y. B., Smith, G. J., et al. (2013). Multilocus sequence analysis of *Treponema denticola* strains of diverse origin. *BMC Microbiology, 13*, 24.

Mutreja, A., Kim, D. W., Thomson, N. R., Connor, T. R., Lee, J. H., Kariuki, S., et al. (2011). Evidence for several waves of global transmission in the seventh cholera pandemic. *Nature, 477*(7365), 462–465.

Priest, F. G., Barker, M., Baillie, L. W., Holmes, E. C., & Maiden, M. C. (2004). Population structure and evolution of the *Bacillus cereus* group. *Journal of Bacteriology, 186*(23), 7959–7970.

Saitou, N., & Nei, M. (1987). The neighbour-joining method: A new method for reconstructing phylogenetic trees. *Molecular Biology and Evolution, 4*(4), 406–425.

Sanger, F., Nicklen, S., & Coulson, A. R. (1977). DNA sequencing with chain-terminating inhibitors. *Proceedings of the National Academy of Sciences of the United States of America, 74*(12), 5463–5467.

Sawabe, T., Ogura, Y., Matsumura, Y., Feng, G., Amin, A. R., Mino, S., et al. (2013). Updating the *Vibrio* clades defined by multilocus sequence phylogeny: Proposal of eight new clades, and the description of *Vibrio tritonius sp. nov. Frontiers in Microbiology, 4*, 414.

Selander, R. K., Caugant, D. A., Ochman, H., Musser, J. M., Gilmour, M. N., & Whittam, T. S. (1986). Methods of multilocus enzyme electrophoresis for bacterial population genetics and systematics. *Applied and Environmental Microbiology, 51*, 837–884.

Tamura, K., Peterson, D., Peterson, N., Stecher, G., Nei, M., & Kumar, S. (2011). MEGA5: Molecular evolutionary genetics analysis using maximum likelihood, evolutionary distance, and maximum parsimony methods. *Molecular Biology and Evolution, 28*(10), 2731–2739.

Thompson, C. C., Chimetto, L., Edwards, R. A., Swings, J., Stackebrandt, E., & Thompson, F. L. (2013). Microbial genomic taxonomy. *BMC Genomics, 14*, 913.

Vamosi, S. M., Heard, S. B., Vamosi, J. C., & Webb, C. O. (2009). Emerging patterns in the comparative analysis of phylogenetic community structure. *Molecular Ecology, 18*(4), 572–592.

Walcher, M., Skvoretz, R., Montgomery-Fullerton, M., Jonas, V., & Brentano, S. (2013). Description of an unusual *N. meningitidis* isolate containing and expressing *N. gonorrhoeae*-specific 16S rRNA gene sequences. *Journal of Clinical Microbiology, 51*(10), 3199–3206.

Walker, T. M., Clp, C. L., Harrell, R. H., Evans, J. T., Kapatai, G., Dedicoat, M. J., et al. (2013). Whole-genome sequencing to delineate *Mycobacterium tuberculosis* outbreaks: A retrospective observational study. *Lancet Infectious Diseases, 13*(2), 137–146.

Woese, C. R. (1987). Bacterial evolution. *Microbiological Reviews, 51*(2), 221–271.

Woese, C. R., Kandler, O., & Wheelis, M. L. (1990). Towards a natural system of organisms—Proposal for the domains *Archaea, Bacteria, and Eucarya. Proceedings of the National Academy of Sciences of the United States of America, 87*(12), 4576–4579.

Zerbino, D. (2010). Using the velvet *de novo* assembler for short-read sequencing technologies. *Current Protocols in Bioinformatics, 11*(5), 1–12.

Zerbino, D. R., & Birney, E. (2008). Velvet: Algorithms for de novo short read assembly using de Bruijn graphs. *Genome Research, 18*(5), 821–829.

Multi-locus Sequence Analysis: Taking Prokaryotic Systematics to the Next Level

11

Xiaoying Rong, Ying Huang[1]

State Key Laboratory of Microbial Resources, Institute of Microbiology, Chinese Academy of Sciences, Beijing, P.R. China
[1]Corresponding author: e-mail address: huangy@im.ac.cn

1 INTRODUCTION

The study of genetic variation among prokaryotes is fundamental for understanding their evolution, and untangling their ecology and evolutionary theory-based systematics. Recently, advances in DNA sequencing technology, the exploration of neglected habitats and the discovery of new species and products have forced microbial systematists to undertake comprehensive analyses of genetic variation within and between species and determine its impact on evolutionary relationships. 16S rRNA gene sequence analyses have enhanced our understanding of prokaryotic diversity and phylogeny, including that of non-culturable microorganisms (Doolittle, 1999; Head, Saunders, & Pickup, 1998; Mason et al., 2009). However, 16S rRNA gene phylogenetic analyses have major drawbacks, notably providing insufficient resolution, especially for distinguishing between closely related species. In addition, there are well-documented cases in which individual prokaryotic organisms have been found to contain divergent 16S rRNA genes, thereby suggesting the potential for horizontal exchange of rRNA genes; such problems pose a challenge for reconstructing the evolutionary history of prokaryotic species and complicate efforts to classify them (Acinas, Marcelino, Klepac-Ceraj, & Polz, 2004; Cilia, Lafay, & Christen, 1996; Michon et al., 2010; Pei et al., 2010; Rokas, Williams, King, & Carroll, 2003).

Multi-locus sequence analysis/typing (MLSA/MLST) is a nucleotide sequence-based approach for the unambiguous characterization of prokaryotes via the Internet, which directly characterizes DNA sequence variations in a set of housekeeping genes and evaluates relationships between strains based on their unique allelic profiles or sequences (Maiden, 2006). The method has been extensively used in the classification and identification of diverse bacteria, to determine the extent of gene exchange within and between species and to establish the relative importance of recombination in population genetics (Doroghazi & Buckley, 2010; Alvarez-Perez, de Vega, & Herrera, 2013; Freel, Millan-Aguinaga, & Jensen, 2013). MLSA is providing new opportunities to evaluate the status of bacterial taxa using patterns of genetic variation.

Methods in Microbiology, Volume 41, ISSN 0580-9517, http://dx.doi.org/10.1016/bs.mim.2014.10.001

Actinobacteria are well known for their ability to produce diverse bioactive metabolites of commercial significance and to produce signalling molecules of ecological value (Bérdy, 2005, 2012; Vetsigian, Jajoo, & Kishony, 2011). Although once considered primarily as soil bacteria, it is now known that actinobacteria are widely distributed in diverse natural habitats where they show a high degree of species and functional diversity (Bull, 2010). Since the discovery of streptomycin from *Streptomyces griseus* (Schatz, Bugie, & Waksman, 1944), actinobacteria, notably streptomycetes, have been isolated from innumerable habitats and screened for bioactive compounds. In the case of the genus *Streptomyces*, such studies have led to the publication of over 600 validly named species that pose significant taxonomic challenges (Kämpfer, 2012; Labeda et al., 2012).

In this chapter, we summarize the practical procedures and the implementation of MLSA to species assignments of representative taxa within the genus *Streptomyces* of the phylum *Actinobacteria*, and how MLSA results are compared with and validated against the current benchmark of DNA–DNA hybridization (DDH) values. We also describe how MLSA could be used to construct a sound framework that bridges taxonomic relationships and ecotype-level functional diversity within prokaryotic species.

2 MULTI-LOCUS SEQUENCE ANALYSIS

2.1 UNDERLYING CONCEPTS

MLSA was developed from the application of the MLST method in order to reconstruct evolutionary relationships between prokaryotes (Sawabe, Kita-Tsukamoto, & Thompson, 2007; Vinuesa et al., 2008; Wagner, Varghese, Hemme, & Wiegel, 2013). Comparative MLST is based on sequencing 450–500 bp fragments of five to seven housekeeping genes that provide information on the spread of nucleotide divergence across the chromosomes of sampled populations. Sequences that differ by even a single nucleotide for each gene are assigned as different alleles, thus making MLSA highly suitable for detecting genetic changes within and between species.

Phylogenetic analysis using multi-locus sequences involves several steps: (1) selection of strains and housekeeping genes, (2) generation of sequences (polymerase chain reaction (PCR) amplification and DNA sequencing), (3) data analysis to identify homologous sites within each gene and (4) MLSA using concatenated sequences or allelic profiles (the latter are usually used for population genetics). In step 1, the most important factors to consider are the clock-like behaviour and phylogenetic range of the gene sequences; the availability of complete genomes makes it easier to establish new MLSA schemes. The amplicon of each gene needs to be based on accurate sequence information from both strands hence high-quality sequencing chromatograms are at a premium. In data analysis, definitive identification of variation is obtained by nucleotide sequence determination of gene fragments, with no weight given to the number of nucleotide differences between alleles when assigning sequence types (STs), as it is not possible to determine whether differences at multiple nucleotide sites are due to multiple point mutations or to single recombination events (Maiden, 2006). Finally, relatedness between strains in MLSA is obtained by comparing concatenated sequences or allelic profiles (also STs). Those undertaking population genetics or phylogenetic adaptation studies may compare expanded candidate functional gene sequences (Tankouo-Sandjong et al., 2007; Parker, Havird, & De La Fuente, 2012; Stefanic et al., 2012).

2.2 SELECTION OF GENE LOCI

The housekeeping genes selected for MLSA analyses should be single copies, orthologous and ubiquitous among all of the sampled strains. They should also be highly conserved, have no linkage disequilibrium on the chromosome but should contain sufficient varying nucleotide positions to accurately establish relationships between closely related strains. To strike a balance among acceptable identification power, time and cost for strain typing, about five to seven housekeeping genes are commonly used. The five housekeeping genes that we have used in our *Streptomyces* MLSA scheme are *atpD* (ATP synthase F1, β-subunit), *gyrB* (DNA gyrase, B subunit), *recA* (recombinase A), *rpoB* (RNA polymerase, β-subunit) and *trpB* (tryptophan synthase, β-subunit); the physical distance between any two of these genes is greater than 30 kb in a single *Streptomyces* genome, thereby avoiding hitchhiking effects (Guo, Zheng, Rong, & Huang, 2008; Rong & Huang, 2012). However, it is not uncommon to use up to 10 housekeeping genes, as exemplified in the case of the genus *Nocardia* where 14 protein-coding genes were examined (Tamura et al., 2012). Thus, both the number and type of housekeeping genes interrogated by MLSA may differ from genus to genus.

2.3 GENERATING SEQUENCES

The PCR is commonly used to generate sequence fragments. Conserved profiles of protein-coding genes provide highly conserved islands that can be used to design amplification and sequencing primers that have broad specificity with respect to phylogenetic diversity. Commonly used primer design tools include Primer Premier (Premier Biosoft International) and OLIGO Primer Analysis software (Molecular Biology Insights), web tools such as Primer3 (http://biotools.umassmed.edu/bioapps/primer3_www.cgi) and BLAST (http://www.ncbi.nlm.nih.gov/tools/primer-blast/) for finding primers specific to PCR templates. DANMAN software (Molecular Biology Insights) can be used to evaluate the resultant primers, e.g. by checking the minimum free energy of the primer structure and if dimers (including self- and cross-dimers), hairpin and GC clamps exist. It is essential that dimers are situated well away from the 3' ends of the primers and continuous cross-dimers between the forward and reverse primers are to be avoided.

The genes and primers used in our streptomycete MLSA scheme are shown in Table 1. The suggested primer set for each gene covers more than the 400 bp internal

Table 1 Genes and Primers Used in the *Streptomyces* MLSA Scheme

Gene	Function	Direction	Primer Sequence (5′–3′)[a]	Position[b]	Chromosomal Location[b]
atpD	Amplification	F	atpDPF-GTCGGCGACTTCACCAAGGGCAAGGTGTTCAACACC	283–318	5842699-5844135
		R	atpDPR-GTGAACTGCTTGGCGACGTGGGTGTTCTGGGACAGGAA	1243–1280	
	Sequencing	F	atpDF-ACCAAGGGCAAGGTGTTCAA	295–314	
		R	atpDR-GCCGGGTAGATGCCCTTCTC	1027–1046	
gyrB	Amplification	F	gyrBPFA-CTCGAGGGTCTGACGCGGTCCGCAAGCGACCCGGTATGTA	121–161	c4265438-4263378
		R	gyrBPRA-GAAGGTCTTCACCTCGGTGTTGCCCAGCTTCGTCTT	1150–1185	
	Sequencing	F	gyrBFA-GCAAGCGACCCGGTATGTAC	143–162	
		R	gyrBRA-GAGGTTGTCGTCCTTCTCGC	1052–1071	
recA	Amplification*	F	recAPFA-GABSCCGCDCTCGCNCAGATYGARCG	31–56	6306654-6307778
		R	recAPRA-GGAAGTTGCGSGCGTTCTCCTTGC	902–925	
	Sequencing	F	recAF-ACAGATTGAACGGCAATTCG	45–64	
		R	recAR-ACCTTGTTCTTGACCACCTT	733–752	
rpoB	Amplification	F	rpoBPF-GAGCGCATGACCACCCAGGACGTCGAGGC	1162–1190	5078061-5081546
		R	rpoBPR-CCTCGTAGTTGTGACCCTCCCACGGCATGA	2126–2155	
	Sequencing	F	rpoBF1-TTCATGGACCAGAACAACC	1273–1291	
		R	rpoBR1-CGTAGTTGTGACCCTCCC	2135–2152	
trpB	Amplification	F	trpBPF-GCGCGAGGACCTGAACCACACCGGCTCACACAAGATCAACA	267–308	c2189306-2188023
		R	trpBPR-TCGATGGCCGGGATGATGCCCTCGGTGCGCGACAGCAGGC	1049–1088	
	Sequencing	F	trpBF-GGCTCACACAAGATCAACAA	289–308	
		R	trpBR-TCGATGGCCGGGATGATGCC	1069–1088	

[a]In order to reduce the impact of the high GC content of streptomycetes, we add 10% DMSO to the amplification reaction. The annealing temperatures for the genes were 60 °C (atpD), 65 °C (gyrB), 61 °C (recA), 63 °C (rpoB) and 65 °C (trpB), respectively.
[b]Positions corresponding to those of Streptomyces coelicolor A3(2).

fragments needed for MLSA sequence determination. We recommend the use of high-quality genomic DNA for PCR amplification to avoid problems caused by DNA templates. Sequencing primers (Table 1) that target internal sites are appropriate for amplified fragments in direct PCR sequencing.

In the era of high-throughput sequencing, MLSA sequence determinations rely on high-throughput MLST with Roche 454 technology (Boers, van der Reijden, & Jansen, 2012). Emulsion PCR uses primers from standardized MLSA schemes based on the housekeeping genes of interest, albeit with universal tails at the 5′ end to allow for the addition of 454-specific sequences and isolate-specific multiplex identifiers in the second PCR round. Third-generation sequencing techniques can now be applied routinely for whole genome sequencing (WGS) as the cost and time to do WGS continues to decline (see chapter "Revolutionizing Prokaryotic Systematics Through Next-Generation Sequencing" by Sangal et al.). This trend is impressively illustrated by the fact that the number of prokaryotic genomes registered in public databases has almost doubled since 2012 (Chun & Rainey, 2014). Software and biotools developed for using WGS data in MLSA are now available, such as PyPop (*Python* for Population Genomics, http://www.pypop.org/) (Lancaster, Nelson, Meyer, Thomson, & Single, 2003), BIGSdb (Bacterial Isolate Genome Sequence Database, http://pubmlst.org/software/database/bigsdb/) (Jolley & Maiden, 2010), ribosomal MLST (rMLST) (Jolley et al., 2012) and the MLST web-based method (www.cbs.dtu.dk/services/MLST) (Larsen et al., 2012).

2.4 DATA ANALYSIS
2.4.1 Properties of loci
2.4.1.1 Sequence alignments
Sequences alignments combined with both prior and subsequent quality checking of the (raw) data for each locus are pre-requisites for MLSA. The Clustal series of programs are the ones most widely used for multiple sequence alignment. The accuracy and speed of multiple alignments can be improved by the use of other programs, including MAFFT, Muscle and T-Coffee, which tend to consider requirements for scalability and accuracy of increasingly large-scale sequence data, influence of functional non-coding RNAs and extract biological knowledge for multiple sequence alignments (Blackburne & Whelan, 2013). MaxAlign software (Gouveia-Oliveira, Sackett, & Pedersen, 2007) can be used to delete unusual sequences from multiple sequence alignments in order to maximize the size of alignment areas, and Gblocks software (Talavera & Castresana, 2007) to select conserved blocks from poorly aligned positions and to saturate multiple substitutions for multiple alignments for MLSA-based phylogenetic analyses.

2.4.1.2 Loci statistics
Sequences of internal gene fragments are used in MLSA studies. It is advisable to avoid hypervariable regions, as they tend to violate overall evolutionary evidence. Statistics for each locus, such as the number and proportion of polymorphic sites,

mean G+C content and the d_N/d_S index (the ratios of non-synonymous to synonymous polymorphisms), can be summarized using START2 (http://pubmlst.org/software/analysis/start2/), MEGA (http://www.megasoftware.net/) and DnaSP (http://www.ub.edu/dnasp/) softwares.

2.4.1.3 Establishing STs

For MLSA sequences used for population genetics, unique sequences at each locus are assigned allele numbers, and then combined into allelic profiles (also STs). The resulting ST data are then ready for determining estimator F_{ST} (fixation index) and I_A^s (index of association), which quantify differentiation and recombination within populations. Strains that have the same alleles at all loci are considered to belong to the same ST; each ST is also given a number. The most commonly used web tool for ST assignments is non-redundant databases (http://pubmlst.org/cgi-bin/mlstanalyse/mlstanalyse.pl?site=pubmlst&page=nrdb), which can search batches of sequences to find those that are identical.

2.4.2 Phylogenetic analysis

2.4.2.1 Models of evolution

Pairwise distances between sequences of each locus can be calculated by using the Jukes–Cantor model, which assumes equal base frequencies and the same probability of any type of base change (Jukes & Cantor, 1969), and the Kimura 2-parameter model (Kimura, 1980), which distinguishes between transitions and transversions. Other more complex models describe the evolution of single sites within a set of sequences. These models include F81 (Felsenstein, 1981), HKY85 (Hasegawa–Kishino–Yano), Tamura–Nei (TN93) and generalized time-reversible, which allow unequal base frequencies and individual substitution rates for all possible changes; the model T92 considers G+C-content bias, as well as different types of transitions and transversions (Rzhetsky & Nei, 1995). The heterogeneity rate can then be parameterized in the substitution models in terms of the gamma distribution (Γ) and the proportion of invariable sites (I). Further details on model selection can be found on Felsenstein's web page (http://evolution.genetics.washington.edu/phylip/software.html#Modelselection).

2.4.2.2 Evaluating phylogenetic congruence

In order to construct multi-locus gene phylogenies, analyses of congruence are needed to determine the underlying histories of each gene. The congruence test also helps to address lateral gene transfer events and verify genes with a common ancestor (Leigh, Lapointe, Lopez, & Bapteste, 2011). The most widely used character-based methods in congruence analyses include the incongruence length difference (Farris, Källersjö, Kluge, & Bult, 1994) implemented in the PAUP* package (Swofford, 1998), the likelihood ratio tests and adapted likelihood-based topology tests. Other topological analyses include maximum agreement subtree and congruence among distance matrices; these tests are generally used in the fields of phylogeography and co-evolution, and are less relevant to phylogenomics and genome evolution than character-based methods (Leigh et al., 2011). Moreover, CONSEL software, developed by Shimodaira and Hasegawa (2001), estimates the site-wise log-likelihood for the trees, where the *p*-value is less biased than the conventional *p*-values of Kishino–Hasegawa, Shimodaira–Hasegawa and weighted Shimodaira–Hasegawa methods, and works in conjunction with several phylogenetic software packages: Molphy (Adachi & Hasegawa, 1996), PAML (Yang, 2007), PAUP* (Swofford, 1998), TREE-PUZZLE (Schmidt, Strimmer, Vingron, & von Haeseler, 2002) and RAxML (Stamatakis, 2006).

2.4.2.3 Construction of phylogenetic trees

The accumulation of nucleotide changes in genes is a relatively slow process; hence, concatenated sequences of bacterial isolates are sufficiently stable over time to be ideal for inferring phylogenetic relationships. To ensure the stability and reliability of phylogenetic relationships between strains based on the MLSA approach, phylogenetic trees are usually constructed using both distance- and character-based methods, including the neighbour-joining (NJ) (Saitou & Nei, 1987), maximum parsimony (MP) (Fitch, 1971; Hartigan, 1973) and maximum likelihood (ML) (Felsenstein, 1981; Kishino & Hasegawa, 1989) algorithms and the Bayesian inference (BI) method (Yang & Rannala, 1997, 2012).

NJ is one of the most frequently used algorithms in MLSA phylogenetic reconstructions, as it can quickly be used to reflect pairwise distances, especially for closely related datasets. However, this method treats all sites as equal, thereby making it more suitable for the analysis of less informative sequences. MP, a non-parametric statistical method used to construct phylogenies, is based on the assumption that tree topologies derived from real alignment column data are a result of a minimum number of evolutionary changes. Parsimony analyses often return a number of unrooted equally probable parsimonious trees that show all possible relationships between taxa. The BI and ML methods use likelihood function and are theoretically immune to long-branch attraction when properly used. As the BI, ML and NJ methods are all parametric statistical methods, they rely on explicit models of character evolution. It has also been shown that for some suboptimal models, these methods can give inconsistent results (Yang & Rannala, 2012).

It is worth noting that phylogenetic estimations can be influenced by branch length, dataset size (both the number of taxa and sites), sequence heterogeneity, evolutionary depth, dataset complexity and the analytical framework. Even if optimal phylogenetic trees are generated successfully, they do not always provide relationships that make sense from a biological viewpoint. This suggests that we should pay attention to phylogenetic estimation as well as algorithmic techniques to obtain optimal solutions.

2.5 COMPARISON WITH OTHER TAXONOMIC METHODS

Since the *ad hoc* committee proposed the combined use of several housekeeping genes for the re-evaluation of the species definition in bacteriology (Stackebrandt et al., 2002), the MLSA approach has shed new light on prokaryotic systematics and phylogeny. Relationships based on 16S rRNA gene sequence data provide an invaluable framework for MLSA studies but do not give sufficient resolution to distinguish between closely related species. In contrast, DDH and DNA-fingerprinting methods can be used to differentiate between closely related species, but both of these approaches are labour-intensive, do not yield cumulative data and are difficult to apply to large numbers of strains. MLSA provides an attractive alternative to DDH by providing high-quality, reproducible data for establishing relationships within and between species. MLSA schemes are also distinct from DNA barcoding, a taxonomic method that uses short genetic markers to recognize known or cryptic species and is based on the fact that mitochondrial DNAs or some parts of ribosomal DNA cistrons have relatively fast mutation rates.

Now that WGS has become more affordable as it is being increasingly used in the classification and identification of prokaryotes and in the generation of phylogenies (Konstantinidis & Tiedje, 2005; Ventura et al., 2007; Zhi, Zhao, Li, & Zhao, 2012), though challenges remain in the use of such data (Leigh et al., 2011). At present, the widespread use of whole-genome sequence data is restricted as such data are only available for 1725 out of nearly 11,000 archaeal and bacterial type strains (Chun & Rainey, 2014). However, this situation is being addressed by the 1000 microbial genomes (KMG-I) project (Genomic Encyclopedia of Type Strains, Phase I) (Kyrpides et al., 2014) which is now underway.

2.6 MLSAs: ADVANTAGES AND DISADVANTAGES

The most important advantage of the MLSA method is that it provides information on the nucleotide divergence of internal fragments of housekeeping genes that can be used to compare data from different laboratories and to construct cumulative and reproducible datasets. As MLSA data provide neutral variations of core constitutive genes across the genome, multi-locus phylogenetic analyses enhance our insight into microbial diversity and taxonomic structure at the genomic level. Compared to single gene-based phylogenies, MLSA gives the phylogenetic resolution needed for species delineation, especially for closely related species, and avoids possible misleading results that attribute to sequences based on single genes that may be affected by recombination (Didelot & Falush, 2007).

MLSA, as mentioned earlier, is a scalable method that can be used to combine advances in high-throughput sequencing techniques and bioinformatics with population genetics in order to infer evolutionary histories among bacteria. MLSA studies are increasingly based on the use of whole-genome core gene sets and functional genes (Caro-Quintero & Konstantinidis, 2012; Didelot, Meric, Falush, & Darling, 2012; Stefanic et al., 2012). The programs ClonalFrame and Structure, initially developed for analyzing multi-locus sequence data, are now widely used to determine population structures of bacteria and to provide genome-based frameworks for the delineation of species (Pritchard, Stephens, & Donnelly, 2000; Didelot & Maiden, 2010; Didelot et al., 2012).

However, depending on which theory-based species concept is used, MLSA faces difficulties as clusters identified at the species level lack coherent phenotypic and genetic demarcations. Positively selected functional genes used to assess micro-diversity present in natural habitats are more likely to confound the results of evolutionary parameters (Ferreira et al., 2012), thereby making it difficult to link species diversity based on neutral housekeeping genes to the diversity of ecophysiological characteristics and adaptiveness. Moreover, it has been shown that the different frequency and relative weight of point mutations and recombination inferred when using dissimilar MLSA loci, strains, or analytical methodologies, also make it difficult to set boundaries between bacteria based on genetic diversity (Didelot & Maiden, 2010; Ferreira et al., 2012). In addition, although multi-locus sequence data for population/group strains can be obtained in routine laboratories, there are a limited number of reference sequences in public databases compared to the number of commonly used rRNA gene sequences.

2.7 MLSA DATABASES

The most widely used MLST databases hosted on web servers are MLST.net (http://www.mlst.net/) and pubMLST.org (http://pubmlst.org), which contain reference allele sequences and STs for a number of species. These websites are based on the use of a variety of analytical software tools that allow users to query deposited allele sequences and STs. The *Streptomyces* MLSA database (http://pubmlst.org/streptomyces/), as other single MLSA database systems, contains two linked databases: one for allelic profiles and sequences and the other for isolate information. At present, this database contains information on 141 *Streptomyces* species and 135 STs (almost all of the sequences are for type strains). Recently, an eMLSA.net database (http://www.emlsa.net/) was implemented in the MLSA website to facilitate the assignment of strains to species clusters via the Internet. The pilot eMLSA.net database is only available for members of the *Streptococcus viridans* group.

3 APPLICATION OF MLSAs IN PROKARYOTIC SYSTEMATICS

3.1 THE GENUS *STREPTOMYCES*

In our most recent studies, the MLSA scheme has shown great promise for refining relationships between closely related *Streptomyces* species. The inter- and intra-specific relationships of streptomycetes assigned to the *Streptomyces albidoflavus*, *Streptomyces griseus* and *Streptomyces hygroscopicus* 16S rRNA gene clades, as well as to the *Streptomyces pratensis* phylogroup, have been established based on the use of the five housekeeping gene sequences (*atpD*, *gyrB*, *recA*, *rpoB* and *trpB*) and DDH data (Rong, Guo, & Huang, 2009; Rong, Liu, Ruan, & Huang, 2010; Guo et al., 2008; Rong et al., 2013; Rong & Huang, 2010, 2012). A strong correlation was found between the MLSA and DDH data, as well as with associated phenotypic characteristics of clusters corresponding to individual species. These results provide further evidence that this MLSA scheme provides a reliable and effective way of classifying and identifying *Streptomyces* species.

3.1.1 *The* Streptomyces *MLSA scheme*

Our MLSA scheme is based on 141 *Streptomyces* strains and the five housekeeping genes mentioned earlier; the representatives of the various taxa are distributed across the *Streptomyces* 16S rRNA gene phylogenetic tree generated by Labeda et al. (2012). PCR amplification and sequencing of the housekeeping genes were performed using the primers shown in Table 1. The purified PCR products were directly sequenced on both strands using an automated DNA sequencer (Applied Biosystems); only sequence chromatograms with clean backgrounds were used. In the final analysis, alignments of 495 bp of *atpD*, 419 bp of *gyrB* (with gaps), 504 bp of *recA*, 540 bp of *rpoB* and 567 bp of *trpB* (with gaps) were obtained (Table 2). In addition, almost complete 16S rRNA gene sequences of the strains were either generated in-house or obtained from GenBank.

The features for each locus (MEGA or START2 programs), including the number of polymorphic sites, STs, mean G+C content and index of selection ($d_N - d_S$), have shown that protein-coding alleles have a higher percentage of variable sites and a wider range of distances that can be used to fine tune relationships between closely related strains. Among the five protein-coding genes, *atpD*, *recA* and *rpoB* are more conserved than the *gyrB* and *trpB* genes. Sequence diversities of the five gene loci were reflected in the corresponding amino acids, where AtpD and RecA were the most conserved. The alignment datasets encompass 16S rRNA gene sequences, the five protein-encoding gene sequences and various concatenations of these genes; the concatenated sequences were obtained by joining head-to-tail in-frame. The phylogenetic analyses were based on the NJ, MP and ML tree-making algorithms implemented in the MEGA package (http://www.megasoftware.net/), with the option of complete deletion of gaps; the topologies of the resultant trees were evaluated in bootstrap analyses based on 1000 resamplings (Felsenstein, 1985). The Kimura 2-parameter model (Kimura, 1980) was chosen as the substitution model for NJ tree construction; and the P-distance model for calculation of amino acid sequence distances. The best-fit model chosen for ML tree construction was performed with parameters selected by ModelTest (Posada & Crandall, 1998). The ML tree was also inferred by using either the PHYLIP package version 3.68 (DNAML program) (Felsenstein, 2008) or RAxML software (Stamatakis, 2006) with 1000 replicates.

The phylogenetic trees based on the individual protein-coding loci were generally congruent and showed much greater resolution when compared with corresponding trees, thereby showing that multiple sequences of the protein-coding genes can be used for integrated analyses (trees not shown). However, the single genes did not have enough resolution to discriminate between all of the *Streptomyces* species, and there were some poorly resolved relationships and disagreements between the trees suggesting the occurrence of recombination events, e.g. the positions of *Streptomyces libani* subsp. *rufus* and *Streptomyces ramulosus* in the *atpD S. hygroscopicus* gene tree were quite different from the positions of these taxa in the four corresponding single-gene trees. It was also shown that bootstrap support for the individual gene trees was generally low, especially for the deeper nodes. Single-gene

Table 2 Properties of Loci

Locus	Allele Length (bp)	No. of Alleles	No. of Polymorphic Sites	Polymorphic Sites (%)	Mean G+C Content (mol%)	Distance Range[a]	Mean K2P Distance[b]	$d_N - d_S$
atpD	495	107	145	29.3	64.6	0–0.170	0.089	−8.494
gyrB	420	109	241	57.4	67.4	0–0.270	0.154	−7.364
recA	504	108	178	35.3	69.2	0–0.178	0.094	−10.799
rpoB	540	111	211	39.1	67.4	0–0.183	0.106	−7.728
trpB	567	124	229	40.4	72.3	0–0.239	0.109	−839
Five-gene	2526	133	1005	39.8	68.3	0–0.175	0.109	−22.359
16S rRNA	1367	95	165	12.1	58.9	0–0.059	0.026	/

[a]Pairwise distances calculated by using the Kimura 2-parameter substitution model.
[b]Distances calculated by using the Kimura 2-parameter substitution model.

phylogenies, therefore, give misleading results thereby justifying further MLSA studies. Since the five protein-coding genes are widely distributed and each has a single copy within the genome, MLSA can reduce the impact of horizontal gene transfer (HGT) and reflect the evolutionary history of recognized *Streptomyces* species at the genome level.

The relationships between the strains assigned to the various *Streptomyces* 16S rRNA gene clades is based on the concatenated alignment of the five protein-coding loci totalling 2526 nt, with evolutionary distances ranging from 0 to over 0.175. The results from the studies on members of the *S. albidoflavus*, *S. griseus* and *S. hygroscopicus* 16S rRNA gene clades clearly show that the MLSA phylogeny shows much higher resolution and a more stable topological structure than the corresponding 16S rRNA gene phylogeny (Figures 1 and 2). Nevertheless, the topologies of the two trees are similar as the same clusters are found in each of the trees. Moreover, the MLSA scheme was found to be more suitable for discriminating between strains with >99% 16S rRNA gene sequence similarities, typified by the strains classified in the *S. albidoflavus* and *S. hygrocopicus* clades, and the upper section of the *S. griseus* clade.

MLSA is an attractive method as it avoids the vagaries of single-gene tree phylogenics, thereby providing a sound framework for establishing relationships between closely related strains. We have also explored the effects of classifying our *Streptomyces* strains based on sequences derived from either three or four of the protein-coding genes. A tree based on the three most conserved genes (*atpD–recA–rpoB*) gave good discriminatory power with respect to strains classified in the *S. griseus* and *S. hygrocopicus* 16S rRNA gene clades. Particularly, good resolution was found in trees based on *atpD–rpoB–gyrB* and *gyrB–recA–rpoB* concatenated sequences when compared with corresponding three-gene datasets; hence, such trees can be recommended for routine use (Rong & Huang, 2010, 2012).

3.1.2 DNA:DNA hybridization and MLSAs

Extensive comparisons were made between DDH and MLSA evolutionary distances in all of the *Streptomyces* 16S rRNA clades and the phylogroup using a microplate method and biotinylated probe DNA, as described in our previous studies (Rong et al., 2013; Rong & Huang, 2010, 2012). The MLSA evolutionary distances between representative strain pairs were calculated and ranged from 0 to 0.114. DNA hybridizations were performed at 53 °C overnight in $2 \times$ SSC containing $5 \times$ Denhardt's solution, 50% formamide and 100 μg denatured salmon sperm DNA ml^{-1}. Reciprocal hybridizations (i.e. $A \times B$ and $B \times A$) were performed in triplicate for each pair of strains and hybridization values expressed as a mean of the corresponding reciprocal values; most strain pairs were examined in different hybridization experiments.

The MLSA evolutionary distances and corresponding DDH values for pairs of strains classified in the various *Streptomyces* taxa are shown in Table 3. A strong correlation was found between the two sets of values; strain pairs with MLSA

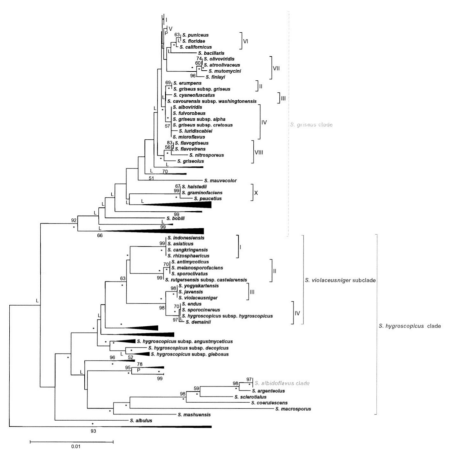

FIGURE 1

Neighbour-joining tree based on 16S rRNA gene sequences showing relationships within and between 142 strains classified in the *S. albidoflavus*, *S. griseus* and *S. hygroscopicus* clades, as well as the *S. pratensis* phylogroup. Parts of the strain clusters have been compressed. P and L indicate branches of the tree that were also determined using the MP and ML algorithms. Asterisks indicate branches of the tree that were supported by all three tree-making algorithms. Bar equals 1% difference in nucleotide sequences.

distances ≤0.007, corresponded to DDH values which ranged from 73.4% to 92.9% (*S. albidoflavus* clade), 75.9% to 89.6% (*S. griseus* clade) and 74.9% to 96.1% (*S. hygroscopicus* clade), while for the strain pairs with MLSA distances >0.007, the corresponding DDH values were well below the 70% (20.9%–57.9%) cut-off recommended for the delineation of species (Wayne et al., 1987).

All of the data obtained from the *Streptomyces* clades, as well as from *S. pratensis* phylogroup and related species were combined for a more comprehensive MLSA–DDH relationship analysis in order to establish an MLSA distance cut-off

FIGURE 2

Neighbour-joining tree based on concatenated five-gene sequences of five housekeeping genes (*atpD, gyrB, recA, rpoB* and *trpB*) showing relationships within and between 142 strains classified in the *S. albidoflavus, S. griseus* and *S. hygroscopicus* 16S rRNA gene clades, as well as in the *S. pratensis* phylogroup. Some clusters are compressed. P and L indicate branches of the tree that were also determined using MP and ML algorithms. Asterisks indicate branches of the tree that were supported by all three tree-making algorithms. Bar equals 1% difference in nucleotide sequences.

point for *Streptomyces* species. The correlation coefficients (r^2) between MLSA distance (range 0–0.015) and DDH values were calculated in a linear regression analysis (Figure 3). A strong correlation was found between the MLSA and DDH data with the 70% DDH cut-off point corresponding to a five-gene MLSA distance of

Table 3 DNA:DNA Hybridization and Multi-Locus Sequence Analysis (MLSA) Evolutionary Distance Values Among Strains and Related Species in the *Streptomyces* Clades and the *Streptomyces pratensis* Phylogroup

Strain Pair	DDH Value (%)	MLSA Evolutionary Distance
S. albidoflavus clade		
S. albidoflavus DSM 40455[T]	S. canescens CGMCC 4.1681[T]	90.3	0.001
S. albidoflavus DSM 40455[T]	S. champavatii CGMCC 4.1615[T]	80.8	0.002
S. albidoflavus DSM 40455[T]	S. coelicolor DSM 40233[T]	78.8	0.002
S. albidoflavus DSM 40455[T]	S. felleus CGMCC 4.1677[T]	87.2	0.001
S. albidoflavus DSM 40455[T]	S. gallaeus CGMCC 4.1320	82.4	0.002
S. albidoflavus DSM 40455[T]	S. globisporus subsp. caucasicus NBRC 100770[T]	92.9	0.000
S. albidoflavus DSM 40455[T]	S. griseus subsp. solvifaciens CGMCC 4.1845[T]	85.7	0.001
S. albidoflavus DSM 40455[T]	S. limosus NBRC 12790[T]	87.7	0.001
S. albidoflavus DSM 40455[T]	S. odorifer NBRC 13365[T]	77.7	0.001
S. albidoflavus DSM 40455[T]	S. sampsonii NBRC 13083[T]	78.9	0.002
S. albidoflavus DSM 40455[T]	S. sioyaensis CGMCC 4.1306	73.4	0.002
S. albidoflavus DSM 40455[T]	S. vinaceus CGMCC 4.1305	74.8	0.002
S. griseus subsp. solvifaciens CGMCC 4.1845[T]	S. canescens CGMCC 4.1681[T]	79.4	0.001
S. griseus subsp. solvifaciens CGMCC 4.1845[T]	S. champavatii CGMCC 4.1615[T]	83.6	0.002
S. griseus subsp. solvifaciens CGMCC 4.1845[T]	S. coelicolor DSM 40233[T]	79.8	0.003
S. griseus subsp. solvifaciens CGMCC 4.1845[T]	S. felleus CGMCC 4.1677[T]	75.9	0.002

Continued

Table 3 DNA:DNA Hybridization and Multi-Locus Sequence Analysis (MLSA) Evolutionary Distance Values Among Strains and Related Species in the *Streptomyces* Clades and the *Streptomyces pratensis* Phylogroup—cont'd

Strain Pair		DDH Value (%)	MLSA Evolutionary Distance
S. griseus subsp. solvifaciens CGMCC 4.1845[T]	S. gallaeus CGMCC 4.1320	75.7	0.002
S. canescens CGMCC 4.1681[T]	S. champavatii CGMCC 4.1615[T]	80.7	0.001
S. canescens CGMCC 4.1681[T]	S. felleus CGMCC 4.1677[T]	92.1	0.001
S. canescens CGMCC 4.1681[T]	S. globisporus subsp. caucasicus NBRC 100770[T]	85.0	0.001
S. felleus CGMCC 4.1677[T]	S. limosus NBRC 12790[T]	90.8	0.000
S. argenteolus CGMCC 4.1693[T]	S. albidoflavus DSM 40455[T]	43.7	0.010
S. argenteolus CGMCC 4.1693[T]	S. gallaeus CGMCC 4.1320	42.1	0.009
S. argenteolus CGMCC 4.1693[T]	S. griseus subsp. solvifaciens CGMCC 4.1845[T]	43.1	0.009
S. griseus subsp. griseus CGMCC 4.1419[T]	S. albidoflavus DSM 40455[T]	27.3	0.114
S. griseus subsp. griseus CGMCC 4.1419[T]	S. felleus CGMCC 4.1677[T]	20.9	0.113
S. griseus subsp. griseus CGMCC 4.1419[T]	S. griseus subsp. solvifaciens CGMCC 4.1845[T]	29.2	0.114
S. griseus clade			
S. californicus	S. puniceus	87.7	0.001
S. alboviridis	S. griseus subsp. alpha	83.7	0.002
S. californicus	S. floridae	89.6	0.002
S. floridae	S. puniceus	84.4	0.002
S. flavogriseus	S. flavovirens	88.3/86.6/87.7	0.004
S. acrimycini	S. baarnensis	86.5/80.5/85.9	0.005
S. graminofaciens	S. halstedii	77.3	0.005
S. acrimycini	S. flavofuscus	84.1/83.5/81.9	0.006

S. griseus subsp. alpha	S. luridiscabiei	79.3	0.006
S. cavourensis subsp. washingtonensis	S. cyaneofuscatus	75.9	0.007
S. alboviridis	S. luridiscabiei	76.6/78.6	0.007
S. griseus subsp. griseus	'S. ornatus'	80.5/78.9	0.007
S. argenteolus	S. galilaeus	42.1/48.7	0.009
S. argenteolus	S. griseus subsp. solvifaciens	43.1/47.1	0.009
S. badius	S. sindenensis	38.6/41.5/37.9	0.012
S. hygroscopicus clade			
S. cangkringensis	S. indonesiensis	90.0	0.000
S. asiaticus	S. cangkringensis	96.1	0.001
S. javensis	S. yogyakartensis	93.6/95.9	0.001
S. asiaticus	S. rhizosphaericus	85.4/81.1	0.002
S. antimycoticus	S. sporoclivatus	85.5	0.003
S. antimycoticus	S. castelarensis	83.7/89.1	0.004
S. endus	S. sporocinereus	81.9/77.6	0.004
S. libani subsp. libani	S. nigrescens	86.7/85.5	0.005
S. castelarensis	S. sporoclivatus	82.9	0.006
S. niger	S. olivaceiscleroticus	78.3/74.9	0.007
S. cangkringensis	S. yogyakartensis	52.9/50.1	0.008
S. hygroscopicus subsp. glebosus	S. platensis	51.8/54.0/55.9	0.008
S. asiaticus	S. yogyakartensis	55.0/53.4	0.009
S. cangkringensis	S. javensis	52.3/47.4	0.009
S. javensis	S. rhizosphaericus	46.3/44.8/50.3	0.009
S. melanosporofaciens	S. castelarensis	55.6/50.5/57.9	0.009
S. rhizosphaericus	S. yogyakartensis	48.4/46.6	0.009
S. antimycoticus	S. melanosporofaciens	50.9/51.5/53.2	0.010

Continued

Table 3 DNA:DNA Hybridization and Multi-Locus Sequence Analysis (MLSA) Evolutionary Distance Values Among Strains and Related Species in the *Streptomyces* Clades and the *Streptomyces pratensis* Phylogroup—cont'd

Strain Pair	DDH Value (%)	MLSA Evolutionary Distance
S. abikoensis S. luteoverticillatus	51.3/46.9	0.012
S. melanosporofaciens S. sporoclivatus	50.7/46.0/47.4	0.012
S. purpurogeneiscleroticus S. violens	47.5/49.3	0.012
S. hiroshimensis S. rectiverticillatus	31.1/32.4	0.014
S. ochraceiscleroticus S. purpurogeneiscleroticus	29.5/26.9/30.3	0.014
S. ochraceiscleroticus S. violens	36.5/37.1/38.9	0.015
S. pratensis phylogroup		
Strain ch24^T Strain W25 26	84.3	0.002
Strain ch24^T Strain ch2	83.0/81.2	0.002
Strain ch2 Strain W25 26	85.6/87.9/89.3	0.002
Strain ch24^T Strain CGMCC 4.1868	82.5/88.7/85.0	0.003
Strain ch24^T S. mutomycini CGMCC 4.1747^T	25.5/32.0/30.1	0.037
Strain CGMCC 4.1868 S. mutomycini CGMCC 4.1747^T	43.9/45.1	0.037
Strain CGMCC 4.1868 S. atroolivaceus CGMCC 4.1405^T	20.7	0.041
Strain CGMCC 4.1868 S. caviscabies CGMCC 4.1836^T	23.4/25.3/21.7	0.072
Strain CGMCC 4.1868 S. flavofuscus CGMCC 4.1938^T	26.7/28.0	0.072
Strain ch24^T S. caviscabies CGMCC 4.1836^T	20.3/19.5	0.070
Strain CGMCC 4.1868 S. setonii CGMCC 4.1367^T	16.3/18.4/19.9	0.071
S. caviscabies CGMCC 4.1836^T S. flavofuscus CGMCC 4.1938^T	87.1/82.9/84.3	0.002
S. caviscabies CGMCC 4.1836^T S. setonii CGMCC 4.1367^T	62.2/61.9	0.021
S. mutomycini CGMCC 4.1747^T S. caviscabies CGMCC 4.1836^T	20.9/21.3	0.070

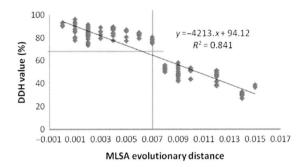

$$y = -4213.x + 94.12$$
$$R^2 = 0.841$$

FIGURE 3

Relationship between five-gene MLSA evolutionary distances and DDH values of *Streptomyces*. The data are from strains assigned to the *S. albidoflavus*, *S. griseus* and *S. hygroscopicus* clades, and from the *S. pratensis* phylogroup. The solid diamonds represent MLSA evolutionary distance (*x*-axis) between two strains plotted against the DDH value (*y*-axis). (See the colour plate.)

0.007; the circumscription of single-gene sequence distances and corresponding DDH values were also determined. In general, poor correlation was found between the evolutionary distances of individual genes and corresponding DDH values (Figure 4), providing further evidence of the limitations of basing phylogenetic relationships between strains on single-gene sequences. It was also encouraging that *Streptomyces* strains assigned to the same genomic species on the basis of five-gene sequence distances ≤ 0.007 shared phenotypic characteristics. Indeed, on this basis, 55 *Streptomyces* species and subspecies classified in the tested clades were considered to be later heterotypic synonyms of 20 genomic species, while two new species, namely, *Streptomyces glebosus* sp. nov., comb. nov and *Streptomyces ossamyceticus* sp. nov., comb. nov. were recognized in the *S. hygroscopicus* 16S rRNA gene clade (Rong et al., 2009; Rong & Huang, 2010, 2012).

The results also showed that the MLSA approach based on five protein-coding genes can be used to clarify relationships between closely related *Streptomyces* species across the range of variation encompassed by the 16S rRNA gene since the *S. albidoflavus*, *S. griseus* and *S. hygroscopicus* clades are well separated in the *Streptomyces* 16S rRNA gene tree generated by Labeda et al. (2012). The correlation found between the MLSA, DDH and phenotypic data shows that our MLSA procedure can be used to clarify relationships between *Streptomyces* species, thereby providing a sound classification for ecological, bioprospecting and biotechnological purposes.

3.2 CLASSIFICATION OF THE *S. PRATENSIS* PHYLOGROUP

Members of the *S. pratensis* phylogroup were found to share identical 16S rRNA gene sequences with eight other strains classified in cluster I of the *S. griseus* 16S rRNA gene clade (Figure 5A) but were consistently recovered as a distant branch

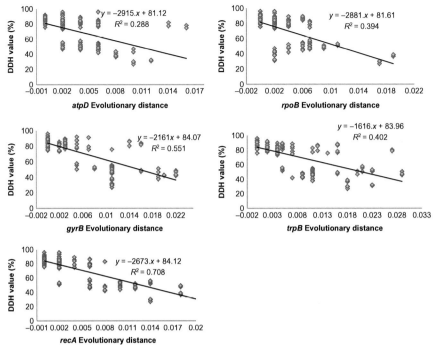

FIGURE 4

Relationships between evolutionary distances of individual genes and DDH values of strains classified in the *S. albidoflavus*, *S. griseus* and *S. hygroscopicus* 16S rRNA gene clades and in the *S. pratensis* phylogroup. Each solid diamond represents the MLSA evolutionary distances (*x*-axis) between two strains plotted against the DDH value (*y*-axis). (See the colour plate.)

in the phylogeny based on the protein-coding housekeeping genes where they were loosely associated with *Streptomyces atroolivaceus* and *Streptomyces mutomycini* in cluster VII of the *S. griseus* clade (Figure 5B). Additional studies based on the sequences of the five protein-coding housekeeping genes (*atpD*, *gyrB*, *recA*, *rpoB* and *trpB*), DNA relatedness and corresponding properties were carried out to establish the value of our MLSA procedure for distinguishing between *Streptomyces* species with identical 16S rRNA gene sequences. It can be seen from Figure 5 that the *S. pratensis* phylogroup strains (i.e., ATCC 33331, CGMCC 4.1868, Cald 193, ch2, ch24^T and W25 26) formed a distinct cluster in the MLSA phylogeny based on an analysis of the concatenated five-gene sequences and three tree-making algorithms.

The members of the *S. pratensis* phylogroup showed MLSA sequence divergences ranging from 0% to 0.4%, values well below the 0.7% cut-off point for the delineation of *Streptomyces* species, a result in line with corresponding DDH data and data based on a broad range of biochemical and physiological characteristics, as

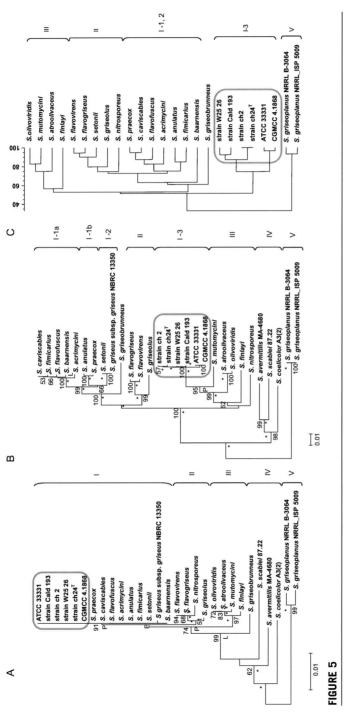

FIGURE 5

Neighbour-joining trees based on 16S rRNA gene sequences (A); five concatenated gene sequences (B), showing relationships between the *S. pratensis* phylogroup and related species and within the phylogroup; and (C) a dendrogram based on a numerical analysis of 49 biochemical and physiological characteristics.

Data from Rong et al. (2013).

shown in Figure 5C. Consequently, we proposed that the members of this phylogroup should be classified as *S. pratensis* sp. nov. (Rong et al., 2013). These results clearly show that, unlike the corresponding 16S rRNA sequence approach, the MLSA procedure not only provides sufficient resolution to distinguish between closely related *Streptomyces* species but can also be used to recognize new species whereas associated 16S rRNA gene sequence data may give erroneous results.

3.3 MLSA OF PHYTOPATHOGENIC *STREPTOMYCES* SPECIES

Phytopathogenicity is rare in the genus *Streptomyces*. The 10 *Streptomyces* species considered to be phytopathogenic belong to seven clades based on the 16S rRNA gene sequence study of Labeda et al. (2012). The type strains of the 10 phytopathogenic *Streptomyces* species, namely, *Streptomyces acidiscabies*, *Streptomyces europaeiscabiei*, *Streptomyces ipomoeae*, *Streptomyces luridiscabiei*, *Streptomyces niveiscabiei*, *Streptomyces puniciscabiei*, *Streptomyces reticuliscabiei*, *Streptomyces scabiei*, *Streptomyces stelliscabiei*, *Strepotomyces turgidiscabies* and six uncharacterized phytopathogenic *Streptomyces* isolates and the type strains of 52 phytogenetically related taxa were the subject of an MLSA phylogeny based on four protein-coding genes (*atpD*, *recA*, *rpoB* and *trpB*) (Labeda, 2011). Good congruence was found between the phylogenetic relationships of the individual housekeeping genes and the concatenated sequences between them and the 16S rRNA gene sequence data. However, the phytopathogenic species were more clearly delineated in the MLSA analysis with individual taxa supported by higher bootstrap values.

The function of most of the *Streptomyces* pathogenome has still to be resolved but it does include genes spread through HGT that encode secreted proteins, transcriptional regulators and membrane transporters, as well as for secondary metabolite biosynthesis (Loria, Kers, & Joshi, 2006; Bignell, Huguet-Tapia, Joshi, Pettis, & Loria, 2010). *S. acidiscabies* and *S. turgidiscabies* are thought to be newly emergent pathogens with a limited geographical distribution. The MLSA data showed that these organisms were located far from *S. scabiei*, the oldest plant pathogen. The results of the studies on the phytopathogenic *Streptomyces* provide further evidence of the value of MLSA for the rapid and accurate classification of *Streptomyces* strains at the species level.

3.4 MLSA: ACTINOBACTERIA

Actinobacteria show tremendous diversity in terms of their morphological, metabolic and ecophysiological properties (Bull, 2010; Stefanic et al., 2012), as exemplified by members of the family *Micromonosporaceae* (Genilloud, 2012). Members of the genera *Micromonospora*, *Salinispora* and *Verrucosispora*, which form extensively branched substrate hyphae, but rarely show sparse aerial hyphae, have received a lot of attention as a promising source of novel biologically important compounds. *Salinispora* strains are obligate marine organisms that are a rich source of novel antibiotics (Jensen, Gontang, Mafnas, Mincer, & Fenical,

2005; Bucarey, Penn, Paul, Fenical, & Jensen, 2012), *Verrucosispora maris*, an isolate from a marine sediment, produces the antibiotics abyssomicin and proximicin, while *Micromonospora* strains were recently found to inhabit the intracellular tissues of nitrogen-fixing nodules of *Pisum sativum* (Carro, Sproer, Alonso, & Trujillo, 2012).

The three closely related species that currently comprise the genus *Salinispora*, namely, *Salinispora arenicola*, *Salinispora pacifica* and *Salinispora tropica*, have been analysed using a multi-locus sequence scheme that targeted the five genes used in our *Streptomyces* studies (Freel et al., 2013); the results underpinned the taxonomic integrity of these species. However, a homologous recombination event was detected in a portion of the *rpoB* gene sequence which seemed to be linked to rifamycin resistance. Subsequent, more detailed population studies showed that *S. arenicola*, *S. pacifica* and *S. tropica* show different levels of nucleotide site divergence; recombination was found to play a central role in the evolution of *S. arenicola* and *S. tropica* while *S. pacifica* might be an amalgam of ecotypes or newly diverging species.

Carro et al. (2012) undertook a phylogenetic analysis of *Micromonospora* strains isolated from the rhizosphere and from nitrogen-fixing nodules of *P. sativum*, based on 16S rRNA gene sequences (1384 bp) and the sequences of four protein-coding genes, namely, *gyrB* (950 bp), *rpoB* (575 bp), *atpD* (873 bp) and *recA* (514 bp). They were able to show that the MLSA procedure can be used to unravel species diversity within the genus *Micromonospora* with sequence similarities <98.5%, indicating that strains belonged to different species. Similarly, an MLSA scheme based on concatenated sequences of *gyrB–rpoB–recA–relA–atpD* (4099 nt) provided an effective way of distinguishing *Kribbella* species, members of the family *Nocardiodaceae* (Curtis & Meyers, 2012).

Two different MLSA schemes have been introduced to clarify relationships between closely related mycobacteria. To this end, eight housekeeping genes (*argH*, *cya*, *glpK*, *gnd*, *murC*, *pgm*, *pta* and *purH*) were used to distinguish between strains of *Mycobacterium abscessus* (*sensu lato*) (Macheras et al., 2011). This scheme was refined by Lu et al. (2012) who used seven highly discriminative gene loci (*recX*, *rpsL*, *rmlC*, *rpmG1*, *mprA*, *gcvH* and *ideR*) to unravel relationships between *Mycobacterium tuberculosis* strains.

It can be concluded from studies such as those outlined above that MLSA can be used to unravel relationships within actinobacterial genera and can be applied to clarify relationships within taxonomically complex taxa, such as *Kocuria* and *Micrococcus*. It can also be used to provide a hierarchical phylogenetic classification of actinobacteria.

4 DETECTION OF ECOTYPES BASED ON MLSAs

It is essential to consider both fundamental and applied aspects of *Streptomyces* systematics, given the importance of these organisms as a source of diverse bioactive metabolites of commercial significance. It is particularly important to understand whether and to what extent the refined taxonomic diversity of streptomycetes is correlated to their functional and ecological diversity. To this end, six endophytic *S. griseus* isolates and two reference strains known to have antagonistic activities and identical 16S rRNA gene sequences were the subject of an MLSA study based on housekeeping and KS-AT genes, respectively, as well as of KS-AT fingerprinting and antimicrobial assays (Rong et al., 2010). Good congruence was found between phylogenies based on the different datasets (Figure 6). Strains FXJ70, FXJ124 and FXJ175, which were isolated from medicinal plants collected from the same county, shared identical sequences for the five genes, and consistently clustered together irrespective of the dataset (Figure 6B–D). Consequently, these strains can be considered to form an ecotype within *S. griseus*, as do each of the five remaining strains included in the study. Similar results were obtained in an MLSA study carried out by Antony-Babu, Stach, and Goodfellow (2008) who found that six alkaliphilic *Streptomyces* strains isolated from four locations within a 60-m transect across a beach and dune sand system were genotypically and phenotypically distinct. The six isolates, which shared almost identical 16S rRNA gene sequences with one another and with representatives of *S. griseus*, were considered to be ecovars of this species. The results also showed that infrasubspecific diversity can be detected over small spatial scales.

The ability of MLSA to recognize ecotypes within the genus *Bacillus* isolated from 'Evolution Canyon' of Israel has been achieved by ecotype simulation (Koeppel et al., 2008). Two sets of genes, *gapA*, *rpoB* and *uvrA*, and *gapA*, *gyrA* and *rpoB* (excluding recombinant sequences), were used to detect ecological

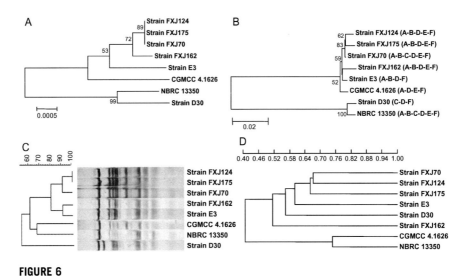

FIGURE 6

Natural diversity of endophytic *S. griseus* isolates based on MLSA phylogenies based on five housekeeping genes (A), KS-AT genes (B), (C) KS-AT fingerprinting and (D) antimicrobial assays.

Data from Rong et al. (2010).

diversity within the *Bacillus simplex* (Haifa, Israel) and *Bacillus subtilis–Bacillus licheniformis* clades that encompassed isolates from the Negev Desert. In each case, the south-facing slope ecotypes shared greater growth rates at stressful high temperatures and a high proportion of isomethyl-branched fatty acids. The concatenated sequence-based divergence can describe diversity within an ecotype. In *Bacillus*, the increased sequence diversity of >98–99% (inflection point) reflects the ephemeral sequence divergence within populations that have not been purged by periodic selection or drift (Koeppel et al., 2008). These results show that MLSA can reflect the evolutionary history of strains and can assess the net rate of ecotype formation (Stefanic et al., 2012). MLSA can also be used to identify clusters of ecological significance within species subject to frequent recombination events, as exemplified by the recognition of ecotypes within *Neisseria* taxa, based on concatenated sequences of seven loci (*abcZ, adk, aroE, fumC, gdh, pdhC* and *pgm*) (Hanage, 2013; Hanage, Fraser, & Spratt, 2006).

5 MLSA BASED ON WHOLE GENOME SEQUENCES

The availability of whole-genome sequence data opens up the prospect of large-scale MLSA studies based on the sequences of housekeeping genes. Alam, Merlo, Takano, and Breitling (2010) undertook a comprehensive genome-based phylogenic analysis of *Streptomyces* and its relatives based on 155 ortholog protein-coding genes from 45 actinobacterial species and 4 outgroup species. The resultant phylogenomic consensus tree, based on an integrated analysis of gene content, gene order and gene concatenation, was in agreement with the traditional taxonomic subdivisions within the phylum *Actinobacteria* albeit in much sharper focus thereby underpinning the value of this approach in the generation of a fully resolved phylogeny. It has also been shown that MLSA studies based on *in silico* analyses of genome sequences of strains within genera can yield reliable phylogenies based on a few genes, as exemplified by a study on *Nocardia* strains based on 12 genes (*atpD, dnaJ, groL1, groL2, gyrB, recA, rpoA, secA, secY, soda, trpB* and *ychF*) derived from 26 genomes (Tamura et al., 2012). However, at present, the majority of large-scale phylogenomic reconstructions are dependent on the skewed sample of whole genome sequences available in public databases, notably GenBank.

MLSA studies can also be constructed on information derived from whole-genome sequence data, such as those based on ribosomal protein subunit genes (rMLST) (Jolley et al., 2012; Read et al., 2013) that allow strains to be assigned to taxonomic ranks from strains to domains; such studies allow the roles that adaptive and virulence genes play in speciation to be explored (Didelot et al., 2012; Penn & Jensen, 2012; Hernandez-Lopez et al., 2013; Kong et al., 2013; Paul et al., 2013).

MLSA is widely used to type the genetic structure of bacterial populations (Kong et al., 2013; Yahara et al., 2012). Doroghazi and Buckley (2010) used MLSA to quantify intra- and inter-species homologous recombination among streptomycetes and found high levels of gene exchange compared to many other bacterial groups (Vos & Didelot, 2009). The intra-species recombination rate was found to have exceeded the inter-species rate by two orders of magnitude, suggesting that patterns of gene exchange and recombination may shape the evolution of streptomycetes. However, the MLSA scheme targets a limited number of metabolic genes that may not be representative of the entire genome. More recently, the whole-genome sequence data have been used to determine the relative importance of recombination in population genetics and to explore the association between genomic evolution and ecological adaptation (Didelot et al., 2012; Shapiro et al., 2012).

In conclusion, it is clear that multi-locus sequence data have a key role to play not only in prokaryotic systematics and evolution but also in ecology. Indeed, MLSAs are undoubtedly taking prokaryotic systematics to the next level.

REFERENCES

Acinas, S. G., Marcelino, L. A., Klepac-Ceraj, V., & Polz, M. F. (2004). Divergence and redundancy of 16S rRNA sequences in genomes with multiple *rrn* operons. *Journal of Bacteriology*, *186*, 2629–2635.

Adachi, J., & Hasegawa, M. (1996). *MOLPHY version 2.3: Programs for molecular phylogenetics based on maximum likelihood.* Tokyo: Institute of Statistical Mathematics.

Alam, M. T., Merlo, M. E., Takano, E., & Breitling, R. (2010). Genome-based phylogenetic analysis of *Streptomyces* and its relatives. *Molecular Phylogenetics and Evolution*, *54*, 763–772.

Alvarez-Perez, S., de Vega, C., & Herrera, C. M. (2013). Multilocus sequence analysis of nectar pseudomonads reveals high genetic diversity and contrasting recombination patterns. *PLoS One*, *8*, e75797.

Antony-Babu, S., Stach, J. E., & Goodfellow, M. (2008). Genetic and phenotypic evidence for *Streptomyces griseus* ecovars isolated from a beach and dune sand system. *Antonie van Leeuwenhoek*, *94*, 63–74.

Bérdy, J. (2005). Bioactive microbial metabolites. *The Journal of Antibiotics*, *58*, 1–26.

Bérdy, J. (2012). Thoughts and facts about antibiotics: Where we are now and where we are heading. *The Journal of Antibiotics*, *65*, 385–395.

Bignell, D. R., Huguet-Tapia, J. C., Joshi, M. V., Pettis, G. S., & Loria, R. (2010). What does it take to be a plant pathogen: Genomic insights from *Streptomyces* species. *Antonie van Leeuwenhoek*, *98*, 179–194.

Blackburne, B. P., & Whelan, S. (2013). Class of multiple sequence alignment algorithm affects genomic analysis. *Molecular Biology and Evolution*, *30*, 642–653.

Boers, S. A., van der Reijden, W. A., & Jansen, R. (2012). High-throughput multilocus sequence typing: Bringing molecular typing to the next level. *PLoS One*, *7*, e39630.

Bucarey, S. A., Penn, K., Paul, L., Fenical, W., & Jensen, P. R. (2012). Genetic complementation of the obligate marine actinobacterium *Salinispora tropica* with the large mechanosensitive channel gene *mscL* rescues cells from osmotic downshock. *Applied and Environmental Microbiology*, *78*, 4175–4182.

Bull, A. T. (2010). Actinobacteria of the extremobiosphere. In K. Horikoshi, G. Antranikian, A. T. Bull, F. T. Robb, & K. O. Stelter (Eds.), *Extremophiles handbook* (pp. 1204–1231). Berlin: Springer-Verlag GmbH.

Caro-Quintero, A., & Konstantinidis, K. T. (2012). Bacterial species may exist, metagenomics reveal. *Environmental Microbiology*, *14*, 347–355.

Carro, L., Sproer, C., Alonso, P., & Trujillo, M. E. (2012). Diversity of *Micromonospora* strains isolated from nitrogen fixing nodules and rhizosphere of *Pisum sativum* analyzed by multilocus sequence analysis. *Systematic and Applied Microbiology*, *35*, 73–80.

Chun, J., & Rainey, F. A. (2014). Integrating genomics into the taxonomy and systematics of the *Bacteria* and *Archaea*. *International Journal of Systematic and Evolutionary Microbiology*, *64*, 316–324.

Cilia, V., Lafay, B., & Christen, R. (1996). Sequence heterogeneities among 16S ribosomal RNA sequences, and their effect on phylogenetic analyses at the species level. *Molecular Biology and Evolution*, *13*, 451–461.

Curtis, S. M., & Meyers, P. R. (2012). Multilocus sequence analysis of the actinobacterial genus *Kribbella*. *Systematic and Applied Microbiology*, *35*, 441–446.

Didelot, X., & Falush, D. (2007). Inference of bacterial microevolution using multilocus sequence data. *Genetics*, *175*, 1251–1266.

Didelot, X., & Maiden, M. C. (2010). Impact of recombination on bacterial evolution. *Trends in Microbiology*, *18*, 315–322.

Didelot, X., Meric, G., Falush, D., & Darling, A. E. (2012). Impact of homologous and non-homologous recombination in the genomic evolution of *Escherichia coli*. *BMC Genomics*, *13*, 256.

Doolittle, W. F. (1999). Phylogenetic classification and the universal tree. *Science*, *284*, 2124–2129.

Doroghazi, J. R., & Buckley, D. H. (2010). Widespread homologous recombination within and between *Streptomyces* species. *The ISME Journal*, *4*, 1136–1143.

Farris, J. S., Källersjö, M., Kluge, A. G., & Bult, C. (1994). Testing significance of incongruence. *Cladistics*, *10*, 315–319.

Felsenstein, J. (1981). Evolutionary trees from DNA sequences: A maximum likelihood approach. *Journal of Molecular Evolution*, *17*, 368–376.

Felsenstein, J. (1985). Confidence limits on phylogenies: An approach using the bootstrap. *Evolution*, *39*, 783–791.

Felsenstein, J. (2008). *PHYLIP (phylogeny inference package). Version 3.68*. Seattle, USA: Department of Genome Sciences, University of Washington.

Ferreira, R., Borges, V., Nunes, A., Nogueira, P. J., Borrego, M. J., & Gomes, J. P. (2012). Impact of loci nature on estimating recombination and mutation rates in *Chlamydia trachomatis*. *G3: Genes, Genomes, Genetics*, *2*, 761–768.

Fitch, W. M. (1971). Toward defining the course of evolution: Minimum change for a specified tree topology. *Systematic Zoology*, *20*, 406–416.

Freel, K. C., Millan-Aguinaga, N., & Jensen, P. R. (2013). Multilocus sequence typing reveals evidence of homologous recombination linked to antibiotic resistance in the genus *Salinispora*. *Applied and Environmental Microbiology*, *79*, 5997–6005.

Genilloud, O. (2012). Family I. *Micromonosporaceae*. In M. Goodfellow, P. Kämpfer, H. J. Busse, M. E. Trujillo, K. Suzuki, W. Ludwig, & W. B. Whitman (Eds.), *The actinobacteria: Vol. 5. Bergey's manual of systematic bacteriology* (2nd ed., pp. 1035–1137). New York: Springer.

Gouveia-Oliveira, R., Sackett, P. W., & Pedersen, A. G. (2007). MaxAlign: Maximizing usable data in an alignment. *BMC Bioinformatics*, *8*, 312.

Guo, Y. P., Zheng, W., Rong, X. Y., & Huang, Y. (2008). A multilocus phylogeny of the *Streptomyces griseus* 16S rRNA gene clade: Use of multilocus sequence analysis for streptomycete systematics. *International Journal of Systematic and Evolutionary Microbiology*, *58*, 149–159.

Hanage, W. P. (2013). Fuzzy species revisited. *BMC Biology*, *11*, 41.

Hanage, W. P., Fraser, C., & Spratt, B. G. (2006). Sequences, sequence clusters and bacterial species. *Philosophical Transactions of the Royal Society of London*, *361*, 1917–1927.

Hartigan, J. A. (1973). Minimum evolution fits to a given tree. *Biometrics*, *29*, 53–65.

Head, I. M., Saunders, J. R., & Pickup, R. W. (1998). Microbial evolution, diversity, and ecology: A decade of ribosomal RNA analysis of uncultivated microorganisms. *Microbial Ecology*, *35*, 1–21.

Hernandez-Lopez, A., Chabrol, O., Royer-Carenzi, M., Merhej, V., Pontarotti, P., & Raoult, D. (2013). To tree or not to tree? Genome-wide quantification of recombination and reticulate evolution during the diversification of strict intracellular bacteria. *Genome Biology and Evolution*, *5*, 2305–2317.

Jensen, P. R., Gontang, E., Mafnas, C., Mincer, T. J., & Fenical, W. (2005). Culturable marine actinomycete diversity from tropical Pacific Ocean sediments. *Environmental Microbiology*, *7*, 1039–1048.

Jolley, K. A., Bliss, C. M., Bennett, J. S., Bratcher, H. B., Brehony, C., Colles, F. M., et al. (2012). Ribosomal multilocus sequence typing: Universal characterization of bacteria from domain to strain. *Microbiology*, *158*, 1005–1015.

Jolley, K. A., & Maiden, M. C. (2010). BIGSdb: Scalable analysis of bacterial genome variation at the population level. *BMC Bioinformatics*, *11*, 595.

Jukes, T. H., & Cantor, C. R. (1969). Evolution of protein molecules. In H. N. Munro (Ed.), *Mammalian protein metabolism: Vol. III*, (pp. 21–132). New York: Academic Press.

Kämpfer, P. (2012). Genus 1. *Streptomyces* Waksman & Henrici, 1943, 339[AL] emend. Witt and Stackebrandt 1990, 370 emend. Wellington, Stackebrandt, Sanders, Wolstrup and Jorgensen 1992, 159. In M. Goodfellow, P. Kämpfer, H.-J. Busse, M. E. Trujillo, K.-I. Suzuki, W. Ludwig, & W. B. Whitman (Eds.), *The actinobacteria: Vol. 5. Bergey's manual of systematic bacteriology* (2nd ed., pp. 1455–1767). New York: Springer.

Kimura, M. (1980). A simple method for estimating evolutionary rates of base substitutions through comparative studies of nucleotide sequences. *Journal of Molecular Evolution*, *16*, 111–120.

Kishino, H., & Hasegawa, M. (1989). Evaluation of the maximum likelihood estimate of the evolutionary tree topologies from DNA sequence data, and the branching order in hominoidea. *Journal of Molecular Evolution*, *29*, 170–179.

Koeppel, A., Perry, E. B., Sikorski, J., Krizanc, D., Warner, A., Ward, D. M., et al. (2008). Identifying the fundamental units of bacterial diversity: A paradigm shift to incorporate ecology into bacterial systematics. *Proceedings of the National Academy of Sciences*, *105*, 2504–2509.

Kong, Y., Ma, J. H., Warren, K., Tsang, R. S., Low, D. E., Jamieson, F. B., et al. (2013). Homologous recombination drives both sequence diversity and gene content variation in *Neisseria meningitidis*. *Genome Biology and Evolution*, *5*, 1611–1627.

Konstantinidis, K. T., & Tiedje, J. M. (2005). Towards a genome-based taxonomy for prokaryotes. *Journal of Bacteriology*, *187*, 6258–6264.

Kyrpides, N. C., Woyke, T., Eisen, J. A., Garrity, G., Lilburn, T. G., Beck, B. J., et al. (2014). Genomic Encyclopedia of Type Strains, Phase I: The one thousand microbial genomes (KMG-I) project. *Standards in Genomic Sciences*, *9*, 1278–1284.

Labeda, D. P. (2011). Multilocus sequence analysis of phytopathogenic species of the genus *Streptomyces*. *International Journal of Systematic and Evolutionary Microbiology*, *61*, 2525–2531.

Labeda, D. P., Goodfellow, M., Brown, R., Ward, A. C., Lanoot, B., Vanncanneyt, M., et al. (2012). Phylogenetic study of the species within the family *Streptomycetaceae*. *Antonie van Leeuwenhoek*, *101*, 73–104.

Lancaster, A., Nelson, M. P., Meyer, D., Thomson, G., & Single, R. M. (2003). *PyPop*: A software framework for population genomics: Analyzing large-scale multi-locus genotype data. *Pacific Symposium on Biocomputing*, *8*, 514–525.

Larsen, M. V., Cosentino, S., Rasmussen, S., Friis, C., Hasman, H., Marvig, R. L., et al. (2012). Multilocus sequence typing of total-genome-sequenced bacteria. *Journal of Clinical Microbiology*, *50*, 1355–1361.

Leigh, J. W., Lapointe, F. J., Lopez, P., & Bapteste, E. (2011). Evaluating phylogenetic congruence in the post-genomic era. *Genome Biology and Evolution*, *3*, 571–587.

Loria, R., Kers, J., & Joshi, M. (2006). Evolution of plant pathogenicity in *Streptomyces*. *Annual Review of Phytopathology*, *44*, 469–487.

Lu, B., Dong, H. Y., Zhao, X. Q., Liu, Z. G., Liu, H. C., Zhang, Y. Y., et al. (2012). A new multilocus sequence analysis scheme for *Mycobacterium tuberculosis*. *Biomedical and Environmental Sciences*, *25*, 620–629.

Macheras, E., Roux, A. L., Bastian, S., Leao, S. C., Palaci, M., Sivadon-Tardy, V., et al. (2011). Multilocus sequence analysis and *rpoB* sequencing of *Mycobacterium abscessus* (*sensu lato*) strains. *Journal of Clinical Microbiology*, *49*, 491–499.

Maiden, M. C. (2006). Multilocus sequence typing of bacteria. *Annual Review of Microbiology*, *60*, 561–588.

Mason, O. U., Di Meo-Savoie, C. A., Van Nostrand, J. D., Zhou, J., Fisk, M. R., & Giovannoni, S. J. (2009). Prokaryotic diversity, distribution, and insights into their role in biogeochemical cycling in marine basalts. *The ISME Journal*, *3*, 231–242.

Michon, A. L., Aujoulat, F., Roudiere, L., Soulier, O., Zorgniotti, I., Jumas-Bilak, E., et al. (2010). Intragenomic and intraspecific heterogeneity in *rrs* may surpass interspecific variability in a natural population of *Veillonella*. *Microbiology*, *156*, 2080–2091.

Parker, J. K., Havird, J. C., & De La Fuente, L. (2012). Differentiation of *Xylella fastidiosa* strains *via* multilocus sequence analysis of environmentally mediated genes (MLSA-E). *Applied and Environmental Microbiology*, *78*, 1385–1396.

Paul, S., Linardopoulou, E. V., Billig, M., Tchesnokova, V., Price, L. B., Johnson, J. R., et al. (2013). Role of homologous recombination in adaptive diversification of extraintestinal *Escherichia coli*. *Journal of Bacteriology*, *195*, 231–242.

Pei, A. Y., Oberdorf, W. E., Nossa, C. W., Agarwal, A., Chokshi, P., Gerz, E. A., et al. (2010). Diversity of 16S rRNA genes within individual prokaryotic genomes. *Applied and Environmental Microbiology*, *76*, 3886–3897.

Penn, K., & Jensen, P. R. (2012). Comparative genomics reveals evidence of marine adaptation in *Salinispora* species. *BMC Genomics*, *13*, 86.

Posada, D., & Crandall, K. A. (1998). MODELTEST: Testing the model of DNA substitution. *Bioinformatics*, *14*, 817–818.

Pritchard, J. K., Stephens, M., & Donnelly, P. (2000). Inference of population structure using multilocus genotype data. *Genetics*, *155*, 945–959.

Read, D. S., Woodcock, D. J., Strachan, N. J., Forbes, K. J., Colles, F. M., Maiden, M. C., et al. (2013). Evidence for phenotypic plasticity among multihost *Campylobacter jejuni* and *C. coli* lineages, obtained using ribosomal multilocus sequence typing and Raman spectroscopy. *Applied and Environmental Microbiology*, *79*, 965–973.

Rokas, A., Williams, B. L., King, N., & Carroll, S. B. (2003). Genome-scale approaches to resolving incongruence in molecular phylogenies. *Nature*, *425*, 798–804.

Rong, X. Y., Doroghazi, J. R., Cheng, K., Zhang, L. M., Buckley, D. H., & Huang, Y. (2013). Classification of *Streptomyces* phylogroup pratensis (Doroghazi and Buckley, 2010) based on genetic and phenotypic evidence, and proposal of *Streptomyces pratensis* sp. nov. *Systematic and Applied Microbiology, 36,* 401–407.

Rong, X. Y., Guo, Y. P., & Huang, Y. (2009). Proposal to reclassify the *Streptomyces albidoflavus* clade on the basis of multilocus sequence analysis and DNA-DNA hybridization, and taxonomic elucidation of *Streptomyces griseus* subsp. *solvifaciens*. *Systematic and Applied Microbiology, 32,* 314–322.

Rong, X. Y., & Huang, Y. (2010). Taxonomic evaluation of the *Streptomyces griseus* clade using multilocus sequence analysis and DNA-DNA hybridization, with proposal to combine 29 species and three subspecies as 11 genomic species. *International Journal of Systematic and Evolutionary Microbiology, 60,* 696–703.

Rong, X. Y., & Huang, Y. (2012). Taxonomic evaluation of the *Streptomyces hygroscopicus* clade using multilocus sequence analysis and DNA-DNA hybridization, validating the MLSA scheme for systematics of the whole genus. *Systematic and Applied Microbiology, 35,* 7–18.

Rong, X. Y., Liu, N., Ruan, J. S., & Huang, Y. (2010). Multilocus sequence analysis of *Streptomyces griseus* isolates delineating intraspecific diversity in terms of both taxonomy and biosynthetic potential. *Antonie van Leeuwenhoek, 98,* 237–248.

Rzhetsky, A., & Nei, M. (1995). Tests of applicability of several substitution models for DNA sequence data. *Molecular Biology and Evolution, 12,* 131–151.

Saitou, N., & Nei, M. (1987). The neighbor-joining method: A new method for reconstructing phylogenetic trees. *Molecular Biology and Evolution, 4,* 406–425.

Sawabe, T., Kita-Tsukamoto, K., & Thompson, F. L. (2007). Inferring the evolutionary history of vibrios by means of multilocus sequence analysis. *Journal of Bacteriology, 189,* 7932–7936.

Schatz, A., Bugie, E., & Waksman, S. A. (1944). Streptomycin, a substance exhibiting antibiotic activity against gram-positive and gram-negative bacteria. *Proceedings of the Society for Experimental Biology and Medicine, 55,* 66–69.

Schmidt, H. A., Strimmer, K., Vingron, M., & von Haeseler, A. (2002). TREE-PUZZLE: Maximum likelihood phylogenetic analysis using quartets and parallel computing. *Bioinformatics, 18,* 502–504.

Shapiro, B. J., Friedman, J., Cordero, O. X., Preheim, S. P., Timberlake, S. C., Szabó, G., et al. (2012). Population genomics of early events in the ecological differentiation of bacteria. *Science, 336,* 48–51.

Shimodaira, H., & Hasegawa, M. (2001). CONSEL: For assessing the confidence of phylogenetic tree selection. *Bioinformatics, 17,* 1246–1247.

Stackebrandt, E., Frederiksen, W., Garrity, G. M., Grimont, P. A., Kampfer, P., Maiden, M. C., et al. (2002). Report of the *ad hoc* committee for the re-evaluation of the species definition in bacteriology. *International Journal of Systematic and Evolutionary Microbiology, 52,* 1043–1047.

Stamatakis, A. (2006). RAxML-VI-HPC: Maximum likelihood-based phylogenetic analyses with thousands of taxa and mixed models. *Bioinformatics, 22,* 2688–2690.

Stefanic, P., Decorosi, F., Viti, C., Petito, J., Cohan, F. M., & Mandic-Mulec, I. (2012). The quorum sensing diversity within and between ecotypes of *Bacillus subtilis*. *Environmental Microbiology, 14,* 1378–1389.

Swofford, D. L. (1998). *PAUP*. Phylogenetic analysis using parsimony (*and other methods). Version 4.* Sunderland, Massachusetts: Sinauer Associates.

Talavera, G., & Castresana, J. (2007). Improvement of phylogenies after removing divergent and ambiguously aligned blocks from protein sequence alignments. *Systematic Biology*, *56*, 564–577.

Tamura, T., Matsuzawa, T., Oji, S., Ichikawa, N., Hosoyama, A., Katsumata, H., et al. (2012). A genome sequence-based approach to taxonomy of the genus *Nocardia*. *Antonie van Leeuwenhoek*, *102*, 481–491.

Tankouo-Sandjong, B., Sessitsch, A., Liebana, E., Kornschober, C., Allerberger, F., Hachler, H., et al. (2007). MLST-v, multilocus sequence typing based on virulence genes, for molecular typing of *Salmonella enterica* subsp. *enterica* serovars. *Journal of Microbiological Methods*, *69*, 23–36.

Ventura, M., Canchaya, C., Tauch, A., Chandra, G., Fitzgerald, G. F., Chater, K. F., et al. (2007). Genomics of *Actinobacteria*: Tracing the evolutionary history of an ancient phylum. *Microbiology and Molecular Biology Reviews*, *71*, 495–548.

Vetsigian, K., Jajoo, R., & Kishony, R. (2011). Structure and evolution of *Streptomyces* interaction networks in soil and in silico. *PLoS Biology*, *9*, e1001184.

Vinuesa, P., Rojas-Jimenez, K., Contreras-Moreira, B., Mahna, S. K., Prasad, B. N., Moe, H., et al. (2008). Multilocus sequence analysis for assessment of the biogeography and evolutionary genetics of four *Bradyrhizobium* species that nodulate soybeans on the asiatic continent. *Applied and Environmental Microbiology*, *74*, 6987–6996.

Vos, M., & Didelot, X. (2009). A comparison of homologous recombination rates in bacteria and archaea. *The ISME Journal*, *3*, 199–208.

Wagner, I. D., Varghese, L. B., Hemme, C. L., & Wiegel, J. (2013). Multilocus sequence analysis of *Thermoanaerobacter* isolates reveals recombining, but differentiated, populations from geothermal springs of the Uzon Caldera, Kamchatka, Russia. *Frontiers in Microbiology*, *4*, 169.

Wayne, L. G., Brenner, D. J., Colwell, R. R., Grimont, P. A. D., Kandler, O., Krichevsky, M. I., et al. (1987). Report of the *ad hoc* committee on reconciliation of approaches to bacterial systematics. *International Journal of Systematic Bacteriology*, *37*, 463–464.

Yahara, K., Kawai, M., Furuta, Y., Takahashi, N., Handa, N., Tsuru, T., et al. (2012). Genome-wide survey of mutual homologous recombination in a highly sexual bacterial species. *Genome Biology and Evolution*, *4*, 628–640.

Yang, Z. (2007). PAML 4: A program package for phylogenetic analysis by maximum likelihood. *Molecular Biology and Evolution*, *24*, 1586–1591.

Yang, Z., & Rannala, B. (1997). Bayesian phylogenetic inference using DNA sequences: A Markov Chain Monte Carlo method. *Molecular Biology and Evolution*, *14*, 717–724.

Yang, Z., & Rannala, B. (2012). Molecular phylogenetics: Principles and practice. *Nature Reviews. Genetics*, *13*, 303–314.

Zhi, X. Y., Zhao, W., Li, W. J., & Zhao, G. P. (2012). Prokaryotic systematics in the genomics era. *Antonie van Leeuwenhoek*, *101*, 21–34.

Bacterial Typing and Identification By Genomic Analysis of 16S–23S rRNA Intergenic Transcribed Spacer (ITS) Sequences

Volker Gürtler[*,1], **Gangavarapu Subrahmanyam**[‡], **Malathi Shekar**[†], **Biswajit Maiti**[‡], **Indrani Karunasagar**[‡]

School of Applied Sciences, RMIT University, Bundoora Campus, Melbourne, Victoria, Australia
[†]*UNESCO-MIRCEN for Marine Biotechnology, College of Fisheries, Karnataka Veterinary, Animal and Fisheries Sciences University, Mangalore, Karnataka, India*
[‡]*Faculty of Biomedical Science, Nitte University Centre for Science Education and Research, University Enclave, Medical Sciences Complex, Deralakatte, Mangalore, Karnataka, India*
[1]*Corresponding author: e-mail address: volker.gurtler@gmail.com*

Abbreviations

5S–5S	region between the two 5S rRNA genes post-23S rRNA when there are two 5S rRNA genes per *rrn* operon
ITS1	16S rRNA–23S rRNA intergenic transcribed spacer
ITS2	23S–5S rRNA intergenic transcribed spacer (when there are two 5S rRNA genes post-23S rRNA, the spacer is relative to the 5S rRNA immediately adjacent to the 23S rRNA gene)
NCBI	National Centre for Biotechnology Information (USA)
nt	nucleotide
Post-5S rRNA	region 3' of the end of the 5S rRNA gene until the next defined gene
Pre-16S rRNA	region 5' of the beginning of the 16S rRNA gene until the next defined gene
RDP	Ribosomal Database Project
RFLP	restriction fragment length polymorphism
SNP	single-nucleotide polymorphism
WGS	whole-genome sequences

1 INTRODUCTION

The *rrn* operon is one of the most widely studied operons in the bacterial genome having been studied at the level of ribosome structure (Gutell, Lee, & Cannone, 2002; Wimberly et al., 2000), ribosomal RNA (rRNA) gene transcription (Condon, Squires, & Squires, 1995; Evguenieva-Hackenberg, 2005), rRNA gene organisation (Martín, Barreiro, González-Lavado, & Barriuso, 2003), *rrn* operon organisation at the level of the bacterial whole genome (Gurtler & Grando, 2013) and the use of the 16S rRNA gene sequence in evolutionary and phylogenetic studies, pioneered by Woese (1987). Indeed, the 16S rRNA gene sequence database is the cornerstone of evolutionary phylogenetics and bacterial taxonomy (Amann, Ludwig, & Schleifer, 1995; Baker, Smith, & Cowan, 2003; Woese, 1987). The *rrn* operon consists of one 16S rRNA gene, one 23S rRNA gene and one 5S rRNA gene in that order in most alleles and in most bacteria examined to date (Deutscher, 2009; Gurtler & Stanisich, 1996; Martín et al., 2003). As a result of the operon gene organisation, there are intergenic transcribed spacer (ITS) sequences from the end of the 16S rRNA gene to the beginning of the 23S rRNA gene (Gurtler & Stanisich, 1996) and the end of the 23S rRNA gene to the beginning of the 5S rRNA gene (ITS1 and ITS2, respectively). In some species, one or two of the *rrn* operons contains two 5S rRNA genes (see Table 1). Some *rrn* operons have ITS1 sequences that contain transfer RNA (tRNA) genes (Gurtler & Stanisich, 1996).

A diagram of the *rrn* operon that represents the general gene organisation in most bacteria examined to date is shown in Figure 1 (Deutscher, 2009; Gurtler & Stanisich, 1996; Martín et al., 2003). Figure 1 shows that the length of the 16S and 23S rRNA genes and a number of sequence regions within them are remarkably conserved within bacteria (Deutscher, 2009; Gurtler & Stanisich, 1996; Martín et al., 2003; Woese, 1987). However, the sequences between these conserved regions vary at a constant rate, dependent on time and the relatedness of bacterial species (Woese, 1987). To further complicate matters, the number of *rrn* operons within a bacterial genome varies between bacterial species (Table 1; Figure 2) and it has been argued that differences between 16S rRNA genes within the same genome are "unlikely to have a profound effect on the classification of taxa" (Coenye & Vandamme, 2003). The possibility remains that this intra-genomic heterogeneity of the *rrn* operon may affect any or all of the biological processes mentioned above, since only a small proportion of the estimated number of bacterial species (and an even smaller proportion of the strains within some of these species) have had whole-genome sequences (WGS) completely sequenced and analysed at the level of the *rrn* operon (Table 1; Figure 2; Coenye & Vandamme, 2003).

In order to study the large amount of diverse data that is generated when analysing multi-copy *rrn* operons, a user-friendly bioinformatic tool and database were developed and previously reported by Gurtler and Grando (2013) that could store, present and add new data to a growing bacterial *rrn* database. This database has been named RiboTyping (RT) and is accessible via an App that runs on FileMaker Go

Table 1 Intra- and interspecies variation in the number of *rrn* operons present in bacterial chromosomes taken from whole-genome sequence data

Bacterial species	Number of strains	Total rRNAs	Total *rrn* operons	Total 16S	Total 23S	Total 5S
Bacillus anthracis	5	33	11		11	
	1	30	10		10	
Bacillus cereus	4	42	14		14	
	5	39	13		13	
	3	36	12		12	
	1	33	11		11	
Bacillus subtilis	9	30	10	10		10
	1	31	10	10		11
	2	27	9		9	
	1	24	8		8	
Bacillus thuringiensis	1	45	15		15	
	5	42	14		14	
	2	39	13		13	
	1	41+	14		14	
	2	36	12		12	
Campylobacter coli	1	9	3		3	
	1	11	3	4	3	4
*Campylobacter jejuni**	12	9	3		3	
	1	3	1	1	1	1
Clostridium botulinum	2	34	11		11	12
	1	30	10		10	
	7	27	9		9	
	2	24	8		8	
	1	20	3	9	2	9
Clostridium difficile[1]	1	36	12		12	
	3	30	10		10	
	3	27	9		9	
	1	50	8	19	20	11
Edwardsiella tarda	3	25	8	8		9
Enterococcus faecalis	5	12	4	4	4	4
Enterococcus faecium	4	18	6		6	
Escherichia coli	52	22	7		7	8
	1	23	7			9
	4	21	7		7	
	1	19	6	6		7
Listeria monocytogenes	23	18	6		6	
	3	15	5		5	
Methanococcus maripaludis[#]	4	10	3		3	4
	1	7	2		2	3
Mycobacterium gilvum Spyr1	2	6	2		2	
Mycobacterium smegmatis	4	6	2		2	

Continued

Table 1 Intra- and interspecies variation in the number of *rrn* operons present in bacterial chromosomes taken from whole-genome sequence data—cont'd

Bacterial species	Number of strains	Total rRNAs	Total *rrn* operons	Total 16S	Total 23S	Total 5S
Mycobacterium tuberculosis	1	4		1		2
	26	3	1		1	
	3	2		1		0
Pyrococcus furiosus*	2	4	1	1		2
Riemerella anatipestifer	4	9	3		3	
Salmonella bongori	2	22	7	7		8
Salmonella enterica subsp. enterica serovar	1	24	8	8	8	8
	2	21			7	
	1	20		7	6	7
	33	22	7		7	8
	1	19		7	4	8
	1	23		8	7	8
Staphylococcus aureus	7	19	6	6		7
	2	18			6	
	3	17		5		7
	21	16	5	5		6
	10	15			5	
	1	12	4	4		
Streptococcus pneumoniae	24	12	4		3	
Vibrio cholerae	5	25	8	8		9
	3	22	7	7		8
Vibrio parahaemolyticus	2	34	11	11		12
Vibrio vulnificus	3	28	9	9		10
Yersinia enterocolitica	3	22	7	7		8
	3	22				
Yersinia pestis	1	24	8		8	
	6	22	7	7		8
	3	20		6		
	2	19	6	6		7
	1	18			6	
Yersinia pseudotuberculosis	4	22	7	7		8

*2 more genomes are available, however, 16s rRNA features are not yet annotated; ¥Archaeal species.

Taken from NCBI - Genome using the "Genome Project Report" to identify completed chromosomes (filled circle) and then "Genome Annotation Report" to identify the "Number of rRNA genes" (16S, 23S and 5S) found on each chromosome.

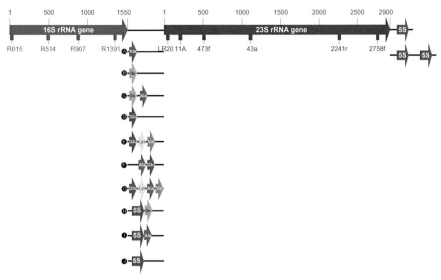

FIGURE 1

Schematic map of the ribosomal RNA operon gene organisation for most bacteria. In most bacteria, the 16S, 23S and 5S rRNA genes are all present in the *rrn* operons, in a single copy per operon, with approximate lengths highly conserved within genomes and species but varying more between species. The variations in intergenic spacers between the 16S and 23S rRNA genes (ITS1) have been grouped according to the type of tRNA gene they contain, with examples of species that have operons with these types of ITS sequences summarised in Gurtler (1999) and further examples for each type given shown in the diagram as follows: (A) *C. difficile* (Gurtler & Grando, 2013); (B) and (C) *Staphylococcus aureus* (Gurtler & Barrie, 1995); (C) and (D) *Escherichia coli* (Condon, Squires, & Squires, 1995) and *Salmonella typhimurium* (Christensen, Moller, Vogensen, & Olsen, 2000; Perez Luz, Rodriguez-Valera, Lan, & Reeves, 1998); (D)–(G) *Vibrio* species (Chun, Huq, & Colwell, 1999; Gonzalez-Escalona, Romero, Guzman, & Espejo, 2006); and (H)–(J) *Butyrivibrio proteoclasticus* (Kelly et al., 2010; Li et al., 2014). The line without tRNA gene boxes at the top refers to many alleles in many species which do not contain any tRNA genes. In most alleles in most bacterial species, there is only one 5S rRNA gene, however, in some bacterial species (Table 1, last column), one operon has two 5S rRNA genes in tandem. The black lines refer to the non-coding regions ITS1 (16S–23S intergenic spacer region) and ITS2 (23S–5S intergenic spacer region). The conserved sequence regions that are marked in the 16S and 23S rRNA genes have been taken from (i) Gurtler and Stanisich (1996) and Gurtler, Wilson, and Mayall (1991) for the 16S rRNA gene and from (ii) Gurtler and Stanisich (1996) and Hunt et al. (2006) for the 23S rRNA gene. They include (i) 16S: R015, (AGAGTTTGATCCTGGCTCAG); R514, (GTGCCAGCAGCCGCGGTAA); R907, (AAACTCAAAGGAATTGACGG); and R1391, (TTGTACACACCGCCCGTC); (ii) 23S: LR20, (GCGGATGCCTTGGCACTAG); 11A, (GGAACTGAAACATCTAAGTA); 473f, (AGTACCGTGAGGGAAAG); 43a, (GGATGTTGGCTTAGAAGCAG); 2241r, (AGTTTGACTGGGGCGGT); and 2758f, (CTGAAAGCATCTAA). (See the colour plate.)

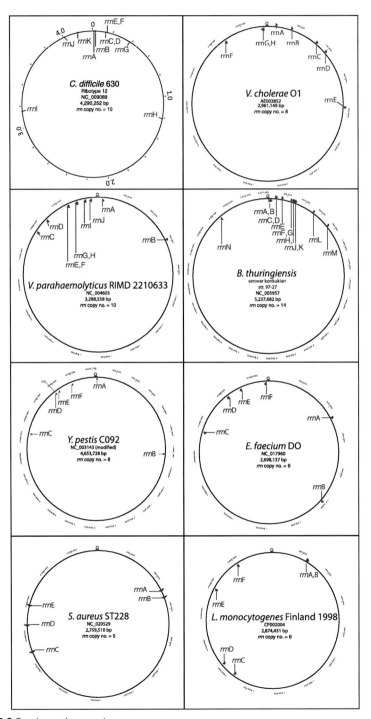

FIGURE 2 See legend on next page.

(Apple iPad, iPhone) and FileMaker 12 (Apple OSX and Microsoft Windows). It was briefly outlined using *Clostridium difficile* as an example in the review by Gurtler and Grando (2013) and its method of construction will be described in detail here. The details of the construction will be described in Section 2. The resultant database will then be described in Section 3. How this database can be applied to the typing of microorganisms will be discussed in Section 4.

2 METHODS

The "RiboTyping App" presented in this chapter addresses fundamental methodological problems faced with searching, analysing, saving, importing/exporting (Figure 3) and presenting large data sets of *rrn* operon DNA sequences (Figures 4–6). A separate but related problem is the fact that each genome has multiple copies of the same *rrn* operon at different locations. These operons have been termed "alleles" because they differ by the presence of single-nucleotide polymorphisms (SNPs) (both inside the operon and flanking the operon), insertions/deletions (indels) and chromosomal location. The principal of these methodological problems are explained in the next sections, expanded in the Appendix (Supplementary materials on http://dx.doi.org/10.1016/bs.mim.2014.07.004) for this section in sufficient detail to be repeated experimentally and solved by the "RiboTyping App".

2.1 SEARCH AND DOWNLOAD BACTERIAL WHOLE-GENOME SEQUENCES

Before downloading from the National Centre for Biotechnology Information (NCBI) genome database (take note of the precautions stated in the Appendix (Supplementary materials on http://dx.doi.org/10.1016/bs.mim.2014.07.004) for this section), the information provided about rRNA operons is a valuable way for determining whether a genome is complete for total rRNA operon analysis. Only complete genome sequences with a full operon set are suitable. The basic numbers obtained from NCBI concerning the number of rRNA genes can be checked using Geneious, firstly by simply counting annotated rRNA genes. The second way is to search for conserved regions in the 16S rRNA and 23S rRNA genes

FIGURE 2

Whole-genome maps for eight bacterial species showing the positions of *rrn* operons. The method used to construct and produce the maps is described in detail in Section 2.11. For all genomes, the origin of replication is shown by "□" and all *rrn* operons have been named in alphabetical order beginning with *rrn*A for the operon closest to the origin. At least one pair of operons are arranged in tandem in *C. difficile* (*rrn*C,D; *rrn*E,F), *Vibrio cholerae* (*rrn*G,H), *Vibrio parahaemolyticus* (*rrn*E,F; *rrn*G,H), *Bacillus thuringiensis* (*rrn*C,D; *rrn*F,G; *rrn*H,I; *rrn*J,K) and *L. monocytogenes* (*rrn*A,B).

FIGURE 3

A schematic map showing the origin of the sequence data, the programs used to analyse the sequence data and the final deposition into FileMaker to construct the RiboTyping database. The method used to obtain "GenBank" whole-genome sequences is described in Section 2.1. These sequences are then imported into Geneious and analysed as described in Sections 2.2–2.7. The graphics files are exported into Illustrator CS5 and then imported into FileMaker according to Section 2.11. The annotations are exported into Excel and then imported into FileMaker according to Section 2.8.

(see Figure 1 and the Appendix [Supplementary materials on http://dx.doi.org/10.1016/bs.mim.2014.07.004] for this section) to detect fragmented or orphaned rRNA genes (Gurtler & Grando, 2013) that had not been annotated.

2.2 ANNOTATION OF *rrn* ALLELES

In order to differentiate between multiple *rrn* alleles in the genomes of bacterial strains, each operon is designated a letter starting with "A" for the operon closest to the origin (*rrn*A) and continuing on in alphabetical order (Figure 2). Even though the order of the genes on chromosomes relative to the *rrn* operons may vary from strain, this method standardises the order in which the alleles are named and makes it possible to determine if recombination has occurred when each operon has different sequence properties. The procedure (see the Appendix [Supplementary materials on http://dx.doi.org/10.1016/bs.mim.2014.07.004] for this section) after the WGS has been imported into Geneious is to add extra annotations to the extra genic regions (pre-16S rRNA, ITS1, ITS2, 5S-5S rRNA and post-5S rRNA).

2.3 EXTRACTION OF THE GENE COMPONENTS (16S, 23S AND 5S) AND EXTRA GENIC REGIONS (ITS1, ITS2, PRE-16S AND POST-5S) THAT MAKE UP *rrn* ALLELES

Separate gene components (16S rRNA gene, 23S rRNA gene and 5S rRNA gene), as well as extra genic regions (ITS1, ITS2, pre-16S rRNA gene and post-5S rRNA gene), can be extracted for separate analysis in order to determine if there are sequence differences between alleles in the discrete regions, keep homologous regions together and to limit the number of gaps introduced into alignments. Once extra annotations have been added for extra genic regions, the procedure for extraction is as outlined in the Appendix (Supplementary materials on http://dx.doi.org/10.1016/bs.mim.2014.07.004) for this section.

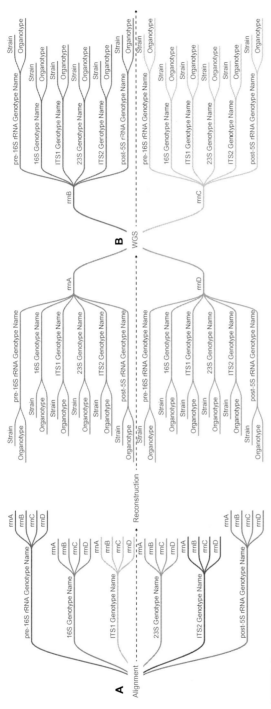

FIGURE 4

Aligning, and then reconstructing the analysed Genotype Names back into their respective operon units from their respective genomes, organotypes and strains. (A) Alignment. Alignments are produced for the sequences of each region (pre-16S rRNA, 16S rRNA, ITS1, 23S rRNA, ITS2 and post-5S rRNA) for all the operons, genomes and strains of a species. This produces alignments with sequences that are sometimes identical within genomes and sometimes between genomes. Other sequences are different between and within genomes and these are labelled "Genotype Names" corresponding to the sequence region. Part of the analysis of the alignments is to summarise the occurence of identical sequences so that they occur only once in the alignment but still remain in the database (these multiples are from different operons, strains, genomes and organotypes). (B) Reconstruction. The aligned sequences now assigned unique "Genotype Names" are now sorted by their respective fields (Genome, operon, strain and organotype) in order to reconstruct them back into their original order. (A) and (B) are two independent figures with the colours not related between figures. (See the colour plate.)

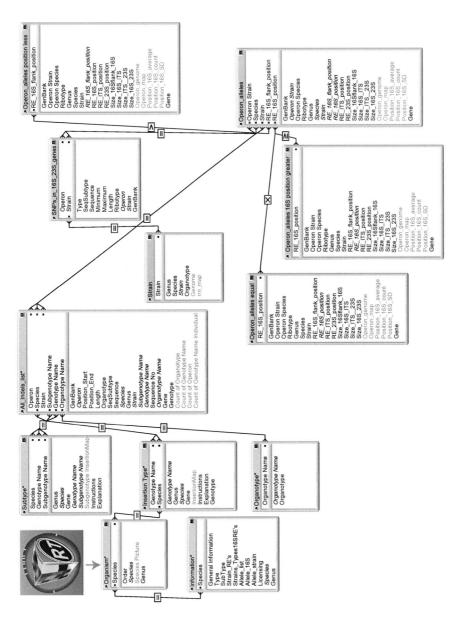

FIGURE 5

RiboTyping database tables, fields and their relationships. Boxes outlined in yellow ("*" for the print version) refer to tables, fields and relationships, respectively, for the entire database, whereas those boxes outlined in grey only refer to the *C. difficile* records. Each box is a named database table (Excel spreadsheet equivalents are listed in Table 2) with fields in bold and summary fields in grey. In each table, the top half of the box shows related fields connected to other tables using the following comparative operators: (=), values in match fields are equal; (<), values in the left match field are less than values in the right match field; (X), all records in the left table are matched to all records in the right table, regardless of the values in the match fields; (□), values in the left match field are greater than or equal to values in the right match field. The FileMaker 12 manual should be consulted for more information.

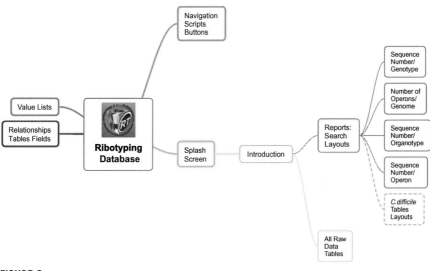

FIGURE 6

The structure of the RiboTyping database. More detail for various components of the database are given in Figure 5 and in Appendix A1–A4 (Supplementary materials on http://dx.doi.org/10.1016/bs.mim.2014.07.004).

2.4 ALIGNMENT OF *rrn* GENE COMPONENTS AND EXTRA GENIC REGIONS

It was found that the best algorithm for performing alignments in Geneious was MUSCLE. The aim is to minimise the number of gaps in the alignment, mostly resulting in moving sequence blocks so that they are contiguous. Mosaic sequence blocks (Sadeghifard, Gurtler, Beer, & Seviour, 2006), often composed of tRNA genes, make this task more obvious. Furthermore, the best alignments are obtained by performing and then editing an alignment on all the allele sequences on individual genomes first, before aligning these multiple alignments (i.e. for 5 single genomes each with 11 *rrn* operons, 5 ITS alignments are produced each with 11 ITS sequences that are aligned to each other to produce a final alignment that compares all 5 genomes containing 55 ITS sequences). Suggestions on how to organise files and folders in Geneious are given in the Appendix (Supplementary materials on http://dx.doi.org/10.1016/bs.mim.2014.07.004) for this section.

2.5 EDITING OF *rrn* GENE COMPONENTS AND EXTRA GENIC REGION ALIGNMENT FILES

The aim of this section is to determine the number of sequence types (field named "Genotype Name" in RiboTyping) present in each alignment of sequences from an *rrn* gene component or an extra genic region (Figure 4A). In many cases, one

Table 2 Origin of data from Geneious into the RiboTyping database

Program origin of field name	Geneious						
Table name and origin of field name		**RiboTyping database**					
	Annotations	**Information**	**All_indels_list**	**Organism**[a]	**Subtype**[a]	**Insertion Type**[a]	**Organotype**[a]
Sequence Name Name			Sequence Name Name				Section 2.8 explains how to extract the information in this field (from the Sequence Name field from Geneious) to the respective fields marked by asterisks, e.g. S01 or 16S123 name of sequence type or SNP
Type			Type				ITS1-long; ITS1-long indel; polymorphism; ITS1-short indel; ITS1-short; indel
Length			Length				Length of nucleotide string – from 1 to 3000 nucleotides
Sequence			Sequence				Nucleotide sequence

GENBANK*				Organotype	GenBank accession number
					16S RNA gene; 23S rRNA gene; intergenic 16S–23S transcribed spacer (ITS1)
GENE*					For example, *rrnA*, *rrnB*; see Figure 2
Serotype*				Organtype name	Designation dependent on the organotype name
Operon*					Serotype, ribotype, biotype
See Figure 5 for all fields	See Figure 5 for all fields	See Figure 5 for all fields	See Figure 5 for all fields	See Figure 5 for all fields	See Figure 5 for all fields

*These tables contain records limited to the number of automatically generated value list members based on the related fields they contain (see Section 2).

"Genotype Name" may be present in the same strain, allele or operon but to completely substantiate this assumption it must be tested. To do this, the alignment must only show those sequences that are unique, they must be given a "Genotype Name" (e.g. for 16S rRNA 16S1, 16S2, etc.) and they must be linked to all the other fields in the database to determine any relationships. The reconstruction of the "Genotype Name" information (Figure 4B) is the second and most important aim of this section. Each gene component and extra genic region is treated separately and separate "Genotype Names" are assigned (e.g. for 16S rRNA 16S1, 16S2, etc.). With all the information linked to each individual sequence, it is then possible to reconstruct the order of the sequences into the Original Strain, Organotype, WGS, Operon Gene Component and Extra Genic Region. How this information is analysed, sorted and presented is the subject of Sections 2.6–2.11. Instructions covering the interactive use of Excel, Geneious and FileMaker for linking the database field "Genotype Name" to the other database fields are explained in detail in the Appendix (Supplementary materials on http://dx.doi.org/10.1016/bs.mim.2014.07.004) for this Section.

2.6 EXPORTING ANNOTATIONS AND ASSOCIATED DATA FROM "GENEIOUS" TO "EXCEL"

This section outlines how the RiboTyping database is constructed and will also be of great benefit to users that require the capability of importing their own data into RiboTyping. In future, the aim is to provide this importing capability directly from an iPad. The large amounts of data that need to be compared within and between genomes means that Geneious is insufficient and that data need to be exported and stored elsewhere for final presentation and analysis. This means that the way that the data is labelled in Geneious is important so that it is easily extracted, imported/exported into Excel/FileMaker, respectively. From Geneious, there is an option for some of the fields to be easily exported directly into Excel, however, some fields need their file names extracted from Geneious then imported into Excel. The Excel formulas for performing these routines are listed in the Appendix (Supplementary materials on http://dx.doi.org/10.1016/bs.mim.2014.07.004) for this section. This explains how to add data into a table (all_indels_list) that contains all the data containing fields (Figure 5). To help with the organisation and analysis of the data, some of the fields and their respective information need to be also added into related tables (Organism, Insertion Type, Subtype, Organotype, Strain, SNPs and Operon_alleles) as shown in Figure 5. The way this is set up in FileMaker and exported using Excel is explained in detail in the Appendix (Supplementary materials on http://dx.doi.org/10.1016/bs.mim.2014.07.004) for this section.

2.7 EXPORTING AND DRAWING GENOME, RESTRICTION AND ALIGNMENT MAPS IN ILLUSTRATOR

Many figures are generated for each collection of strains making it necessary to edit, store and find new ways of presentation. One way in which this can be done is to use the genome field as a container field for WGS with the positions of *rrn* operons marked making it possible to scroll through the records and observe "movie-like" operon position differences. Maps of the ITS can be constructed in a similar way. To do this, it is necessary to export the graphics files from Geneious into Illustrator for editing and then import into RiboTyping. This results in a database that contains diagrams, maps and DNA sequences.

2.8 DESIGN OF EXCEL TABLES FOR FILEMAKER DATABASE CONSTRUCTION

Further to Section 2.6, this section goes into greater detail about how the related tables (Organism, Insertion Type, Subtype, Organotype, Strain, SNPs and Operon_alleles) are constructed (Figure 5). See the Appendix (Supplementary materials on http://dx.doi.org/10.1016/bs.mim.2014.07.004) for this section for more detailed instructions.

2.9 CONSTRUCTION OF FILEMAKER DATABASE FROM EXCEL TABLES

The tables constructed in Section 2.8 make up the skeleton of the RiboTyping database and Figure 6 shows how a RiboTyping database can be constructed using this skeleton with the other tools provided by FileMaker. The value lists are derived from the respective field values making data entry, data sorting and data searching simpler. The relationships can be assigned to specific fields from one table to another table. Navigation, sorting and searching can be done with the help of scripts operated by the user through buttons on the iPad interface. The final reports are displayed on a number of search layouts that can be easily navigated to and from by buttons. See Figure 6 for a list of the report layouts that can be constructed and the Appendix (Supplementary materials on http://dx.doi.org/10.1016/bs.mim.2014.07.004) for this section for more detailed construction instructions.

2.10 DESIGN OF FILEMAKER STATISTICAL AND GRAPHICAL REPORTS

The design of the statistical reports is dictated by the information that is required from the data and is summarised in Figure 6 reports (Sequence Number/Genotype; Number of Operons/Genome; Sequence Number/Organotype and Sequence Number/Operon). The Sequence Number is the total number of sequences found such that in a group of sequences with the same "Genotype Name", the number of sequences found may be much greater than one and may include sequences from different operons, strains, genomes or organotypes. Each layout has a number of search fields for entering all (or only one) search criteria and a different emphasis of fields that have been sorted and presented based on the specified relationships in each layout. These layouts can be constructed according to standard FileMaker procedures and more detail is given in the Appendix (Supplementary materials on http://dx.doi.org/10.1016/bs.mim.2014.07.004) for this section.

2.11 A NEW WAY OF REPORTING LARGE AMOUNTS OF STATISTICAL AND GRAPHICAL INFORMATION IN FILEMAKER GO (SEE APPENDIX B4 FOR INSTRUCTIONS ON HOW TO USE RIBOTYPING, INCLUDED AS APPENDIX B1 (SUPPLEMENTARY MATERIALS ON HTTP://DX.DOI.ORG/ 10.1016/BS.MIM.2014.07.004))

The WGS data imported from GenBank is presented and linked together relationally in the following ways: (i) Maps of operons, genomes and restriction sites from individual strains, alleles and species; (ii) lists of nucleotide (nt) sequences showing differences; and (iii) statistical summary of related fields making it possible to compare alleles between strains and within genomes.

3 RESULTS

3.1 SEQUENCE ACQUISITION AND PREPARATION

The importance of the points made in the introduction about the genomic location and the number of *rrn* operons per genome (Figures 1 and 2) must be re-emphasised in relation to obtaining and preparing GenBank WGS from PubMed where they have not had ITS or conserved regions annotated. Once the whole-genome sequences are edited according to Section 2.1, the components of each *rrn* operon can then be extracted and aligned.

3.2 EXTRACTION OF *rrn* OPERON GENES AND SEQUENCE ALIGNMENT

It is important that each *rrn* operon is named consistently for each genome so that comparisons can be made between genomes for operons that reside at similar or even identical genome positions. Once named, the 16S rRNA, ITS and 23S rRNA components can be extracted for three respective alignments. It is at this stage that the naming of each operon becomes critical because the order of phylogenetic clusters will be different for each of the three alignments after a phylogenetic tree has been constructed. These phylogenetic trees form the basis of determining sequence types for each of the three components (16S rRNA, ITS and 23S rRNA). Once the alignments have been made for each of the three *rrn* components, the constructed phylogenetic trees identify identical sequences. The importance of this is that these identical sequences can then be grouped into sequence types called "Genotype Name" in the database. Using the sorting facility of the RiboTyping database an alignment can be more easily constructed which is only based on the single sequence types with a unique "Genotype Name" (see Gurtler & Grando, 2013; Figures 2 and 3). The "Genotype Name" field then allows the sequence records to be relationally sorted by their sequence identity. This will vary between the three *rrn* components and so once the components have been extracted and imported into the RiboTyping database, it is important to have an accurate way of reconstructing the connections of the components with their respective metadata. By using related fields, the RiboTyping database does this in an efficient and accurate way.

3.3 ANNOTATION OF SEQUENCE DIFFERENCES IN THE *rrn* OPERONS

Once these simplified alignments have been constructed for each of the *rrn* components, the sequence differences between sequence types with a unique "Genotype Name" can be annotated in Geneious. This involves annotating every sequence with the respective "Genotype Name". Furthermore, to study differences at different Genotype levels, the database is also divided up at the level of Genotype to study different genetic levels (e.g. Whole gene, ITS1-long, ITS1-short, polymorphism and other Genotypes: see Figure 5 and Appendix 1A [Supplementary materials on http://dx.doi.org/10.1016/bs.mim.2014.07.004]). This makes it possible to divide the database up further so that individual polymorphisms can be extracted (see *C. difficile* records in Appendix B1 [Supplementary materials on http://dx.doi.org/10.1016/bs.mim.2014.07.004] RT Database). However, to do this each individual polymorphism needs to be annotated in Geneious on the respective *rrn* component alignment. In Figure 5, the tables in "*" show the *C. difficile* records that were used to define the polymorphisms in the whole *rrn* operon.

The *rrn* operons are also annotated according to their position from the origin of replication (Figure 2) such that the operon closest to the origin is named *rrn*A, then the next *rrn*B and the next *rrn*C and so on. This has to be done for each genome so that nomenclature is consistent. Annotation for the ITS1 sequence is also added for each *rrn* operon. This means that when each *rrn* component is extracted, metadata associated with each component is still able to identify which allele it originated from. Furthermore, when graphical presentations (see Section 3.4) of genomes are made they can be associated with their respective strains easily.

3.4 GRAPHICAL PRESENTATION OF ALIGNMENTS, OPERON RESTRICTION MAPS AND WHOLE-GENOME MAPS

A unique feature of the RiboTyping database is the ability to store and present graphical presentations of the data that have been constructed in other programs. One of the major difficulties of presenting and analysing the data produced from comparisons of *rrn* operons is the large number of similar diagrams that could be easily viewed by superimposing them on top of each other, to produce a "movie file" (see Appendix B1 (Supplementary materials on http://dx.doi.org/10.1016/bs.mim.2014.07.004) RT Database for numerous examples where sequence alignments have been overlaid on top of each other). These "movie files" are constructed by adding into the "Insertion Type" table (see Figure 5 fields in grey) a separate alignment indel graphic file to each "InsertionMap" field of each record with a unique "Genotype Name". Similarly, a separate alignment SNP graphic file to each "Subgenotype Insertion Map" field of each record with a unique "Subgenotype

Name" can be added to the "Subtype" table and a separate WGS map and *rrn* map graphic file to each "Genome" and "*rrn*_map" field respectively of each record with a unique "Strain". All these fields can be accessed on layouts navigated from buttons found on the original "Information" and "Type Map_Cd" layout pages from which the RiboTyping database first opens. From the "Type Map_Cd" layout page, specific finds and sorts can be accessed using the black buttons and user-defined finds and sorts can be accessed using the red buttons.

3.5 STATISTICAL PRESENTATION OF ANNOTATION AND METADATA, LINKING ALIGNMENTS, WHOLE GENOMES, *rrn* OPERON ALLELES, STRAINS AND ORGANOTYPES

Relationships, sub-summary reports and charts in FileMaker can be used to compare various metadata parameters, the details of which are shown in Appendix A2 (Supplementary materials on http://dx.doi.org/10.1016/bs.mim.2014.07.004). As described in Section 3.4, there are layouts with find requests and sorts already defined (black buttons) and layouts that can have find requests user-defined (red buttons). Both red and black buttons include four layout types: (i) Number of operons per genome; (ii) number of sequences per "Genotype Name"; (iii) number of sequences per "Organotype" (e.g. for *Listeria monocytogenes*, it is the number of sequences with a specific "Genotype Name", sorted and then grouped in each graph by "serotype"); (iv) number of sequences per Operon. With the black buttons that have specific find requests only, in this example *L. monocytogenes* was included in the report. However, navigation back to "Type Map_Cd" layout page will restore the previous find and sort requests until new ones are made. By navigating to the respective report activated by the red button, customising the find request on this page and leaving the sort setting the same, it is possible to enter new find requests for the same four sort/chart criteria stated above by using a combination of the red and black buttons.

Currently, the RiboTyping database is capable of generating complex relational reports for a number of criteria about the number of sequences present per "Genotype Name", "Organotype" and "Operon". The RiboTyping database can also determine the number of operons per genome. Accordingly the RiboTyping database can give information about which operon and strain in which these sequence differences occur and the frequency they occur. A list of the sequences is generated and all the information is accessible from one page, much in graphical format.

4 DISCUSSION

The RiboTyping database described in this chapter makes the novel contribution of combining and extending the features of previous databases to analyse and present multiple copies of bacterial *rrn* operons from complete WGS. The databases previously available for *rrn* operon analysis and presentation have not taken into account the number of copies per genome of the sequence in the database and were primarily for alignment and analysis of separate 16S rRNA gene sequences, including the Ribosomal Database Project (RDP) for collating 16S rRNA gene sequences for bacterial taxonomy and 28S rRNA gene sequences for fungal taxonomy (Cole et al., 2009), SILVA for aligning 16S and 23S ribosomal RNA sequence data (Pruesse et al., 2007), Greengenes for identifying chimeric 16S rRNA genes (Ashelford, Chuzhanova, Fry, Jones, & Weightman, 2005; DeSantis et al., 2006), RIDOM for 16S ribosomal differentiation of medical microorganisms (Harmsen, Rothganger, Frosch, & Albert, 2002) and the 16SpathDB 2.0 database of 16S rRNA genes from 689 whole bacterial genomes (Teng et al., 2014). The Ribosomal RNA Database (*rrn*DB) does catalogue the number of 16S rRNA, 23S rRNA, 5S rRNA, tRNA and ITS regions that are present in specific bacterial genomes (Lee, Bussema, & Schmidt, 2009). However, it does not perform alignments or any other analyses on these sequences. In addition, there are a number of databases that analyse and align ITS sequences including the Ribosomal Intergenic Spacer Sequence Collection (RISSC) which is a collection of ITS sequences not from WGS (Garcia-Martinez, Bescos, Rodriguez-Sala, & Rodriguez-Valera, 2001), IWoCS (D'Auria, Pushker, & Rodriguez-Valera, 2006) and which analyses mosaic sequences in bacterial ITS sequence regions, and two studies that have compiled sequence difference information in database formats for mycobacteria (Gurtler, Harford, Bywater, & Mayall, 2006; Mohamed et al., 2005). Of these, the first is a collection of RFLPs and the second is a collection of SNPs. The general DNA analysis and database program Geneious has access to many of the features of the above *rrn* analysis programs and has been used in the development of RiboTyping. However, it does not have the capability of presenting and storing graphical data or relating information from multicopy genes. This database will be useful for analysing the data that target specific *rrn* operons such as the method recently developed by Zhang et al. (2014) to PCR walk along specific *rrn* operons. The novel contribution of the RiboTyping database outlined in this chapter is that it combines many of the features of the above *rrn* analysis programs into an App that is much more interactive and user-friendly. RiboTyping also has the advantages of storing and presenting graphical data but more significantly it relates information from multi-copy genes, something other programs have not addressed.

REFERENCES

Amann, R. I., Ludwig, W., & Schleifer, K. H. (1995). Phylogenetic identification and in situ detection of individual microbial cells without cultivation. *Microbiological Reviews, 59*, 143–169.

Ashelford, K. E., Chuzhanova, N. A., Fry, J. C., Jones, A. J., & Weightman, A. J. (2005). At least 1 in 20 16S rRNA sequence records currently held in public repositories is estimated to contain substantial anomalies. *Applied and Environmental Microbiology, 71*, 7724–7736.

Baker, G. C., Smith, J. J., & Cowan, D. A. (2003). Review and re-analysis of domain-specific 16S primers. *Journal of Microbiological Methods, 55*, 541–555.

Christensen, H., Moller, P. L., Vogensen, F. K., & Olsen, J. E. (2000). Sequence variation of the 16S to 23S rRNA spacer region in Salmonella enterica. *Research in Microbiology, 151,* 37–42.

Chun, J., Huq, A., & Colwell, R. R. (1999). Analysis of 16S-23S rRNA intergenic spacer regions of Vibrio cholerae and Vibrio mimicus. *Applied and Environmental Microbiology, 65,* 2202–2208.

Coenye, T., & Vandamme, P. (2003). Intragenomic heterogeneity between multiple 16S ribosomal RNA operons in sequenced bacterial genomes. *FEMS Microbiology Letters, 228,* 45–49.

Cole, J. R., Wang, Q., Cardenas, E., Fish, J., Chai, B., Farris, R. J., et al. (2009). The Ribosomal Database Project: Improved alignments and new tools for rRNA analysis. *Nucleic Acids Research, 37,* D141–D145.

Condon, C., Squires, C., & Squires, C. L. (1995). Control of rRNA transcription in Escherichia coli. *Microbiological Reviews, 59,* 623–645.

D'Auria, G., Pushker, R., & Rodriguez-Valera, F. (2006). IWoCS: Analyzing ribosomal intergenic transcribed spacers configuration and taxonomic relationships. *Bioinformatics, 22,* 527–531.

DeSantis, T. Z., Hugenholtz, P., Larsen, N., Rojas, M., Brodie, E. L., Keller, K., et al. (2006). Greengenes, a chimera-checked 16S rRNA gene database and workbench compatible with ARB. *Applied and Environmental Microbiology, 72,* 5069–5072.

Deutscher, M. P. (2009). Chapter 9 Maturation and degradation of ribosomal RNA in bacteria. *Progress in Molecular Biology and Translational Science, 85,* 369–391.

Drummond, A. J., Ashton, B., Buxton, S., Cheung, M., Cooper, A., Duran, C., Field, M., Heled, J., Kearse, M., Markowitz, S., Moir, R., Stones-Havas, S., Sturrock, S., Thierer, T., Wilson, A., 2013. *Geneious v6.1.6.* http://www.geneious.com/.

Evguenieva-Hackenberg, E. (2005). Bacterial ribosomal RNA in pieces. *Molecular Microbiology, 57,* 318–325.

Garcia-Martinez, J., Bescos, I., Rodriguez-Sala, J. J., & Rodriguez-Valera, F. (2001). RISSC: A novel database for ribosomal 16S-23S RNA genes spacer regions. *Nucleic Acids Research, 29,* 178–180.

Gonzalez-Escalona, N., Romero, J., Guzman, C. A., & Espejo, R. T. (2006). Variation in the 16S-23S rRNA intergenic spacer regions in Vibrio parahaemolyticus strains are due to indels nearby their tRNAGlu. *FEMS Microbiology Letters, 256,* 38–43.

Gurtler, V. (1999). The role of recombination and mutation in 16S-23S rDNA spacer rearrangements. *Gene, 238,* 241–252.

Gurtler, V., & Barrie, H. D. (1995). Typing of Staphylococcus aureus strains by PCR-amplification of variable-length 16S-23S rDNA spacer regions: Characterization of spacer sequences. *Microbiology (Reading, England), 141*(Pt. 5), 1255–1265.

Gurtler, V., & Grando, D. (2013). New opportunities for improved ribotyping of C. difficile clinical isolates by exploring their genomes. *Journal of Microbiological Methods, 93,* 257–272.

Gurtler, V., Harford, C., Bywater, J., & Mayall, B. C. (2006). Direct identification of slowly growing Mycobacterium species by analysis of the intergenic 16S-23S rDNA spacer region (ISR) using a GelCompar II database containing sequence based optimization for restriction fragment site polymorphisms (RFLPs) for 12 enzymes. *Journal of Microbiological Methods, 64,* 185–199.

Gurtler, V., & Stanisich, V. A. (1996). New approaches to typing and identification of bacteria using the 16S-23S rDNA spacer region. *Microbiology (Reading, England), 142*(Pt. 1), 3–16.

Gurtler, V., Wilson, V. A., & Mayall, B. C. (1991). Classification of medically important clostridia using restriction endonuclease site differences of PCR-amplified 16S rDNA. *Journal of General Microbiology, 137*, 2673–2679.

Gutell, R. R., Lee, J. C., & Cannone, J. J. (2002). The accuracy of ribosomal RNA comparative structure models. *Current Opinion in Structural Biology, 12*, 301–310.

Harmsen, D., Rothganger, J., Frosch, M., & Albert, J. (2002). RIDOM: Ribosomal Differentiation of Medical Micro-organisms Database. *Nucleic Acids Research, 30*, 416–417.

Hunt, D. E., Klepac-Ceraj, V., Acinas, S. G., Gautier, C., Bertilsson, S., & Polz, M. F. (2006). Evaluation of 23S rRNA PCR primers for use in phylogenetic studies of bacterial diversity. *Applied and Environmental Microbiology, 72*, 2221–2225.

Kelly, W. J., Leahy, S. C., Altermann, E., Yeoman, C. J., Dunne, J. C., Kong, Z., et al. (2010). The glycobiome of the rumen bacterium Butyrivibrio proteoclasticus B316(T) highlights adaptation to a polysaccharide-rich environment. *PLoS One, 5*, e11942.

Lee, Z. M., Bussema, C., 3rd., & Schmidt, T. M. (2009). rrnDB: Documenting the number of rRNA and tRNA genes in bacteria and archaea. *Nucleic Acids Research, 37*, D489–D493.

Li, D., Leahy, S., Henderson, G., Kelly, W., Cookson, A., Attwood, G., et al. (2014). Atypical bacterial rRNA operon structure is prevalent within the Lachnospiraceae, and use of the 16S-23S rRNA internal transcribed spacer region for the rapid identification of ruminal Butyrivibrio and Pseudobutyrivibrio strains. *Annals of Microbiology*. http://dx.doi.org/10.1007/s13213-014-0806-2.

Martín, J. F., Barreiro, C., González-Lavado, E., & Barriuso, M. (2003). Ribosomal RNA and ribosomal proteins in corynebacteria. *Journal of Biotechnology, 104*, 41–53.

Mohamed, A. M., Kuyper, D. J., Iwen, P. C., Ali, H. H., Bastola, D. R., & Hinrichs, S. H. (2005). Computational approach involving use of the internal transcribed spacer 1 region for identification of Mycobacterium species. *Journal of Clinical Microbiology, 43*, 3811–3817.

Perez Luz, S., Rodriguez-Valera, F., Lan, R., & Reeves, P. R. (1998). Variation of the ribosomal operon 16S-23S gene spacer region in representatives of Salmonella enterica subspecies. *Journal of Bacteriology, 180*, 2144–2151.

Pruesse, E., Quast, C., Knittel, K., Fuchs, B. M., Ludwig, W., Peplies, J., et al. (2007). SILVA: A comprehensive online resource for quality checked and aligned ribosomal RNA sequence data compatible with ARB. *Nucleic Acids Research, 35*, 7188–7196.

Sadeghifard, N., Gurtler, V., Beer, M., & Seviour, R. J. (2006). The mosaic nature of intergenic 16S-23S rRNA spacer regions suggests rRNA operon copy number variation in Clostridium difficile strains. *Applied and Environmental Microbiology, 72*, 7311–7323.

Teng, J. L., Ho, T. C., Yeung, R. S., Wong, A. Y., Wang, H., Chen, C., et al. (2014). Evaluation of 16SpathDB 2.0, an automated 16S rRNA gene sequence database, using 689 complete bacterial genomes. *Diagnostic Microbiology and Infectious Disease, 78*, 105–115.

Wimberly, B. T., Brodersen, D. V., Clemons, W. M., Morgan-Warren, R. J., Carter, A. P., Vonhrheln, C., et al. (2000). Structure of the 30S ribosomal subunit. *Nature, 407*, 327–339.

Woese, C. R. (1987). Bacterial evolution. *Microbiological Reviews, 51*, 221–271.

Zhang, H., You, C., Ren, J., Xu, D., Han, M., & Liao, W. (2014). A simple one-step PCR walking method and its application of bacterial rRNA for sequencing identification. *Current Microbiology, 68*, 486–494.

GLOSSARY

Browse mode   The FileMaker mode in which you enter and edit information in fields

Button   A layout object that performs a specified script in Browse or Find modes

Chimera   DNA damage or degradation caused by recombination of homologous but not identical DNA sequences (Ashelford et al., 2005)

Fields   A collection of data pertaining to a subject, such as customers or stock prices. A database file contains one or more tables, which consist of fields and records

FileMaker   Database construction program (FileMaker, Inc.)

Find mode   The FileMaker mode in which you specify criteria for finding a subset of records

Geneious   Geneious 6 DNA analysis program (Drummond et al., 2013)

Illustrator CS5   Vector-based illustration program (Adobe)

Indel   Insertion or deletion of nucleotide sequence from 1 nt to 100–200 nt

Metadata   XML name for "fields"

Mosaic   Presence or absence of sequence blocks (indels) up to 100–200 nt in length

Records   One set of fields in a database table. Each record contains data about a single activity, individual, subject or transaction

Related field   For relational databases, a field in one table that is related to a field in another table. If a relationship is defined between two tables, data in fields in one table can be accessed from the other table

Relationships   Relationships provide access to data from one table to another that can join one or more records in one table to one or more records in another table

Sub-summary parts   Use summary parts to view and display information from one or more records. You place a summary field in a summary part to display a summary of information for each group of records sorted on the break field. You can add one or more sub-summaries above (leading) or below (trailing) the body

Summary field   A field that contains the result of a summary calculation of values across a group of records

Table   A collection of data pertaining to a subject, such as customers or stock prices. A database file contains one or more tables, which consist of fields and records

MALDI-TOF Mass Spectrometry Applied to Classification and Identification of Bacteria

13

Peter Schumann[*,1], **Thomas Maier**[†]

Leibniz Institute DSMZ-German Collection of Microorganisms and Cell Cultures, Braunschweig, Germany

[†]*Bruker Daltonics, Bremen, Germany*

[1]*Corresponding author: e-mail address: psc@dsmz.de*

1 INTRODUCTION

The idea of classification and identification of bacteria by mass spectrometry (MS) dates back to the 1970s (Anhalt & Fenselau, 1975; Meuzelaar & Kistemaker, 1973). Pyrolysis MS of intact bacterial cells (Gutteridge, 1988) resulted in a bewildering multitude of fragment peaks, the evaluation of which was a challenge for any data processing equipment available at that time. The peak patterns depended strongly on cultivation conditions and on instrument parameters fluctuating with time (source-ageing effect), the latter affecting the peak intensities in a way that meant identification libraries could be used only for a given time span. The introduction of matrix-assisted laser desorption/ionisation time-of-flight (MALDI-TOF) MS for the analysis of large biomolecules by Karas and Hillenkamp (1988) and Tanaka et al. (1988) was the breakthrough for the analysis of complex biological systems such as bacterial cells (Easterling, Colangelo, Scott, & Amster, 1998), because this technique inhibits the fragmentation of the analytes and provides clearly identifiable spectra of molecular ions of intact cellular macromolecules.

Since the first pioneering attempts for classification and identification of intact microorganisms by MALDI-TOF MS (Claydon, Davey, Edwards-Jones, & Gordon, 1996; Holland et al., 1996; Krishnamurthy & Ross, 1996), this approach has gained increasing application for bacteria, archaea, yeasts and fungi. Because the spectra of bacterial cells are dominated by the peaks of ribosomal proteins and these are rather conservative molecules in the course of evolution, spectra are characteristic of a given taxon and are only marginally susceptible to changes in cultivation conditions (Maier & Kostrzewa, 2007; Ryzhov & Fenselau, 2001; Teramoto, Sato, Sun, Torimura, & Tao, 2007a). MALDI-TOF MS for identification of

Methods in Microbiology, Volume 41, ISSN 0580-9517, http://dx.doi.org/10.1016/bs.mim.2014.06.002

potentially pathogenic microorganisms rapidly entered the market of medical diagnostics, not least due to the progress in maintaining a stable instrument performance and to the availability of spectra databases covering the area of medically relevant bacteria. Because of its success in medical diagnostics, the introduction of MALDI-TOF MS is now considered an "ongoing revolution in bacteriology" (Seng et al., 2009).

Searching the literature database PubMed (United States National Library of Medicine, NLM) for the terms "Identification", "Bacteria" and "MALDI" in combination resulted in 1522 references including 65 review articles at the time of writing. This high impact on the scientific literature demonstrates the broad application of MALDI-TOF MS to bacterial cells not only in clinical diagnosis but also in quality control of food production and in the pharmaceutical and biotechnological industries, as well as in ecological and environmental research and bacterial systematics. In these applied areas, MALDI-TOF MS has proven its superiority over classical phenotypic methods and is even challenging technologies based on nucleic acid sequences by its exceptionally favourable ratio of analytical turn-around time to costs. While many publications deal with the optimisation of sample preparation and comparison of different protocols (e.g. Alatoom, Cunningham, Ihde, Mandrekar, & Patel, 2011; Drevinek, Dresler, Klimentova, Pisa, & Hubalek, 2012; Dridi & Drancourt, 2011; Juiz et al., 2012; Liu, Du, Wang, & Yang, 2007; Šedo, Sedláček & Zdráhal, 2011; Šedo, Voráč & Zdráhal, 2011; Vaidyanathan, Winder, Wade, Kell, & Goodacre, 2002), there are only very few articles about the optimisation of instrumental conditions (e.g. Freiwald & Sauer, 2009), the evaluation of the quality of spectra and the discriminatory power of MALDI-TOF MS for classification, identification and strain typing (e.g. Sandrin, Goldstein, & Schumaker, 2013; Šedo, Sedláček, et al., 2011; Seng et al., 2010).

2 SAMPLE PREPARATION

The way samples are prepared for MALDI-TOF MS measurements of bacteria is crucial for determining the number of peaks obtained, the signal-to-noise (S/N) ratio and the discriminatory power of the spectra as well as for the reproducibility, reliability, turn-around time and cost of the analyses. There are two general strategies for sample preparation:

I. Intact cells are transferred from agar plates or centrifugal pellets directly to the target (see Section 2.3) or

II. Proteins are extracted from cells and subjected after separation of the cells or as crude lysate to MALDI-TOF MS analysis (see Section 2.4).

2.1 CULTIVATION OF BACTERIA

MALDI-TOF mass spectra of bacteria are of highest taxonomic significance and reproducibility when dominated by the peaks of ribosomal proteins. Ribosomal protein patterns are rather robust towards changes in cultivation conditions

(Maier & Kostrzewa, 2007). However, cells should be harvested after cultivation under optimal conditions and in a growth phase that ensures the highest yield of ribosomes. The number of ribosomes was found to be maximal in the mid-log-phase of the slow growing oligotroph *Sphingomonas alaskensis* but decreased rapidly by 90% during the late log phase (Fegatella, Lim, Kjelleberg, & Cavicchioli, 1998). Deviation from optimal cultivation conditions may give rise to the formation of stress proteins that may dominate the MALDI-TOF mass spectra as can be seen, e.g., from the proteomic response to a downshift of the cultivation temperature (Kim, Han, Lee, & Lee, 2005). Storage of bacterial cultures at temperatures markedly below the optimal cultivation temperature (e.g. in a refrigerator) before sample preparation should be avoided because altered mass spectra may result that do not match the entry of the identification database for a given strain. Storage of an agar plate with *Pseudonocardia autotrophica* DSM 43082 (GYM *Streptomyces* agar, www.dsmz.de, cultivated for six days at 28 °C) in the refrigerator (8 °C) gave rise to additional peaks superimposing on those of ribosomal proteins after 3 days (Figure 1E). These peaks did not occur when a second plate of the strain was stored at 23 °C for 8 days (Figure 1G). The series of equidistant peaks (distance approx. 142 *m/z*) in the range of approx. 3000–6000 *m/z* might be caused by proteins formed by the cells when coping with the reduced temperature. A similar series of equidistant peaks (distance approx. 130 m/z) was reported by Vargha, Takats, Konopka, and Nakatsu (2006) for an *Arthrobacter* strain and discussed as a temporary growth-stage related effect during the coccus-rod-coccus cycle typical of this genus.

Changes in the physiological state of the cell, as in the case of sporulation of *Bacillus* species, may drastically alter MALDI-TOF mass spectra. With the appearance of the first spores of *Bacillus thuringiensis* DSM 2046[T] after cultivation on FP agar at 28 °C for 48 h, spore marker peaks typical of the *Bacillus cereus* group (Lasch et al., 2009) occur in the mass range of 6600–7090 *m/z*. These spore marker peaks originate from small acid-soluble proteins (Castanha, Fox, & Fox, 2006) and predominate in the mass spectrum after cultivation for 7 days when vegetative cells are lysed and the culture consists mainly of spores (Figure 2). Successful identification by MALDI-TOF MS requires that the strains under study are analysed in the same spectrum-relevant physiological state for which the database entries were generated.

2.2 SELECTION OF THE MATRIX AND COMPOSITION OF THE MATRIX SOLUTION

The selection of the matrix and the composition of its solution have to match the analytical purpose and are crucial for the quality of mass spectra. Common matrices for MALDI-TOF MS analyses of bacteria are alpha-cyano-4-hydroxycinnamic acid (HCCA), sinapinic acid (SA), 2,5-dihydroxybenzoic acid (DHB), ferulic acid, 2-mercaptobenzothiazole (MBT), 5-chloro-2-mercaptobenzothiazole (CMBT) and 2-(4-hydroxyphenylazo)benzoic acid. According to a recent review (Šedo, Sedláček, et al., 2011), HCCA was used as matrix in almost half of the studies reviewed, followed by SA (25%) and DHB (10%). HCCA is superior to many other

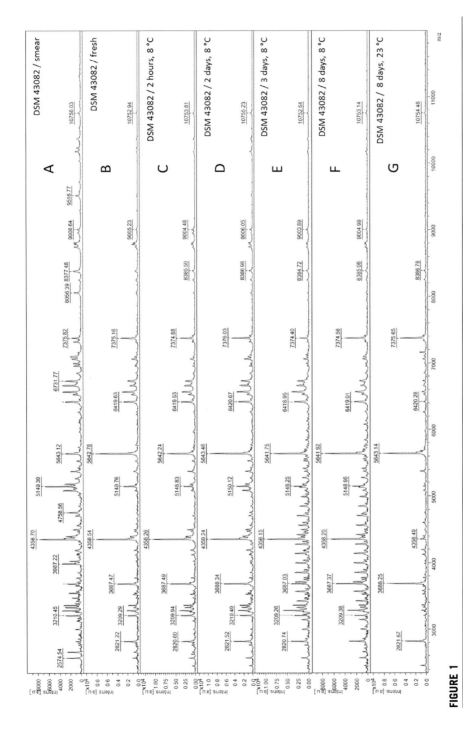

FIGURE 1

MALDI-TOF mass spectra of *Pseudonocardia autotrophica* DSM 43082 after cultivation on GYM *Streptomyces* agar (www.dsmz.de) for 6 days at 28 °C. HCCA was used as matrix. (A) Spectrum obtained after direct colony transfer (Protocol 2a) without further storage. (B–G) Spectra obtained after ethanol/formic acid extraction (Protocol 3). (B) Biomass taken from the plate without further storage. (C) After storage of the plate for 2 h at 8 °C. (D) After storage of the plate for 2 days at 8 °C. (E) After storage of the plate for 3 days at 8 °C. (F) After storage of the plate for 8 days at 8 °C. (G) After storage of a second plate for 8 days at 23 °C.

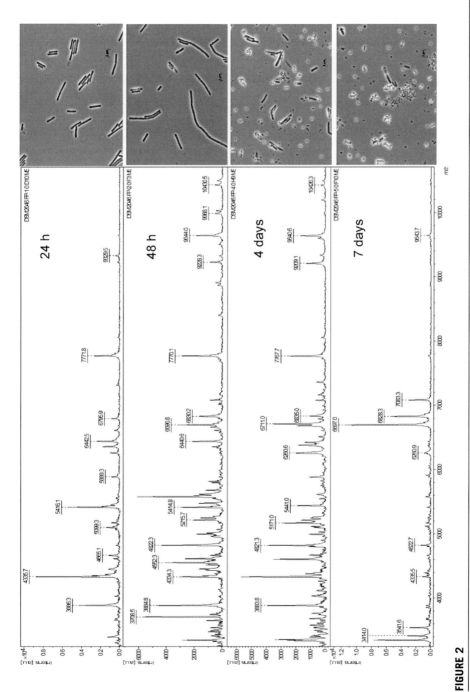

FIGURE 2

Changes in the MALDI-TOF mass spectra (3300–10600 m/z) of *Bacillus thuringiensis* DSM 2046[T] (cultivated on FP agar at 28 °C) depending on the degree of sporulation (phase-contrast images; bar, 5 µm).

matrices due to its better sensitivity, higher intensity and higher number of signals in the lower mass range. The lack of signals in the higher mass range might be due to the formation of double-charged ions. A further advantage is crystallisation as a homogeneous layer is beneficial for automated MALDI-TOF MS measurements. Disadvantages of HCCA are the lower signal resolution, peak broadening and absence of signals in the high-mass range when compared to ferulic acid and SA (Šedo, Sedláček, et al., 2011). Ferulic acid is selected for molecules of high molecular weight. However, the long crystal needles render the sample/matrix mixtures rather inhomogeneous. MBT and CMBT are preferred when analyses are focused on lipid components.

Among organic solvents, acetonitrile (ACN) is preferred because of its high solubilising capability for both matrices and proteins. A high shot-to-shot reproducibility can be achieved by a quick co-crystallisation of the sample/matrix mixture giving rise to the formation of many small crystals. A slow crystallisation, on the other hand, is favourable for the ionisation of large biomolecules (Šedo, Sedláček, et al., 2011). The speed of crystallisation can be influenced via the volatility of the solvent and the ratio of ACN to water in the solvent system. A ratio of approximately 70% ACN to 30% water resulted in the highest number of peaks (Vargha et al., 2006). The addition of an acidic component such as trifluoroacetic acid (TFA) to peptides and proteins is beneficial for MALDI-TOF MS measurements in the positive ion mode. The TFA concentration of 0.1% (v/v) commonly used in proteomic applications appears to be too low for the direct measurement of microbial samples. The decrease in the pH value caused by the lysis of intact cells may affect the quality of spectra drastically and needs to be compensated by TFA concentrations of 1% (v/v) or higher. TFA concentrations as high as 40% have been used for the extraction of acid-soluble proteins from spores of bacilli (Dickinson et al., 2004).

A solution of 10 mg/ml of HCCA in 50% (v/v) aqueous ACN containing 2.5% (v/v) TFA turned out to be useful in broad diagnostic routine applications (Protocol 1).

Protocol 1: Preparation of the HCCA matrix solution in 50% (v/v) aqueous ACN containing 2.5% (v/v) TFA

1. Place 10 mg HCCA in an Eppendorf tube and add a mixture of 475 μl deionised water, 25 μl TFA and 500 μl ACN. Alternatively, 1 ml LC-MS CHROMASOLV® (Fluka #19182) as a ready-to-use solution of the same composition can be used.
2. Vortex the suspension for several minutes at approximately 23 °C until the solution becomes clear.
3. The matrix solution can be used for 1 week when stored at ambient temperature.

Systematic studies revealed that overlaying the whole cells or their extracts after air drying by the matrix solution gives better results than pre-mixing cells or extracts with the matrix solution prior to the application on the target (Keys et al., 2004; Vaidyanathan et al., 2002).

2.3 DIRECT COLONY TRANSFER

The fastest, easiest and cheapest way of obtaining MALDI-TOF mass spectra of intact bacterial cells is the direct transfer of biomass from agar plates or centrifugal pellets of cultures in liquid media to spots of the stainless steel target and covering by a matrix layer (Protocol 2a).

Protocol 2a: Direct transfer of bacterial biomass to stainless steel targets

1. Transfer a small, barely visible amount of biomass directly to the sample spot of a stainless steel MALDI-TOF MS target. A thin film should be obtained by spreading the cells evenly on the whole area of the spot by using a sterile plastic pipette tip or inoculation loop. To become acquainted with using the appropriate amount of cells, please refer to Figure 3.

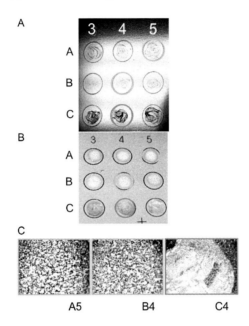

FIGURE 3

Evaluation of the correct amount of biological material for direct application (Protocol 2a). (A) MALDI-TOF MS target after application of different amounts of biological material and before matrix application. Row A, appropriate amount of biological material; row B, lower but still sufficient amount of biological material; row C, too much biological material. Green (suitable amount) and red (too high amount) frames mark the same target positions which are shown in panel (B) after matrix addition and visualised in panel (C) after crystal formation. (B) The same MALDI target as shown in panel (A) after matrix application and drying. (C) Crystal formation as observed by the "Camera Image View" of the software package flexControl (Bruker Daltonics): spots A5 and B4 demonstrate optimal crystal formation; C4 is an example for suboptimal crystal formation due to excessive biological material applied in the direct transfer step. (See the colour plate.)

2. Overlay the biological material with 1 μl of the HCCA matrix solution prepared according to the instructions in Protocol 1.
3. Allow to dry in a fume hood prior to the insertion of the target into the mass spectrometer.

The quality of spectra may be improved by adding formic acid prior to step 2 (Protocol 2b). The prolonged contact with concentrated acid supports the release of proteins and gives rise to a better success rate of automated MALDI-TOF MS measurements for Gram positive bacteria and yeasts. For Gram negative bacteria, this step is commonly omitted in order to keep the hands-on time per sample short.

Protocol 2b: Extended direct transfer of bacterial biomass to stainless steel targets

1. Step 1 of Protocol 2a.
2. Overlay the biomass film with 1 μl of 70% formic acid and allow for drying at room temperature (RT).
3. Overlay the biomass after formic acid treatment with 1 μl of the HCCA matrix solution prepared according to the instructions in Protocol 1.
4. Allow to dry in a fume hood prior to the insertion of the target into the mass spectrometer.

2.4 EXTRACTION METHODS

In early applications of MALDI-TOF MS for bacterial identification, most workers used direct colony transfer techniques. Only when spectra of insufficient intensity and resolution, or no spectra at all were obtained, were extraction methods applied as a last resort. However, recent comparisons of different sample preparation techniques (Alatoom et al., 2011; Fournier et al., 2012; Saffert et al., 2011; Schulthess et al., 2013) came unequivocally to the conclusion that methods aimed at the extraction of proteins prior to the measurement resulted in more high-level identifications when compared to the direct colony transfer. In a study of 238 strains representing 34 species of anaerobic bacteria, 207 strains were identified by using the direct colony method while 218 strains were identified after application of an extraction procedure (Fournier et al., 2012). In addition, the score values indicating the reliability of species identification could be significantly increased by prior extraction of proteins in this study. Spectra obtained by an extraction protocol usually show the same diagnostic peaks, less minor peaks of ambiguous origin and a better S/N ratio than those obtained by direct colony transfer (compare Figure 1A and B). Despite the fact that extraction of proteins requires additional inconvenient steps for sample preparation, results in a longer turnaround time of analyses and may expose workers to additional chemical hazards, extraction procedures are on the rise in MALDI-TOF MS identification of bacteria due to their aforementioned advantages (see also Croxatto, Prod'hom, & Greub, 2012).

Though proteins can be released by the application of surfactants, enzymatic and mechanical procedures (Giebel, Fredenberg, & Sandrin, 2008; Meetani & Voorhees, 2005; Teramoto et al., 2007a), ultra sonication (Easterling et al., 1998) and corona plasma discharge (Birmingham et al., 1999), the extraction of bacterial cells with solvent systems containing TFA or formic acid turned out to be convenient and to give rise to high-quality MALDI-TOF mass spectra (Schulthess et al., 2013).

The matrix solvent is the only cell lysing and protein extracting agent when using direct transfer (Protocol 2a) and it interacts with the biomass only until it has dried out. Especially for organisms with capsules or a robust cell wall the quality of MALDI-TOF mass spectra can be significantly improved if proteins are released from the cells prior to measurements. During the short steps of extraction protocols several additional factors contribute to the improvement of MALDI-TOF MS measurements: Cells are washed and agar media components like salt, carbohydrates or peptides are removed. In addition, more biological material can be used and concentrated acids and ACN can be applied for destruction of cell walls and protein release. The interaction time with the solvents can be carefully controlled, too. Finally a homogeneous cell-free extract results for application to the target.

Among the solvent-based extraction protocols, the so-called "ethanol-formic acid extraction" has found the most widespread application (Protocol 3). After step 4 of Protocol 3, inactivated samples can be shipped at ambient temperature between laboratories or be stored for months at $-20\,°C$ before finalising the sample preparation and subsequent measurement.

Protocol 3: Sample preparation by ethanol–formic acid extraction for MALDI-TOF MS-based identification of bacteria

1. Pipette 300 µl deionised water into an Eppendorf tube.
2. Place biological material (from one single colony up to 5–10 mg) into the tube.
3. Mix thoroughly. Pipetting or vortexing should be appropriate. For microorganisms which are difficult to suspend, a micro-pistil (Eppendorf 0030 120.973) can be used.
4. Add 900 µl absolute ethanol and mix thoroughly.
5. Centrifuge at approx. $15.800 \times g$ for 2 min, decant the supernatant, centrifuge again and remove all the residual ethanol by carefully pipetting it off to waste without disturbing the pellet.
6. Dry the ethanol-pellet for several minutes at RT.
7. Add 70% formic acid (1–80 µl) to the pellet and mix very well by pipetting and/or by vortexing.
   - If only a little material (one single colony) was used, the volume of formic acid must be adjusted accordingly (refer to Table 1).
8. Add pure ACN (1–80 µl) and mix carefully.
   - Likewise, if only little material (one single colony) was used, the volume of ACN must be adjusted as well (refer to Table 1).
9. Centrifuge at approx. $15.800 \times g$ for 2 min until all the material is collected neatly in a pellet.

Table 1 Estimates of Volumes of Formic Acid and Acetonitrile used in Protocol 3 Depending on the Amount of Biomass

	Small Colony	Big Colony	1 μl Loop	10 μl Loop
Formic acid 70%	1–5 μl	5–15 μl	10–40 μl	30–80 μl
Acetonitrile	1–5 μl	5–15 μl	10–40 μl	30–80 μl

10. Pipette 1 μl of supernatant from the previous step onto a target plate and allow it to dry in air at normal RT.
- It is recommended to work rapidly in order to avoid oxidation and acylation of the protein extract which may give rise to artefact peaks.

11. Overlay with 1 μl of matrix solution and allow it to dry in air (in a fume hood) at ambient temperature.
- It is not recommended to speed up the drying process in steps 10 and 11 by heating.

2.5 ORGANISM-SPECIFIC SAMPLE PREPARATION

Though Protocol 3 is applicable to sample preparation from the vast majority of bacteria and allows spectra of high quality to be recorded, certain groups of organisms may cause difficulties in obtaining spectra of sufficient intensities. The experience of users and manufacturers of MALDI-TOF MS identification systems have resulted in suggestions for optimisation of the standard protocol for specific groups of bacteria.

Potentially harmful bacteria, in particular, those affiliated to biological safety level 3 (BSL-3) and classified as group A or group B bioterrorism agents, need to be inactivated prior to handling outside BSL-3 facilities. Suitable methods for efficient inactivation while maintaining the capability for successful MALDI-TOF MS-based identification have been reviewed previously (Drevinek et al., 2012; Dridi & Drancourt, 2011). Protocol 3 was found to be appropriate for inactivation of viable cells but not of resistant spores. In order to remove spores from the extracts, the inclusion of centrifugal filtration steps (pore size 0.1 μm) was suggested (Drevinek et al., 2012; Dybwad, van der Laaken, Blatny, & Paauw, 2013). A procedure starting with suspending the biomass in 80% (v/v) TFA was suggested as an alternative to Protocol 3 for reliable inactivation of spores and highly pathogenic bacteria by Freiwald and Sauer (2009) and Lasch et al. (2008).

Many filamentous actinomycetes may cause problems due to the strong aggregation of mycelia and adherence to the agar surface. Cultivation on cellulose nitrate membrane filters (pore size 45 μm) laid on the agar medium and grinding of the mycelium scraped from the filters with microtube pestles in water prior to standardised sample preparation allowed successful analyses of streptomycetes and related organisms (Schumann & Pukall, 2013). Cultivation of filamentous organisms in liquid media is an alternative remedy for this problem and results additionally in a more homogeneous biomass. Experience suggests that mycelial biomass may strongly absorb aqueous ethanol and requires more intensive drying at step 6 of Protocol 3, i.e., by application of centrifugal evaporators.

In several cases where bacteria tend to give poor spectra, it has been found to be advantageous to wash the biomass intensively with deionised water and aqueous ethanol prior to step 5 of Protocol 3 in order to get rid of salts and other polar substances that might affect the mass spectrometric ionisation process. It is suggested to suspend 10–15 mg biomass in 1 ml water by agitation on a vortex mixer for 2 min. After centrifugation for 2 min, the pellet should be subjected to the same washing procedure again. The pellet obtained by spinning down for 2 min is carefully re-suspended in 100 µl of water before adding 200 µl of water and 900 µl of ethanol and mixing with a vortex mixer. Also this washing step needs to be repeated before the sample can be processed by Protocol 3, starting at step 5.

Actinomycetes containing mycolic acids (e.g. members of the families *Mycobacteriaceae* and *Nocardiaceae*) tend to give only poor spectra possibly due to their thick lipid cell envelope layers. Additional remedial measures, such as water washing, mechanical disintegration and heat treatment for inactivation of pathogens, supplementing ethanol–formic acid extraction render the recording of mass spectra of mycobacteria possible (Protocol 4). Whether or not Protocol 4 is superior to the usual ethanol–formic acid extraction (Protocol 3) for other mycolata (e.g. mycolic-acid containing corynebacteria and members of the genera *Dietzia, Gordonia, Millisia, Nocardia, Rhodococcus, Segniliparus, Skermania, Smaragdicoccus, Tomitella, Tsukamurella, Williamsia*) and is a remedy for unsatisfactory results obtained by Protocol 3 needs to be tested in each individual case. Application of the bead preparation protocol appears to be unnecessary for *Nocardia* strains and did not improve the quality of their spectra when compared to the results obtained by Protocol 3 according to the experience of the authors.

Protocol 4: Bead preparation protocol for MALDI-TOF MS sample preparation of mycobacteria

A1—Solid medium samples

1. Pipette 300 µl deionised water (or HPLC or MS grade water) into a 1.5 ml Eppendorf tube.
2. Transfer mycobacteria biomass into the tube. (Avoid collecting medium! Try to get one to three 10 µl inoculation loops of biomass. As estimate of the amount of biomass: 2 µl water in an Eppendorf tube represents a small pellet, 5 µl of water represents a pellet of optimal size).
3. Heat by boiling for 30 min in a water bath (95 °C or better 99 °C in a thermomixer may also be suitable).

Warning: The lids of reaction tubes may burst during heat inactivation. Take precautions to avoid burn injury. This applies to all heat inactivation steps of this protocol.

A2—Liquid medium samples, e.g., MGIT^{TM} medium

1. Using a disposable Pasteur pipette, collect 1.2 ml liquid medium from the bottom of the cultivation tube (where biomass has deposited or will settle within 5–10 min) in a 1.5 ml Eppendorf tube. Centrifuge for 2 min at 15,800–20,000 × g, then remove medium carefully by pipetting.
2. Add 300 μl deionised water.
3. Heat by boiling for 30 min in a water bath or by using a thermomixer.

B—Follow-up for both solid and liquid medium samples

1. Add 900 μl ethanol and mix thoroughly.
2. Centrifuge for 2 min at 15,800–20,000 × g and decant supernatant, centrifuge again and remove residual 75% (v/v) ethanol carefully by pipetting.
3. Allow the pellet to dry at RT for a few minutes.
4. Add beads (0.5 mm Zirconia/Silica beads, BioSpec Products, Catalogue no. 11079105z or similar). Quantity: volume of approximately 100 μl.
5. Add pure ACN (in general 10–50 μl, depending on size of the pellet; if unsure use 20 μl).
6. Vortex mix at maximum speed for 1 min.
7. Add 70% formic acid (same volume as ACN) and mix by vortexing for approx. 5 s.
8. Centrifuge at 15,800–20,000 × g for 2 min.
9. Place 1 μl of supernatant on a MALDI target plate and allow to dry at RT.
10. Overlay with 1 μl of HCCA solution immediately after drying.

2.6 MALDI-TOF MS-BASED IDENTIFICATION OF BACTERIA IN COMPLEX BIOLOGICAL MATRICES

2.6.1 Identification of bacteria in positive blood cultures

Bloodstream infections are associated with high morbidity and mortality (Kumar et al., 2006). For septic patients, an immediate start of intravenous antibiotics has great priority. Blood cultures (BC) are one important diagnostic tool. However, not only timely administration but also the appropriate choice of antimicrobial medication determines the survival rates of patients with septic shock. To speed up the diagnostic process for bacteraemia, several protocols have been established for the direct identification of bacteria from positive BC bottles by MALDI-TOF MS. However, blood and media components interfere with many rapid testing methods, requiring BC to be processed in order to separate the microorganisms (which represent the minority in terms of biomass in BC) from interfering substances.

Protocol 5: Application of the Bruker MALDI Sepsityper Kit

(Schubert et al., 2011)

1. Transfer 1 ml culture fluids drawn from positive BC bottles to a 1.5-ml Eppendorf tube.
2. Add 200 μl of the Sepsityper lysis buffer and mix the sample using a vortex mixer for 10 s.

3. Centrifuge for 1 min at 17,900 × g using a centrifuge (e.g. Eppendorf 5417, Germany).
4. Discard supernatant.
5. Resuspend the pellet in 1 ml of Sepsityper washing buffer.
6. Centrifuge again for 1 min at 17,900 × g.
7. Discard supernatant.
8. Suspend the pellet in 300 µl of distilled water and continue according to Protocol 3.

Protocol 6: Differential centrifugation—"Vacutainer" *(Gray, Thomas, Olma, Iredell, & Chen, 2013; Moussaoui et al., 2010)*
1. Recover approx. 1.5 ml from positive BC vials and inject it with a syringe into a gel separator tube (e.g. Clot Activator and Gel BD Vacutainer tubes [Becton Dickinson] or ZSerumSeptClotActivator [Greiner Bio One, Courtaboeuf, France]).
2. Centrifuge the tubes at 500 × g for 10 min at RT to separate blood cells at the bottom of the gel from plasma, cell debris and bacteria at the surface of the gel.
3. Suspend the surface bacterial sediment in 1.5 ml of sterile water and transfer it to an Eppendorf tube.
4. Centrifuge at 300 × g for 1 min at RT for complete removal of cell debris still present.
5. Transfer 1 ml of the supernatant to another microtube to collect bacteria after a 10,000 × g centrifugation for 2 min at RT.
6. Suspend the pellet in 300 µl of distilled water and continue according to Protocol 3.

Protocol 7: Processing of positive BC containing charcoal
(Wüppenhorst et al., 2012)
1. Collect 5 ml of positive BC broth.
2. Centrifuge 10 min at 400 × g (charcoal and blood cells are removed).
3. Transfer 1 ml of the supernatant to another microtube.
4. Add 200 µl 5% saponin[1] for cell lysis (Sigma-Aldrich, St. Louis, MO, USA).
5. Incubate the mixture for 5 min at RT.
6. Remove remaining charcoal and blood cells with a centrifugation step using a SigmaPrep™ spin column (800 µl, Sigma-Aldrich, St. Louis, MO, USA).
7. Centrifuge for 2 min at 330 × g, collecting the filtrate.
8. Centrifuge the filtrate for 2 min at 15,800 × g.
9. Remove supernatant carefully and wash the pellet in 1 ml of ultra-pure water.
10. Centrifuge again for 1 min at 15,800 × g.
11. Discard the supernatant.
12. Suspend the pellet in 300 µl of distilled water and continue according to Protocol 3.

---

[1]Several authors report on the use of alternative detergents (Ferroni et al., 2010; Meex et al., 2012; Saffert et al., 2011).

Protocol 8: Lysis-filtration method for VITEK MS systems
(Fothergill et al., 2013)

1. Briefly vent the positive blood bottles by using a venting device (e.g. Steri/VENT device, Sterimatic Worldwide Ltd.), allowing filtered air to be drawn into the bottle head space under low pressure. Remove the venting device and invert the bottles several times.
2. Add 2 ml of broth to 1.0 ml of lysis buffer (0.6% polyoxyethylene [10] oleyl ether [Brij 97] in 0.4 M 3-[cyclohexylamino]-1-propanesulfonic acid [CAPS]).
3. Vortex the samples for 5 s and incubate them for 2–4 min at RT.
4. Pass lysates in a constant stream through a 25-mm filter of 0.45 μm pore size (shiny side down) for 40 s. Collect all effluents in the filter manifold vacuum apparatus and discard them at the end.
5. Up to three samples can be processed simultaneously.
6. After pulling the lysate completely into the filter, wash the remaining microbial cells on the membrane three times with wash buffer (20 mM sodium phosphate, 0.05% Brij 97 and 0.45% NaCl).
7. Wash the samples thereafter three times with deionised water.
8. Remove the microorganisms from the surface by firmly scraping the membrane with a polyester microswab (Texwipe CleanTips swabs; catalogue no. TX754B; Kernersville, NC).
9. Hold the swab nearly vertically and slightly tilted away from the user. Use a downward force sufficient to almost tear the membrane to free the microorganisms from the membrane. Make strokes across the membrane in a nearly overlapping pattern over the entire area where the lysate was applied.
10. Apply the organisms directly to disposable VITEK MS target plates.
11. Hold the swab in the same orientation as it was during organism collection and blot (not wipe) the cells firmly onto a spot on the MALDI target plate.
12. Immediately following application, cover the microorganisms with 1 μl of HCCA solution.

For a more detailed description, please refer to the supplemental material of Fothergill et al. (2013).

2.6.2 Examples of MALDI-TOF MS applications for direct bacterial identification in complex matrices

Urinary tract pathogens were successfully identified directly from 260 urine samples by MALDI-TOF MS, while the sample preparation took only approx. 6 min longer than for routine identification of pure cultures (Ferreira et al., 2010). Direct identification processing methods for MALDI-TOF MS detection of microbes in urinary tract infection were discussed by DeMarco and Ford (2013). Microscopic magnetic beads coated with antibodies were used by Ochoa and Harrington (2005) for selective pre-concentration of enterohemorrhagic *Escherichia coli* O157:H7 strains from ground beef prior to identification by MALDI-TOF MS. Parisi et al. (2008) used MALDI-TOF MS for detection of *E. coli* O157:H7 and *Yersinia enterocolitica* directly in spiked beef samples. Barreiro et al. (2012) detected intact proteins originating from bacteria relevant for the diagnosis of mastitis and monitoring of dairy products in experimentally contaminated milk by MALDI-TOF MS.

3 OPTIMISATION OF MEASUREMENT CONDITIONS

The standard settings of a MALDI-TOF mass spectrometer in the appropriate mass range (e.g. 2000–20,000 m/z) based on the manufacturer's specifications are commonly sufficient for identification. No further optimisation by users is necessary since IVD-CE labelled products even inhibit any user interaction. Nevertheless the need to create the user's own MALDI-TOF MS-based reference libraries exists due to the incomplete coverage of certain taxa by commercial databases.

The ability of users to analyse and control the optimal instrument settings is a precondition for reference generation. A minimal knowledge about the technical background is necessary for optimal spectra measurement and reference generation. An "optimal mass spectrum" means:

- Many peaks (up to 150 peaks are possible)
- High S/N ratio (i.e. low noise and peaks with high intensities)
- High resolution of peaks over the whole mass range
- Correct calibration

Prior to the creation of any identification database, the taxonomic identification of all candidates for reference strains is mandatory. It is required to use a set of methods (biochemical/physiological tests, nucleic acid sequencing, chemotaxonomic analyses, clinical studies, etc.) to identify a strain "polyphasically" and unambiguously prior to its introduction as a reference. Even when designed for the same purpose (identification of bacteria), a database of bacterial MALDI-TOF mass spectra is exposed to additional factors of influence when compared to 16S rRNA gene sequence databases. Sequencing of the 16S rRNA gene is more or less "digital" (i.e. it was either successful or unsuccessful) and the result is independent from cultivation or other biological influences. The quality of MALDI-TOF mass spectra of bacteria, on the other hand, may be graduated in a broad range and depends on biological and instrumental factors as well as on a variety of other influences that need to be controlled.

A "mass spectrum" covers a certain mass range and contains a certain number of peaks. A "peak" is represented by its mass per charge ratio (m/z; its value on the X-axis), its intensity (its value on the Y-axis) and its resolution (depending on the peak width). The MALDI-TOF MS measurement of mass spectra depends on several technical instrument settings such as:

- Laser power
- Acceleration voltage
- Detector voltage
- Calibration constants
- Delayed ion extraction

The following features are used to describe a MALDI-TOF mass spectrum or peaks within a mass spectrum:

- Mass range
- Mass accuracy
- Presence or absence of certain peaks
- Intensity (absolute and relative intensities)
- Intensity distribution
- Total ion count
- Number of laser shots
- Resolution
- S/N ratio

Software for spectral data analysis (quality control or peak picking) uses several tools such as:

- Smoothing
- Baseline correction
- Recalibration

The achievable quality of a mass spectrum depends strongly on the taxonomic group of organisms studied. For example, it is much easier to obtain "high-quality" mass spectra from fast growing Gram negative rods than from slow growing mycobacteria. Nevertheless, it is possible to create acceptable mass spectra from nearly any kind of organism. It is necessary but also possible to realise the optimum for a particular group of organisms.

To assess the optimal instrument setting an external standard can be used. The Bruker Bacterial Test Standard (Bruker Part Number #255343) is such a standard with a concentration of proteins optimal for checking the instrument settings. The protein concentration of this standard is close to the detection limit and contains substances covering the whole mass range of bacterial MALDI-TOF mass spectra.

3.1 MALDI-TOF MS LIBRARY CREATION

The following essential phases should be considered for reference generation and control measures must be established at all steps in the creation of libraries in order to ensure their reliability. All these factors have to be controlled and standardised (using standard operating procedures, standard parameters, etc.):

- Pre-analytical phase (PA-1)
  - Correct taxonomical pre-identification
  - Cultivation conditions (low influence)
  - Age of culture (low influence)
  - Vegetative cells/spores
- Sample preparation (SP-2)
  - MALDI target plate preparation (direct transfer/extended direct transfer/ extraction)
  - Matrix and solvent

- • Analyte concentration and analyte: matrix ratio
- • Crystallisation conditions (temperature, humidity)
- • Measurement (M-3)
  - • Instrument settings
  - • Correct calibration
  - • Reliable methods for automatic spectra acquisition
- • Data analysis and library calculation (DA-4)
  - • Data processing (spectra assessment and reference generation)
- • Reliability check (RC-5)
  - • New reference entries can be compared to already stored reference entries

MALDI-TOF MS measurements have to meet requirements at different levels, depending on whether they are intended for routine identification or for the generation of databases (Table 2).

Table 2 Requirements of MALDI-TOF MS Measurements Depending on Its Applications

Requirements of MALDI-TOF MS Measurements	
Routine ID	**Library Construction**
• Fast measurement with sufficient spectra quality • No/few interaction with acquisition software • Fully automated data processing • Mainly automated quality control at data acquisition and interpretation	• High spectra quality • Quality > speed • More interaction with acquisition software • Semi-automated data processing • Automated, but also significant manual/visual quality control at data acquisition and interpretation

4 APPLICATION OF MALDI-TOF MS FOR CLASSIFICATION AND IDENTIFICATION

4.1 SOFTWARE USED FOR TAXONOMIC EVALUATION OF MALDI-TOF MASS SPECTRA

The comparison of MALDI-TOF mass spectral datasets of prokaryotes with respect to the mass signals and intensities of the peaks requires dedicated software tools. The commercial MALDI-TOF MS identification systems VITEK MS (bioMérieux), Andromas (Andromas SAS) and MALDI Biotyper (Bruker Daltonics) contain integrated proprietary software packages. The BioNumerics software (Applied Maths, Belgium) has been used for the evaluation of bacterial mass spectra recorded with different instruments (Farfour et al., 2012; Ghyselinck, Van Hoorde, Hoste, Heylen, & De Vos, 2011; Saffert et al., 2011; Teramoto et al., 2007b). Appropriate functions of the statistical tools Matlab (The MathWorks, Inc.) and the R package (http://www.r-project.org) have also been applied to the evaluation of the similarity of bacterial MALDI-TOF mass spectra (Freiwald & Sauer, 2009; Gibb & Strimmer, 2012; Hettick et al., 2006; Sauer et al., 2008).

4.2 CLASSIFICATION AND TAXONOMIC RESOLUTION

Despite the increasing acceptance of MALDI-TOF MS in microbiology, there seems still to be a lack of clarity as to which taxonomic questions this technique may be applied to. In order to define its optimal field of application, the discriminatory power of MALDI-TOF MS is compared in the following examples with the taxonomic resolution of established tools for classification, identification and typing of bacteria.

4.2.1 Classification of genera within a family

Because ribosomal proteins dominate mass spectra (Ryzhov & Fenselau, 2001) and because structures of ribosomal components are conservative enough to mirror the phylogeny of organisms (Winker & Woese, 1991), it can be assumed that MALDI-TOF mass spectra may contain information for inferring the phylogenetic relationship of bacteria. However, the clustering of type strains of type species of genera of the family *Microbacteriaceae* in a dendrogram generated on the basis of their MALDI-TOF mass spectra does not agree with the topology of the tree calculated from their 16S rRNA gene sequences (Figure 4). The only exceptions, where the same bifurcations occur in both dendrograms, are the pairs *Rhodoglobus vestalii* DSM 21947[T]/*Salinibacterium amurskyense* DSM 16400[T] and *Microterricola viridarii* DSM 21772[T]/*Phycicola gilvus* DSM 18319[T]. Both pairs belong to those showing the highest binary 16S rRNA gene sequence similarities (98.2% and 99.2%, respectively, as determined by the EzTaxon server 2.1; Chun et al., 2007) among the 28 type strains included in this study. However, the pair *Yonghaparkia alkaliphila* DSM 19663[T]/*Microcella putealis* DSM 19627[T] which also shows 98.2% 16S rRNA gene sequence similarity did not form a bifurcation in the MALDI-TOF MS-based dendrogram. These findings suggest that the clustering of strains by MALDI-TOF MS may possibly correspond to results of sequence analyses only for the most closely related strains, i.e., ones sharing at least approx. 98% 16S rRNA gene sequence similarity.

4.2.2 Classification of species within a genus

The example of the genus *Arthrobacter* was chosen for evaluating the capability of MALDI-TOF MS for displaying intrageneric relationships of species. The genus *Arthrobacter* is a heterogeneous conglomeration of approx. 80 species and may require dissection into several novel genera. Candidates for membership in these new genera form clusters within the phylogenetic tree of the genus *Arthrobacter* and are rather consistent in their peptidoglycan structure and menaquinone composition (Busse, Wieser, & Buczolits, 2012). A dendrogram generated from their MALDI-TOF mass spectra (Figure 5) shows that the representatives of subclade I and rRNA clusters 2 and 3 (Busse et al., 2012) with members displaying >97% 16S rRNA gene

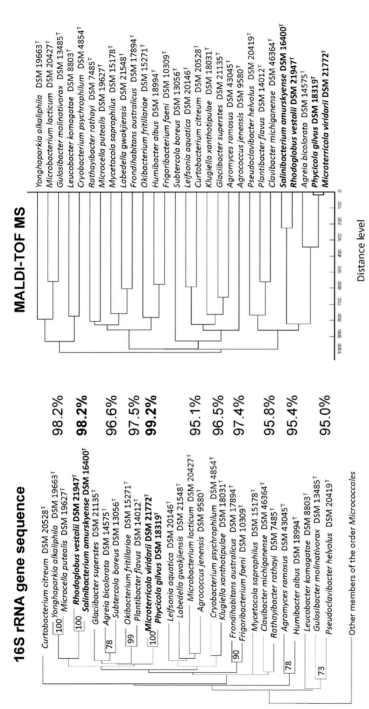

FIGURE 4

Two dendrograms showing the relationship of type strains of type species of selected genera belonging to the family *Microbacteriaceae* as revealed by comparisons of 16S rRNA gene sequences (left) and MALDI-TOF mass spectra (right). Percentage binary sequence similarities of bifurcations and bootstrap values are indicated in the dendrogram based on 16S rRNA gene sequences. Bifurcations occurring in both dendrograms are labelled in bold.

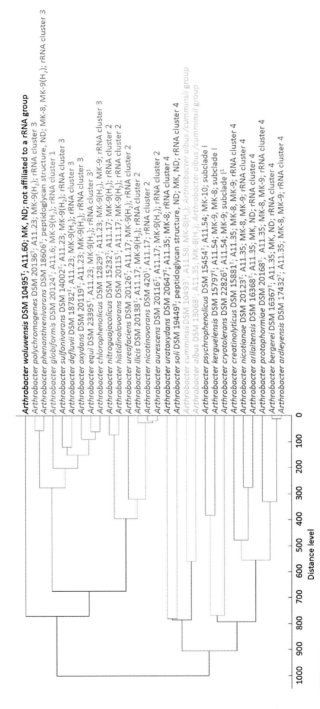

FIGURE 5

Score-oriented dendrogram based on MALDI-TOF mass spectra of selected type strains of the genus *Arthrobacter*. Peptidoglycan structure (abbreviated according to Schumann, 2011), major menaquinone(s) (MK) and rRNA group are indicated. ND, not determined. [1]Species affiliated to the rRNA group on the basis of 16S rRNA gene sequence analyses and chemotaxonomic evidence after the publication of the article of Busse et al. (2012). (See the colour plate.)

Data from Busse et al. (2012).

sequence similarity, and the pair *Arthrobacter cumminsii/Arthrobacter albus* (99.1% 16S rRNA gene sequence similarity) form coherent mass spectral clusters, too. *Arthrobacter woluwensis* represents a separate lineage within the MALDI-TOF MS dendrogram and differs also in its 16S rRNA gene sequence from all *Arthrobacter* species except *Arthrobacter nasiphocae* (Busse et al., 2012). However, in contrast to the 16S rRNA gene sequence analysis, *Arthrobacter globiformis* (rRNA cluster 1) falls within the mass spectral clade of the members of rRNA cluster 3. rRNA cluster 4, with members showing only 95.5% 16S rRNA gene sequence similarity, splits into two MALDI-TOF MS subclusters (Figure 5). These results show that MALDI-TOF MS has a potential to give an insight into the phylogenetic structure of a genus and confirm the agreement with 16S rRNA clustering for strains with similarities above 97% (see also Sauer et al., 2008).

4.2.3 Classification of strains within a species
Strains that share more than 97% (Stackebrandt & Goebel, 1994), 98.5% (Stackebrandt & Ebers, 2006) or 98.65% (Kim, Oh, Park, & Chun, 2014) 16S rRNA gene sequence similarity need to be examined for their species' status by DNA–DNA hybridisation where a value >70% binding indicates membership in the same genomospecies (Wayne et al., 1987). Due to its taxonomic resolution being limited to the range of approx. 97–99% 16S rRNA gene sequence similarity, MALDI-TOF MS is best suited for classification and identification at the species level (De Bruyne et al., 2011; Meetani & Voorhees, 2005; Teramoto et al., 2007a; Welker & Moore, 2011).

Molecular techniques often meet serious problems in the differentiation of species that were defined on the basis of a very limited number of discriminatory phenotypic traits, like, e.g., *B. cereus/B. thuringiensis*, *E. coli/Shigella* spp., *Streptococcus mitis/Streptococcus pneumoniae*. Here taxonomic unifications might be appropriate and it is not surprising that routinely applied MALDI-TOF MS also comes to its limits when attempting to resolve these groups (Denapaite et al., 2010; Lan, Alles, Donohoe, Martinez, & Reeves, 2004; Zheng et al., 2013).

4.2.4 Resolution at the level of subspecies
Strains belonging to the same species tend to display very subtle mass spectral differences that cause difficulties in their discrimination (Rezzonico, Vogel, Duffy, & Tonolla, 2010; Sandrin et al., 2013). Ruiz-Moyano, Tao, Underwood, and Mills (2012) reported the successful differentiation of representatives of *Bifidobacterium animalis* subspecies *animalis* and *Bifidobacterium animalis* subspecies *lactis* by sets of subspecies-specific mass spectrometric biomarker peaks while these two subspecies could also be distinguished by sequencing of the genes *tuf* and *atpD*, as well as by single-nucleotide polymorphisms. Tanigawa, Kawabata, and Watanabe (2010) discriminated strains of *Lactococcus lactis* subspecies *lactis* and *Lactococcus lactis* subspecies *cremoris* by MALDI-TOF MS and by genotypic and phenotypic methods. However, *Campylobacter sputorum* subspecies *sputorum* DSM 10535[T] (from the human oral cavity) and *Campylobacter sputorum* subspecies *bubulus* DSM 5363[T] (from bull sperm), showed 100% 16S rRNA gene sequence similarity and could not

be distinguished by MALDI-TOF MS (P. Schumann, unpublished). These examples demonstrate that MALDI-TOF MS is capable of discriminating subspecies that are genetically different. However, if subspecies are defined, e.g., by the hosts from which the isolates originated or by only a single phenotypic trait, their molecular and MALDI-TOF MS-based differentiation may be challenging or even impossible.

4.2.5 Differentiation of bacterial strains

Early studies were dedicated to the question of whether or not MALDI-TOF MS was capable of discriminating at the strain level (Arnold & Reilly, 1998; Dickinson et al., 2004; Siegrist et al., 2007; Vargha et al., 2006) because the typing of strains is of crucial relevance, for example, for epidemiological studies, detection of strains resistant to antibiotics, quality control of culture collections and microbial source tracking in the food industry. Judging the resolution of strains by MALDI-TOF MS requires the comparison of results of comprehensive studies to those of established methods for strain typing such as, e.g., serotyping, pulsed-field gel electrophoresis (PFGE), repetitive extragenic palindromic PCR, multilocus sequence typing, housekeeping gene sequence analysis and ribotyping. Schumann and Pukall (2013) showed that automated ribotyping appears to be superior to routinely applied MALDI-TOF MS in its discriminatory power at the strains level for *Campylobacter jejuni* and *E. coli* strains. The EHEC strains DSM 19206 and DSM 15856 could not be distinguished from non-pathogenic *E. coli* strains by MALDI-TOF MS in this study. However, the discriminatory power of MALDI-TOF MS can be enhanced by screening for discriminatory mass spectrometric features. Mazzeo et al. (2006) claimed to recognise members of the serotype EHEC O157:H7 by the presence of a peak at 9740 *m/z* combined with the lack of a peak at 9060 *m/z* typical of other *E. coli* strains. Also in other studies where spectra were screened for strain-specific peaks as biomarkers (Ruelle, Moualij, Zorzi, Ledent, & Pauw, 2004), subtle differences between spectra were detected by a sensitive correlation analysis (Arnold & Reilly, 1998) or the weight of biomarker peaks was enhanced in comparison to non-specific signals (Dieckmann, Helmuth, Erhard, & Malorny, 2008; Sauer et al., 2008). A resolution of MALDI-TOF MS as high as that of PFGE was demonstrated by Fujinami et al. (2011) in epidemiological studies of *Legionella* strains, while MALDI-TOF MS analyses took only few hours the PFGE runs took several days. A discriminatory power of MALDI-TOF MS comparable to that of serotyping was achieved for *Salmonella enterica* serovars and pathogenic *Y. enterocolitica* strains by Dieckmann and Malorny (2011) and Stephan et al. (2011), respectively. The perspectives and limitations of MALDI-TOF MS for discerning bacterial strains are discussed in several reviews (e.g. Arnold, Karty, & Reilly, 2006; Lartigue, 2013; Sandrin et al., 2013).

While only limited standardisation of cultivation conditions and sample preparation is needed for identification at the species level, rigorously standardised protocols are crucial if the aim is the differentiation of bacterial strains by MALDI-TOF MS. Alteration of growth parameters may result in the fading of strain-specific mass spectral differences in such a way that single strains cannot be recognised anymore

(Rezzonico et al., 2010). The quality of spectra, their reproducibility and mass accuracy have to be much higher for differentiation of strains than for species-level identifications. Cell extract-based sample preparation methods rather than direct colony transfer have been used by many groups when the focus was on strain-specific biomarkers for the building of databases (Sandrin et al., 2013).

Bioinformatic-enabled approaches (Demirev, Ho, Ryzhov, & Fenselau, 1999) for strain differentiation are based on genome sequence data to identify proteins that are discriminatory for bacterial strains (Sandrin et al., 2013). Intact protein markers that turned out to be strain-resolving are identified by comparison of their masses to those of proteins predicted from genome sequences (see Section 4.4 for taxon-specific biomarkers). Dieckmann et al. (2008) used this technique for the strain-level resolution in the genus *Salmonella*. Pre-fractionated proteins can be digested enzymatically to peptides for mass spectrometric identification in another approach (bottom-up) to identification of strains. In a third strategy (top-down) tandem mass spectrometry (MS-MS) is applied to cleave proteins into peptides. The MS-MS spectra are compared to proteome database entries for a given bacterium in order to identify a strain. Bioinformatic-enabled approaches might open up novel possibilities for strain-level profiling when compared to library based approaches but are limited by their requirements for expensive tandem mass spectrometers, sophisticated software packages, more time and labour and the availability of information from genome sequences (Sandrin et al., 2013).

4.2.6 *Optimal field of application of MALDI-TOF MS in bacterial systematics*
While the method appears to be inappropriate for information on the relationship of distantly related genera or even higher taxonomic ranks (see also Welker & Moore, 2011), MALDI-TOF MS seems to be capable of reflecting the intrageneric relationship of species. Differentiation and identification of taxonomically soundly defined bacterial species is doubtlessly the strong point of MALDI-TOF MS where its performance is comparable to that of 16S rRNA gene sequence analyses. A taxonomic resolution high enough for subspecies or even strains cannot usually be achieved by MALDI-TOF MS as applied in routine identification but requires sophisticated measures for increasing the discriminatory power of this technique (see Section 4.2.5).

4.3 IDENTIFICATION
The success of MALDI-TOF MS in identification of bacterial species is impressively documented in the scientific literature. The essential step for species identification is the comparison of the mass spectrum of the strain under study with a database containing reference mass spectra acquired under quality criteria as outlined in Section 3. The automated output should be an identification hit list of proposed taxa with a numerical value indicating the rating of each match and the taxonomic rank (genus or species) to which the result is applicable. The strains on which the reference spectra are based have to be specified in the database in order to give the user the possibility to judge the reliability of the identification hits. The content of the

database should match the purpose of identification. However, even a database designed for medical diagnostics should contain in addition to spectra of representatives of infection outbreaks also those of type strains as "taxonomic marker entries" in order to provide a systematically correct name, e.g., for searching the scientific literature. As mass spectra may be strain-specific (see Section 4.2), a species should be represented at best not only by the type strain but also by additional authentic strains. As most erroneous identifications are due to wrongly classified reference strains or insufficient coverage by the database, the selection of reference strains for building up a database as well as its regular quality control and continuous completion require responsible attention (Lartigue, 2013). As commercial databases are usually built up and driven by the interests of financially strong customers, public customised databases created and shared by users for specific fields of interest should be encouraged by the instrument manufacturers and supported by software development.

4.4 MALDI-TOF MS AND DISCOVERY OF NOVEL ORGANISMS

The increasing application of MALDI-TOF MS in classification and identification of bacteria raises the question about the role of this technique in the discovery of novel organisms and in their discrimination from validly named taxa. Since 2006, 76 novel taxa have been described in the International Journal of Systematic and Evolutionary Microbiology with reference to MALDI-TOF MS data as discriminatory characteristics. Minimal standards for describing new taxa of the suborder *Micrococcineae* (Schumann, Kämpfer, Busse, Evtushenko, & for the Subcommittee on the Taxonomy of the Suborder Micrococcineae of the International Committee on Systematics of Prokaryotes, 2009) and of *Bifidobacterium*, *Lactobacillus* and related genera (Mattarelli et al., 2014) encourage the application of MALDI-TOF MS for delineation of novel organisms. Indeed, it has been suggested recently that this technique be seen as a new tool in polyphasic taxonomy (Ramasamy et al., 2014; Vandamme & Peeters, 2014). Descriptions of novel taxa using MALDI-TOF MS data emphasise in their argumentation the occurrence of sets of discriminatory peaks or distant clustering in dendrograms based on mass spectral similarity. However, formal criteria for the delineation of species and genera by MALDI-TOF MS have not been established. Ramasamy et al. (2014) consider strains as unknown and subject them to subsequent 16S rRNA gene sequence analysis when the score value to entries of the database is lower than 2. This approach depends on the comprehensiveness of the database and appears to be derived from criteria of the MALDI Biotyper software (Bruker Daltonics) where score values between 2.000 and 2.299 stand for a "secure genus, probable species identification", while score values between 1.700 and 1.999 are considered indicative of a "probable genus identification". However, ranges of score values should be interpreted with care and their meaning be understood just as advice as to how to estimate the confidence level of an identification result. For this reason and because the meaning of score values suggested by the MALDI Biotyper software is not universally applicable to all taxonomic groups, these ranges cannot be considered appropriate criteria for decisions in systematics.

The LC-MS/MS identification of housekeeping proteins that give rise to major peaks in MALDI-TOF mass spectra for defining taxon-specific biomarker sets (see Section 4.2.5 for strain-specific biomarkers) is a concept of great promise for application in bacterial systematics (Wynne, Fenselau, Demirev, & Edwards, 2009). Increasing availability of genome sequences offers the possibility for predicting suitable biomarkers for related strains from gene sequences of the most abundant low-molecular-weight proteins identified by shotgun proteomics. The internally calibrated m/z peaks of whole-cell MALDI-TOF mass spectra can be matched with the theoretical molecular weights of these proteins (including consideration of post-translational modifications, e.g., methionine removal and/or acetylation) using their singly and doubly charged molecular ions. For prediction of best suited biomarkers, the occurrence of the corresponding conserved sequences as a single copy in the genomes of other strains of the same taxon needs to be verified. Because related strains may display small variations in the molecular weight of a marker protein due to amino acid substitutions, different m/z values may represent a biomarker. Although the content of even the most comprehensive database cannot cover the overwhelming diversity of microorganisms in environmental samples, this approach was developed and successfully tested for screening new representatives of the genus *Ruegeria* and the *Roseobacter* clade (Christie-Oleza, Miotello, & Armengaud, 2013; Christie-Oleza, Pina-Villalonga, et al., 2013).

CONCLUSIONS AND OUTLOOK

MALDI-TOF MS has proven its applicability under scrutiny for clinical diagnostics and the number of laboratories trusting in this technology has been continuously increasing. It can be expected that in the next 10 years MALDI-TOF MS will largely replace established phenotypic methods, which incur approximately triple costs and typically take more than 30-fold longer in diagnostic routines (Seng et al., 2010). The steadily growing availability of genome sequences coinciding with improved performance and also dropping costs of mass spectrometers and software will give rise to next-generation MALDI-TOF MS applications in bacteriology. Some trends for novel technologies are becoming apparent, e.g., proteome-based approaches (Intelicato-Young & Fox, 2013), *S10*-GERMS method (Tamura, Hotta, & Sato, 2013), SELDI-TOF MS (Dubska et al., 2011), imaging MALDI-TOF MS (Yang et al., 2012) and even beyond strain-level differentiation such as discrimination of metabolic states (Kuehl, Marten, Bischoff, Brenner-Weiss, & Obst, 2011) or tracking the transition from susceptibility to resistance against antibiotics (Kostrzewa, Sparbier, Maier, & Schubert, 2013; Shah et al., 2011). These studies raise expectations for further innovation in MALDI-TOF MS and its novel potential in microbiology.

ACKNOWLEDGMENTS

The authors gratefully acknowledge the skillful assistance of Ulrike Steiner and Claudia Wahrenburg (both DSMZ) in recording MALDI-TOF mass spectra and taking the microphotographs of Figure 2, respectively.

REFERENCES

Alatoom, A. A., Cunningham, S. A., Ihde, S. M., Mandrekar, J., & Patel, R. (2011). Comparison of direct colony method versus extraction method for identification of gram-positive cocci by use of Bruker Biotyper matrix-assisted laser desorption ionization-time of flight mass spectrometry. *Journal of Clinical Microbiology*, *49*, 2868–2873.

Anhalt, J. P., & Fenselau, C. (1975). Identification of bacteria using mass spectrometry. *Analytical Chemistry*, *47*, 219–225.

Arnold, R. J., Karty, J. A., & Reilly, J. P. (2006). Bacterial strain differentiation by mass spectrometry. In C. L. Wilkins & J. O. Lay (Eds.), *Identification of microorganisms by mass spectrometry* (pp. 181–201). Hoboken: John Wiley & Sons, Inc.

Arnold, R. J., & Reilly, J. P. (1998). Fingerprint matching of *E. coli* strains with matrix-assisted laser desorption/ionization time-of-flight mass spectrometry of whole cells using a modified correlation approach. *Rapid Communications in Mass Spectrometry*, *12*, 630–636.

Barreiro, J. R., Braga, P. A., Ferreira, C. R., Kostrzewa, M., Maier, T., Wegemann, B., et al. (2012). Nonculture-based identification of bacteria in milk by protein fingerprinting. *Proteomics*, *12*, 2739–2745.

Birmingham, J., Demirev, P., Ho, Y. P., Thomas, J., Bryden, W., & Fenselau, C. (1999). Corona plasma discharge for rapid analysis of microorganisms by mass spectrometry. *Rapid Communications in Mass Spectrometry*, *13*, 604–606.

Busse, H. J., Wieser, M., & Buczolits, S. (2012). Genus III arthrobacter. In M. Goodfellow, P. Kampfer, H. J. Busse, M. E. Trujillo, K. Suzuki, W. Ludwig, & W. B. Whitman (Eds.), *Bergey's manual of systematic bacteriology: Vol. 5.* (pp. 578–624). New York: Springer.

Castanha, E. R., Fox, A., & Fox, K. F. (2006). Rapid discrimination of *Bacillus anthracis* from other members of the *B. cereus* group by mass and sequence of "intact" small acid soluble proteins (SASPs) using mass spectrometry. *Journal of Microbiological Methods*, *67*, 230–240.

Christie-Oleza, J. A., Miotello, G., & Armengaud, J. (2013). Proteogenomic definition of biomarkers for the large *Roseobacter* clade and application for a quick screening of new environmental isolates. *Journal of Proteome Research*, *12*, 5331–5339.

Christie-Oleza, J. A., Pina-Villalonga, J. M., Guerin, P., Miotello, G., Bosch, R., Nogales, B., et al. (2013). Shotgun nanoLC-MS/MS proteogenomics to document MALDI-TOF biomarkers for screening new members of the Ruegeria genus. *Environmental Microbiology*, *15*, 133–147.

Chun, J., Lee, J. H., Jung, Y., Kim, M., Kim, S., Kim, B. K., et al. (2007). EzTaxon: A web-based tool for the identification of prokaryotes based on 16S ribosomal RNA gene sequences. *International Journal of Systematic and Evolutionary Microbiology*, *57*, 2259–2261.

Claydon, M. A., Davey, S. N., Edwards-Jones, V., & Gordon, D. B. (1996). The rapid identification of intact microorganisms using mass spectrometry. *Nature Biotechnology*, *14*, 1584–1586.

Croxatto, A., Prod'hom, G., & Greub, G. (2012). Applications of MALDI-TOF mass spectrometry in clinical diagnostic microbiology. *FEMS Microbiology Reviews, 36*, 380–407.

De Bruyne, K., Slabbinck, B., Waegeman, W., Vauterin, P., De Baets, B., & Vandamme, P. (2011). Bacterial species identification from MALDI-TOF mass spectra through data analysis and machine learning. *Systematic and Applied Microbiology, 34*, 20–29.

DeMarco, M. L., & Ford, B. A. (2013). Beyond identification: Emerging and future uses for MALDI-TOF mass spectrometry in the clinical microbiology laboratory. *Clinics in Laboratory Medicine, 33*, 611–628.

Demirev, P. A., Ho, Y. P., Ryzhov, V., & Fenselau, C. (1999). Microorganism identification by mass spectrometry and protein database searches. *Analytical Chemistry, 71*, 2732–2738.

Denapaite, D., Bruckner, R., Nuhn, M., Reichmann, P., Henrich, B., Maurer, P., et al. (2010). The genome of Streptococcus mitis B6–What is a commensal? *PLoS One, 5*, e9426.

Dickinson, D. N., La Duc, M. T., Haskins, W. E., Gornushkin, I., Winefordner, J. D., Powell, D. H., et al. (2004). Species differentiation of a diverse suite of *Bacillus* spores by mass spectrometry-based protein profiling. *Applied and Environmental Microbiology, 70*, 475–482.

Dieckmann, R., Helmuth, R., Erhard, M., & Malorny, B. (2008). Rapid classification and identification of salmonellae at the species and subspecies levels by whole-cell matrix-assisted laser desorption ionization-time of flight mass spectrometry. *Applied and Environmental Microbiology, 74*, 7767–7778.

Dieckmann, R., & Malorny, B. (2011). Rapid screening of epidemiologically important *Salmonella enterica* subsp. *enterica* serovars by whole-cell matrix-assisted laser desorption ionization-time of flight mass spectrometry. *Applied and Environmental Microbiology, 77*, 4136–4146.

Drevinek, M., Dresler, J., Klimentova, J., Pisa, L., & Hubalek, M. (2012). Evaluation of sample preparation methods for MALDI-TOF MS identification of highly dangerous bacteria. *Letters in Applied Microbiology, 55*, 40–46.

Dridi, B. B., & Drancourt, M. (2011). Characterization of prokaryotes using MALDI-TOF mass spectrometry. *Methods in Microbiology, 38*, 283–297.

Dubska, L., Pilatova, K., Dolejska, M., Bortlicek, Z., Frostova, T., Literak, I., et al. (2011). Surface-enhanced laser desorption ionization/time-of-flight (SELDI-TOF) mass spectrometry (MS) as a phenotypic method for rapid identification of antibiotic resistance. *Anaerobe, 17*, 444–447.

Dybwad, M., van der Laaken, A. L., Blatny, J. M., & Paauw, A. (2013). Rapid identification of Bacillus anthracis spores in suspicious powder samples by using matrix-assisted laser desorption ionization-time of flight mass spectrometry (MALDI-TOF MS). *Applied and Environmental Microbiology, 79*, 5372–5383.

Easterling, M. L., Colangelo, C. M., Scott, R. A., & Amster, I. J. (1998). Monitoring protein expression in whole bacterial cells with MALDI time-of-flight mass spectrometry. *Analytical Chemistry, 70*, 2704–2709.

Farfour, E., Leto, J., Barritault, M., Barberis, C., Meyer, J., Dauphin, B., et al. (2012). Evaluation of the Andromas matrix-assisted laser desorption ionization-time of flight mass spectrometry system for identification of aerobically growing Gram-positive bacilli. *Journal of Clinical Microbiology, 50*, 2702–2707.

Fegatella, F., Lim, J., Kjelleberg, S., & Cavicchioli, R. (1998). Implications of rRNA operon copy number and ribosome content in the marine oligotrophic ultramicrobacterium *Sphingomonas* sp. strain RB2256. *Applied and Environmental Microbiology, 64*, 4433–4438.

Ferreira, L., Sanchez-Juanes, F., Gonzalez-Avila, M., Cembrero-Fucinos, D., Herrero-Hernandez, A., Gonzalez-Buitrago, J. M., et al. (2010). Direct identification of urinary tract pathogens from urine samples by matrix-assisted laser desorption ionization-time of flight mass spectrometry. *Journal of Clinical Microbiology*, *48*, 2110–2115.

Ferroni, A., Suarez, S., Beretti, J.-L., Dauphin, B., Bille, E., Meyer, J., et al. (2010). Real-time identification of bacteria and Candida species in positive blood culture broths by matrix-assisted laser desorption ionization-time of flight mass spectrometry. *Journal of Clinical Microbiology*, *48*, 1542–1548.

Fothergill, A., Kasinathan, V., Hyman, J., Walsh, J., Drake, T., & Wang, Y. F. (2013). Rapid identification of bacteria and yeasts from positive-blood-culture bottles by using a lysis-filtration method and matrix-assisted laser desorption ionization-time of flight mass spectrum analysis with the SARAMIS database. *Journal of Clinical Microbiology*, *51*, 805–809.

Fournier, R. M., Wallet, F. D. R., Grandbastien, B., Dubreuil, L., Courcol, R., Neut, C., et al. (2012). Chemical extraction versus direct smear for MALDI-TOF mass spectrometry identification of anaerobic bacteria. *Anaerobe*, *18*, 294–297.

Freiwald, A., & Sauer, S. (2009). Phylogenetic classification and identification of bacteria by mass spectrometry. *Nature Protocols*, *4*, 732–742.

Fujinami, Y., Kikkawa, H. S., Kurosaki, Y., Sakurada, K., Yoshino, M., & Yasuda, J. (2011). Rapid discrimination of *Legionella* by matrix-assisted laser desorption ionization time-of-flight mass spectrometry. *Microbiological Research*, *166*, 77–86.

Ghyselinck, J., Van Hoorde, K., Hoste, B., Heylen, K., & De Vos, P. (2011). Evaluation of MALDI-TOF MS as a tool for high-throughput dereplication. *Journal of Microbiological Methods*, *86*, 327–336.

Gibb, S., & Strimmer, K. (2012). MALDIquant: A versatile R package for the analysis of mass spectrometry data. *Bioinformatics*, *28*, 2270–2271.

Giebel, R. A., Fredenberg, W., & Sandrin, T. R. (2008). Characterization of environmental isolates of *Enterococcus* spp. by matrix-assisted laser desorption/ionization time-of-flight mass spectrometry. *Water Research*, *42*, 931–940.

Gray, T. J., Thomas, L., Olma, T., Iredell, J. R., & Chen, S. C. (2013). Rapid identification of Gram-negative organisms from blood culture bottles using a modified extraction method and MALDI-TOF mass spectrometry. *Diagnostic Microbiology and Infectious Disease*, *77*, 110–112.

Gutteridge, C. S. (1988). Characterization of microorganisms by pyrolysis mass spectrometry. *Methods in Microbiology*, *19*, 227–272.

Hettick, J. M., Kashon, M. L., Slaven, J. E., Ma, Y., Simpson, J. P., Siegel, P. D., et al. (2006). Discrimination of intact mycobacteria at the strain level: A combined MALDI-TOF MS and biostatistical analysis. *Proteomics*, *6*, 6416–6425.

Holland, R. D., Wilkes, J. G., Rafii, F., Sutherland, J. B., Persons, C. C., Voorhees, K. J., et al. (1996). Rapid identification of intact whole bacteria based on spectral patterns using matrix-assisted laser desorption/ionization with time-of-flight mass spectrometry. *Rapid Communications in Mass Spectrometry*, *10*, 1227–1232.

Intelicato-Young, J., & Fox, A. (2013). Mass spectrometry and tandem mass spectrometry characterization of protein patterns, protein markers and whole proteomes for pathogenic bacteria. *Journal of Microbiological Methods*, *92*, 381–386.

Juiz, P. M., Almela, M., Melción, C., Campo, I., Esteban, C., Pitart, C., et al. (2012). A comparative study of two different methods of sample preparation for positive blood cultures for the rapid identification of bacteria using MALDI-TOF MS. *European Journal of Clinical Microbiology & Infectious Diseases, 31,* 1353–1358.

Karas, M., & Hillenkamp, F. (1988). Laser desorption ionization of proteins with molecular masses exceeding 10,000 daltons. *Analytical Chemistry, 60,* 2299–2301.

Keys, C. J., Dare, D. J., Sutton, H., Wells, G., Lunt, M., McKenna, T., et al. (2004). Compilation of a MALDI-TOF mass spectral database for the rapid screening and characterisation of bacteria implicated in human infectious diseases. *Infection, Genetics and Evolution, 4,* 221–242.

Kim, Y. H., Han, K. Y., Lee, K., & Lee, J. (2005). Proteome response of *Escherichia coli* fed-batch culture to temperature downshift. *Applied Microbiology and Biotechnology, 68,* 786–793.

Kim, M., Oh, H.-S., Park, S.-C., & Chun, J. (2014). Towards a taxonomic coherence between average nucleotide identity and 16S rRNA gene sequence similarity for species demarcation of prokaryotes. *International Journal of Systematic and Evolutionary Microbiology, 64,* 346–351.

Kostrzewa, M., Sparbier, K., Maier, T., & Schubert, S. (2013). MALDI-TOF MS: An upcoming tool for rapid detection of antibiotic resistance in microorganisms. *Proteomics: Clinical Applications, 7,* 767–778.

Krishnamurthy, T., & Ross, P. L. (1996). Rapid identification of bacteria by direct matrix-assisted laser desorption/ionization mass spectrometric analysis of whole cells. *Rapid Communications in Mass Spectrometry, 10,* 1992–1996.

Kuehl, B., Marten, S. M., Bischoff, Y., Brenner-Weiss, G., & Obst, U. (2011). MALDI-ToF mass spectrometry-multivariate data analysis as a tool for classification of reactivation and non-culturable states of bacteria. *Analytical and Bioanalytical Chemistry, 401,* 1593–1600.

Kumar, A., Roberts, D., Wood, K. E., Light, B., Parrillo, J. E., Sharma, S., et al. (2006). Duration of hypotension before initiation of effective antimicrobial therapy is the critical determinant of survival in human septic shock. *Critical Care Medicine, 34,* 1589–1596.

Lan, R., Alles, M. C., Donohoe, K., Martinez, M. B., & Reeves, P. R. (2004). Molecular evolutionary relationships of enteroinvasive *Escherichia coli* and *Shigella* spp. *Infection and Immunity, 72,* 5080–5088.

Lartigue, M.-F. (2013). Matrix-assisted laser desorption ionization time-of-flight mass spectrometry for bacterial strain characterization. *Infection, Genetics and Evolution, 13,* 230–235.

Lasch, P., Beyer, W., Nattermann, H., Stammler, M., Siegbrecht, E., Grunow, R., et al. (2009). Identification of *Bacillus anthracis* by using matrix-assisted laser desorption ionization-time of flight mass spectrometry and artificial neural networks. *Applied and Environmental Microbiology, 75,* 7229–7242.

Lasch, P., Nattermann, H., Erhard, M., Stammler, M., Grunow, R., Bannert, N., et al. (2008). MALDI-TOF mass spectrometry compatible inactivation method for highly pathogenic microbial cells and spores. *Analytical Chemistry, 80,* 2026–2034.

Liu, H., Du, Z., Wang, J., & Yang, R. (2007). Universal sample preparation method for characterization of bacteria by matrix-assisted laser desorption ionization-time of flight mass spectrometry. *Applied Environmental Microbiology, 73,* 1899–1907.

Maier, T., & Kostrzewa, M. (2007). Fast and reliable MALDI-TOF MS-based microorganism identification. *Chemistry Today, 25,* 68–71.

Mattarelli, P., Holzapfel, W., Franz, C. M. A. P., Endo, A., Felis, G. E., Hammes, W., et al. (2014). Recommended minimal standards for description of new taxa of the genera *Bifidobacterium, Lactobacillus* and related genera. *International Journal of Systematic and Evolutionary Microbiology, 64,* 1434–1451.

Mazzeo, M. F., Sorrentino, A., Gaita, M., Cacace, G., Di Stasio, M., Facchiano, A., et al. (2006). Matrix-assisted laser desorption ionization-time of flight mass spectrometry for the discrimination of food-borne microorganisms. *Applied and Environmental Microbiology*, *72*, 1180–1189.

Meetani, M. A., & Voorhees, K. J. (2005). MALDI mass spectrometry analysis of high molecular weight proteins from whole bacterial cells: Pretreatment of samples with surfactants. *Journal of the American Society for Mass Spectrometry*, *16*, 1422–1426.

Meex, C., Neuville, F., Descy, J., Huynen, P., Hayette, M.-P., De Mol, P., et al. (2012). Direct identification of bacteria from BacT/ALERT anaerobic positive blood cultures by MALDI-TOF MS: MALDI Sepsityper kit versus an in-house saponin method for bacterial extraction. *Journal of Medical Microbiology*, *61*, 1511–1516.

Meuzelaar, H. L. C., & Kistemaker, P. G. (1973). Technique for fast and reproducible fingerprinting of bacteria by pyrolysis mass spectrometry. *Analytical Chemistry*, *45*, 587–590.

Moussaoui, W., Jaulhac, B., Hoffmann, A. M., Ludes, B., Kostrzewa, M., Riegel, P., et al. (2010). Matrix-assisted laser desorption ionization time-of-flight mass spectrometry identifies 90% of bacteria directly from blood culture vials. *Clinical Microbiology and Infection*, *16*, 1631–1638.

Ochoa, M. L., & Harrington, P. B. (2005). Immunomagnetic isolation of enterohemorrhagic *Escherichia coli* O157:H7 from ground beef and identification by matrix-assisted laser desorption/ionization time-of-flight mass spectrometry and database searches. *Analytical Chemistry*, *77*, 5258–5267.

Parisi, D., Magliulo, M., Nanni, P., Casale, M., Forina, M., & Roda, A. (2008). Analysis and classification of bacteria by matrix-assisted laser desorption/ionization time-of-flight mass spectrometry and a chemometric approach. *Analytical and Bioanalytical Chemistry*, *391*, 2127–2134.

Ramasamy, D., Mishra, A. K., Lagier, J. C., Padhmanabhan, R., Rossi, M., Sentausa, E., et al. (2014). A polyphasic strategy incorporating genomic data for the taxonomic description of novel bacterial species. *International Journal of Systematic and Evolutionary Microbiology*, *64*, 384–391.

Rezzonico, F., Vogel, G., Duffy, B., & Tonolla, M. (2010). Application of whole-cell matrix-assisted laser desorption ionization-time of flight mass spectrometry for rapid identification and clustering analysis of *Pantoea* species. *Applied and Environmental Microbiology*, *76*, 4497–4509.

Ruelle, V., Moualij, B. E., Zorzi, W., Ledent, P., & Pauw, E. D. (2004). Rapid identification of environmental bacterial strains by matrix-assisted laser desorption/ionization time-of-flight mass spectrometry. *Rapid Communications in Mass Spectrometry*, *18*, 2013–2019.

Ruiz-Moyano, S., Tao, N., Underwood, M. A., & Mills, D. A. (2012). Rapid discrimination of *Bifidobacterium animalis* subspecies by matrix-assisted laser desorption ionization-time of flight mass spectrometry. *Food Microbiology*, *30*, 432–437.

Ryzhov, V., & Fenselau, C. (2001). Characterization of the protein subset desorbed by MALDI from whole bacterial cells. *Analytical Chemistry*, *73*, 746–750.

Saffert, R. T., Cunningham, S. A., Ihde, S. M., Jobe, K. E., Mandrekar, J., & Patel, R. (2011). Comparison of Bruker Biotyper matrix-assisted laser desorption ionization-time of flight mass spectrometer to BD Phoenix automated microbiology system for identification of Gram-negative bacilli. *Journal of Clinical Microbiology*, *49*, 887–892.

Sandrin, T. R., Goldstein, J. E., & Schumaker, S. (2013). MALDI TOF MS profiling of bacteria at the strain level: A review. *Mass Spectrometry Reviews*, *32*, 188–217.

Sauer, S., Freiwald, A., Maier, T., Kube, M., Reinhardt, R., Kostrzewa, M., et al. (2008). Classification and identification of bacteria by mass spectrometry and computational analysis. *PLoS One, 3*, e2843.

Schubert, S., Weinert, K., Wagner, C., Gunzl, B., Wieser, A., Maier, T., et al. (2011). Novel, improved sample preparation for rapid, direct identification from positive blood cultures using matrix-assisted laser desorption/ionization time-of-flight (MALDI-TOF) mass spectrometry. *Journal of Molecular Diagnostics, 13*, 701–706.

Schulthess, B., Brodner, K., Bloemberg, G. V., Zbinden, R., Böttger, E. C., & Hombach, M. (2013). Identification of Gram-positive cocci by use of matrix-assisted laser desorption ionization-time of flight mass spectrometry: Comparison of different preparation methods and implementation of a practical algorithm for routine diagnostics. *Journal of Clinical Microbiology, 51*, 1834–1840.

Schumann, P. (2011). Peptidoglycan structure. *Methods in Microbiology, 38*, 101–129.

Schumann, P., Kämpfer, P., Busse, H.-J., Evtushenko, L. I., & for the Subcommittee on the Taxonomy of the Suborder Micrococcineae of the International Committee on Systematics of Prokaryotes (2009). Proposed minimal standards for describing new genera and species of the suborder Micrococcineae. *International Journal of Systematic and Evolutionary Microbiology, 59*, 1823–1849.

Schumann, P., & Pukall, R. (2013). The discriminatory power of ribotyping as automatable technique for differentiation of bacteria. *Systematic and Applied Microbiology, 36*, 369–375.

Sedo, O., Sedlacek, I., & Zdrahal, Z. (2011). Sample preparation methods for MALDI-MS profiling of bacteria. *Mass Spectrometry Reviews, 30*, 417–434.

Sedo, O., Vorac, A., & Zdrahal, Z. (2011). Optimization of mass spectral features in MALDI-TOF MS profiling of *Acinetobacter* species. *Systematic and Applied Microbiology, 34*, 30–34.

Seng, P., Drancourt, M., Gouriet, F., La Scola, B., Fournier, P. E., Rolain, J. M., et al. (2009). Ongoing revolution in bacteriology: Routine identification of bacteria by matrix-assisted laser desorption ionization time-of-flight mass spectrometry. *Clinical Infectious Diseases, 49*, 543–551.

Seng, P., Rolain, J. M., Fournier, P. E., La Scola, B., Drancourt, M., & Raoult, D. (2010). MALDI-TOF-mass spectrometry applications in clinical microbiology. *Future Microbiology, 5*, 1733–1754.

Shah, H. N., Rajakaruna, L., Ball, G., Misra, R., Al-Shahib, A., Fang, M., et al. (2011). Tracing the transition of methicillin resistance in sub-populations of *Staphylococcus aureus*, using SELDI-TOF mass spectrometry and artificial neural network analysis. *Systematic and Applied Microbiology, 34*, 81–86.

Siegrist, T. J., Anderson, P. D., Huen, W. H., Kleinheinz, G. T., McDermott, C. M., & Sandrin, T. R. (2007). Discrimination and characterization of environmental strains of *Escherichia coli* by matrix-assisted laser desorption/ionization time-of-flight mass spectrometry (MALDI-TOF-MS). *Journal of Microbiological Methods, 68*, 554–562.

Stackebrandt, E., & Ebers, J. (2006). Taxonomic parameters revisited: Tarnished gold standards. *Microbiology Today, 33*(November), 152–155.

Stackebrandt, E., & Goebel, B. M. (1994). A place for DNA-DNA reassociation and 16SrRNA sequence analysis in the present species definition in bacteriology. *International Journal of Systematic Bacteriology, 44*, 846–849.

Stephan, R., Cernela, N., Ziegler, D., Pfluger, V., Tonolla, M., Ravasi, D., et al. (2011). Rapid species specific identification and subtyping of *Yersinia enterocolitica* by MALDI-TOF mass spectrometry. *Journal of Microbiological Methods, 87*, 150–153.

Tamura, H., Hotta, Y., & Sato, H. (2013). Novel accurate bacterial discrimination by MALDI-time-of-flight MS based on ribosomal proteins coding in S10-spc-alpha operon at strain level S10-GERMS. *Journal of the American Society for Mass Spectrometry, 24,* 1185–1193.

Tanaka, K., Waki, H., Ido, Y., Akita, S., Yoshida, Y., & Matsuo, T. (1988). Protein and polymer analyses up to m/z 100 000 by laser ionization time-of-flight mass spectrometry. *Rapid Communications in Mass Spectrometry, 2,* 151–153.

Tanigawa, K., Kawabata, H., & Watanabe, K. (2010). Identification and typing of *Lactococcus lactis* by matrix-assisted laser desorption ionization-time of flight mass spectrometry. *Applied and Environmental Microbiology, 76,* 4055–4062.

Teramoto, K., Sato, H., Sun, L., Torimura, M., & Tao, H. (2007). A simple intact protein analysis by MALDI-MS for characterization of ribosomal proteins of two genome-sequenced lactic acid bacteria and verification of their amino acid sequences. *Journal of Proteome Research, 6,* 3899–3907.

Teramoto, K., Sato, H., Sun, L., Torimura, M., Tao, H., Yoshikawa, H., et al. (2007). Phylogenetic classification of *Pseudomonas putida* strains by MALDI-MS using ribosomal subunit proteins as biomarkers. *Analytical Chemistry, 79,* 8712–8719.

Vaidyanathan, S., Winder, C. L., Wade, S. C., Kell, D. B., & Goodacre, R. (2002). Sample preparation in matrix-assisted laser desorption/ionization mass spectrometry of whole bacterial cells and the detection of high mass (>20 kDa) proteins. *Rapid Communications in Mass Spectrometry, 16,* 1276–1286.

Vandamme, P., & Peeters, C. (2014). Time to revisit polyphasic taxonomy. *Antonie Van Leeuwenhoek, 106,* 57–65.

Vargha, M., Takats, Z., Konopka, A., & Nakatsu, C. H. (2006). Optimization of MALDI-TOF MS for strain level differentiation of *Arthrobacter* isolates. *Journal of Microbiological Methods, 66,* 399–409.

Wayne, L. G., Brenner, D. J., Colwell, R. R., Grimont, P. A. D., Kandler, O., Krichevsky, M. I., et al. (1987). Report of the ad hoc committee on reconciliation of approaches to bacterial systematics. *International Journal of Systematic Bacteriology, 37,* 463–464.

Welker, M., & Moore, E. R. B. (2011). Applications of whole-cell matrix-assisted laser-desorption/ionization time-of-flight mass spectrometry in systematic microbiology. *Systematic and Applied Microbiology, 34,* 2–11.

Winker, S., & Woese, C. R. (1991). A definition of the domains archaea, bacteria and eucarya in terms of small subunit ribosomal RNA characteristics. *Systematic and Applied Microbiology, 14,* 305–310.

Wüppenhorst, N., Consoir, C., Lorch, D., & Schneider, C. (2012). Direct identification of bacteria from charcoal-containing blood culture bottles using matrix-assisted laser desorption/ionisation time-of-flight mass spectrometry. *European Journal of Clinical Microbiology & Infectious Diseases, 31,* 2843–2850.

Wynne, C., Fenselau, C., Demirev, P. A., & Edwards, N. (2009). Top-down identification of protein biomarkers in bacteria with unsequenced genomes. *Analytical Chemistry, 81,* 9633–9642.

Yang, J. Y., Phelan, V. V., Simkovsky, R., Watrous, J. D., Trial, R. M., Fleming, T. C., et al. (2012). Primer on agar-based microbial imaging mass spectrometry. *Journal of Bacteriology, 194,* 6023–6028.

Zheng, J., Peng, D., Song, X., Ruan, L., Mahillon, J., & Sun, M. (2013). Differentiation of *Bacillus anthracis, B. cereus,* and *B. thuringiensis* on the basis of the csaB gene reflects host source. *Applied and Environmental Microbiology, 79,* 3860–3863.

Continuing Importance of the "Phenotype" in the Genomic Era

14

Peter Kämpfer[1]

Institut für Angewandte Mikrobiologie, Justus-Liebig-Universität Giessen, Heinrich-Buff-Ring 26,
Giessen, Germany
[1]*Corresponding author: e-mail address: peter.kaempfer@umwelt.uni-giessen.de*

1 PHYLOGENY AND GENOTYPE

The goal of establishing an evolutionary-based classification of *Bacteria* and *Archaea* is increasingly being driven by the need to quickly generate whole-genome sequences and the requirement to develop bioinformatic tools able to handle the resultant enormous amounts of data. The development of cost-effective, high-throughput deoxyribonucleic acid (DNA) sequencing, known as next-generation sequencing (NGS), is currently revolutionizing many areas of microbiology, including prokaryotic systematics.

This development began over 30 years ago with the recognition that molecular markers were present in all organisms, particularly in small subunit rRNAs, ssRNAs. The introduction of 16S rRNA sequence data into the classification and identification of prokaryotic micro-organisms can be regarded as a milestone because it became possible to use "one" molecular marker that covered a wide range of taxonomic ranks closely mirroring the whole or a major part of the evolutionary order back to the origin of life and thereby offering the possibility of establishing a hierarchical taxonomic system.

At present, a "phylogenetic analysis" based on 16S rRNA gene sequences and a determination of similarity between sequences stored in databases are routinely carried out as the first step in identifying novel organisms at the genus level and sometimes at the rank of species (Chun & Rainey, 2014; Stackebrandt & Goebel, 1994; Stackebrandt et al., 2002; Tindall et al., 2010).

It is important to note once again that the success of 16S rRNA sequence analyses in the classification of prokaryotes has led to its current undisputed position as the "backbone" of prokaryotic classification and identification at taxonomic ranks ranging from the genus up to the domain level. A very useful and comprehensive review on the important aspects on the information content of the 16S rRNA gene as the central marker molecule, its resolution, other markers, and important information on the criteria for marker selection, sequence alignment and quality problems, different methods of sequence analyses and tree comparisons has been provided by Ludwig (2010).

Despite the unchallenged importance of the 16S rRNA molecule, it has limitations, functional constraints, variable information content, problems of plesiomorphy, multiple copies of 16S rRNA genes with small intragenomic differences (up to 2 or even 5%) and problems with tree reconstructions; these problems have been addressed by Ludwig and Klenk (2001) and Ludwig (2010). It is obvious from such articles that the 16S rRNA molecule is not suitable to clearly differentiate between organisms at the currently defined species level. However, it has also been pointed out that there are not many alternative genetic markers found in all prokaryotic genomes. In this regard, the number of conserved genes present in all prokaryotic organisms has been estimated to be less than 50 (Ciccarelli et al., 2006; Koonin, 2003). Detailed comparative analyses of alternative markers such as the 23S rRNA gene, genes encoding elongation and initiation factors for translation, RNA polymerase subunits, ATPase subunits, DNA gyrases, *recA* and heat-shock proteins, have usually, but not always, been in line with the results of corresponding 16S rRNA gene sequence analyses (Ludwig, 2010).

As mentioned above, it is now general practice that all prokaryotes are allocated to genera and also to higher taxonomic ranks ("phylum", "class", "order", "family", "genus"), up to the domains "*Archaea*" and "*Bacteria*", on the basis of 16S rRNA gene sequence comparisons. Indeed, Wayne et al. (1987) pointed out more than 25 years ago that "... an ideal taxonomy would involve one system (a hierarchical system) and in microbiology, the ultimate ambition would be to establish a system that mirrors the taxonomic relationships as an 'order in nature ...", a vision that is becoming more and more of a reality.

Wayne et al. (1987) also pointed out "that the complete DNA sequence would be the reference standard to determine phylogeny and that phylogeny should determine taxonomy". Wayne and his colleagues did not provide any more details, as to how, and to what extent, the "complete DNA sequence" should be used in taxonomy but, with improvements in technology, NGS is now routinely used as a rapid and cost-effective method to obtain whole-genome sequences of prokaryotic strains (see Chapter 'Revolutionizing Prokaryotic Systematics Through Next-Generation Sequencing' by Vartul Sangal, Leena Nieminen, Nicholas P. Tucker and Paul A. Hoskisson in this volume). In the meantime, more than 12,000 archaeal and bacterial genome sequences are available for comparison, but only around 1800 of them are sequences of type strains, a situation that limits the use of genomic data in comparative taxonomic studies (Chun & Rainey, 2014). However, it can be foreseen that genome sequence data will soon be available for all type strains deposited in service culture collections.

A short history of the development and application of genome sequencing in prokaryotic systematics has been provided by Chun and Rainey (2014) from the first use of the Roche 454 sequencing system, which was based on pyrosequencing (Margulies et al., 2005), followed by Illumina DNA sequencing, initially developed by Solexa and based on bridge amplification and reversible terminators (Bentley,

2006). Now there are numerous NGS technologies available for which the protocols and the performance of instruments are constantly being updated and improved (see chapters 'Revolutionizing Prokaryotic Systematics Through Next-Generation Sequencing' by Sangal et al., and 'Whole-Genome Sequencing for Rapid and Accurate Identification of Bacterial Transmission Pathways' by Harris and Okoro).

When Wayne et al. (1987) wrote that the "complete deoxyribonucleic acid (DNA) sequence would be the reference standard to determine phylogeny and that phylogeny should determine taxonomy", it was not possible to generate large sequence data. Wayne and his colleagues advocated the estimation of genomic relatedness on the basis of experimental DNA–DNA hybridization (DDH) methods, which measure the degree of similarities/relatedness between two different genomes, without having detailed sequence data. The DDH technique, which was introduced to microbiology by Johnson and Ordal (1968), has "served" for many years as the "gold standard" for species delineation of *Archaea* and *Bacteria*, despite obvious problems, such as the "labour-intensive and error-prone nature of this method" as summarized by Rosselló-Móra (2006). Wayne et al. (1987) recommended a DDH value of 70% as a "threshold" for the bacterial species boundary, and with some exceptions, this method is still the gold standard for species delineation. Later, Stackebrandt et al. (2002) voiced the need for alternative genotypic standards.

It had become clear by this time that 16S rRNA gene sequence data alone did not provide sufficient resolution for species delineation. Indeed, detailed comparisons between 16S rRNA gene sequence similarity data and DDH values had already revealed that 16S rRNA gene sequence similarities of 97.0% and below were never accompanied by DDH values >70% (Stackebrandt & Goebel, 1994). This observation led to the view that DDH studies (which are material and time consuming) were only required for pairs of strains showing 97% or more 16S rRNA gene sequence similarity in cases involving the delineation of novel species (Tindall et al., 2010). In a detailed statistical based study, Meier-Kolthoff, Auch, Klenk, and Göker (2013) concluded that a threshold between 98.2% and 99.0% appeared to be more appropriate cut-off values, depending on the group under study.

It was suggested almost 10 years ago that DDH be replaced by comparisons of whole-genome sequence data, the latter being seen as a form of digital, *in silico* DDH (Konstantinidis & Tiedje, 2005); several such methods have been proposed (Auch et al., 2010; Meier-Kolthoff, Auch, et al., 2013; Meier-Kolthoff, Klenk, & Göker, 2014; Richter & Rosselló-Móra, 2009). An average nucleotide identity (ANI) was first introduced as a mean of establishing values between multiple sets of orthologous regions shared by two genomes (Konstantinidis & Tiedje, 2005). Later, Goris et al. (2007) updated this protocol by artificially cutting the query genome sequence into fragments of 1020 bp, thereby simulating the DNA fragmentation step in DDH experiments. At present, ANI calculations based on the BLASTN method (ANIb) are used more often in comparison to the ANI based on the MUMmer algorithm (ANIm) (Kurtz et al. 2004; as used by Richter & Rosselló-Móra, 2009). Comparative studies between ANI and DDH values revealed that ANI values of

95–96% are comparable to the 70% DDH threshold level (see chapter 'Whole-Genome Analyses:Average Nucleotide Identity' by Arahal, in this volume). A detailed summary of the indices used in ANI studies has been provided by Chun and Rainey (2014) who coined the term "overall genome relatedness indices (OGRI)" for all these indices which utilize whole-genome sequences, but not individual gene sequences or a set of individual gene sequences. With the increasing number of total or partial genome sequences, OGRI will provide simple, reproducible and objective data for comparing any two genomes and will certainly replace the cumbersome DDH methods in the near future.

However, as some OGRI cannot provide detailed information for phylogenetic analyses, investigations have been carried out on comparison of orthologous gene sequences. In this area, new developments will certainly lead to further improvements (see chapter '16S rRNA Gene-Based Identification of *Bacteria* and *Archaea* using the EzTaxon Server' by Kim and Chun, in this volume). There is also no doubt that whole-genome sequencing will increasingly become a cornerstone in identification and epidemiological typing.

One approach that is presently widely used in molecular epidemiology is multi-locus sequencing typing (MLST) (Maiden et al., 1998; Sullivan et al., 2005). In such studies, gene sequences of 8–12 selected, more or less conserved genes are determined, and each sequence is grouped into a sequence type (ST) reflecting its sequence. MLST has become established as a useful and very powerful typing method for differentiating between closely related strains within species (see chapters 'Multi-Locus Sequence Typing and the Gene-by-Gene Approach to Bacterial Classification Analysis of Population Variation' by Cody et al., and 'Multi-locus Sequence Analysis: Taking Prokaryotic Systematics to the Next Level' by Rong and Huang, in this volume). It can be regarded as a way of providing "intermediate resolution" between 16S rRNA gene and genome-based approaches and thereby provides "phylogenetic" resolution at the species level (Cole et al., 2010; Gevers et al., 2005; Konstantinidis & Tiedje, 2007). Many of the bacterial MLST Web sites that have been set up for pathogenic bacteria over the last decade are available online (http://www.mlst.net/ and http://www.pubmlst.org); comprehensive databases, for instance, are available for *Campylobacter jejuni* and *Neisseria meningitidis*. "Phylogenetic" calculations using sequence information generated in MLST schemes (referred to as MLSA) can be used for prokaryotic classification and identification. This offers the introduction of insights available from population genetics and phylogenetic approaches to bacterial systematics. However, there are also problems with MLSA, some of which have been addressed previously (Kämpfer & Glaeser, 2012). Thus, discrepancies in the topology of single gene trees have been observed; from this, it is obvious that individual makers have only a limited information content and, in addition, may have preserved information from different evolutionary time spans (Ludwig, 2010; Ludwig & Klenk, 2001).

A common strategy designed to overcome the problem of conflicting branching patterns of individual genes is to use concatenated sequences of protein-coding genes and at the same time include 16S rRNA gene sequence data. For example, this strategy was used by Serrano et al. (2010) who applied MLSA to resolve taxonomic conflicts in the genus *Marichromatium* (anoxygenic phototrophic bacteria of the family Chromatiaceae). However, this approach may also cause problems as the resulting differences may not reflect the result of very different genetic (evolutionary) processes. In a simulation study (Kubatko & Degnan, 2007), concatenated sequences of a number of protein-coding genes were found to give a high level of discord among individual gene trees and this led to statistically inconsistent estimations of relatedness. These authors also noted that the use of bootstrap values to measure support for inferred phylogenies may lead to moderate to strong support for an incorrect tree. Vinuesa (2010) pointed out that "concatenation approaches estimate a phylogeny that reflects some average of gene trees but do not explicitly estimate species trees", adding that "the latter are the trees that really matter to systematists". A critical comparison of different tree topologies is important, especially to determine genes that may be affected by homologous recombination or lateral gene transfer (LGT) (Bapteste & Boucher, 2008; Bapteste et al., 2009; Dagan, Artzy-Rrandup, & Martin, 2008; Williams, Andam, & Gogarten, 2010; Young, 2001) and (at least in certain genera) widespread recombination events (Doroghazi & Buckley, 2010). These events present a general problem in MLSA studies and are dependent on the frequency of recombination for different genes; different groups of organisms are affected to different extents by these phenomena.

The most critical aspect of MLSA studies is the selection of genes and primers. As pointed out by Gevers et al. (2005), large-scale studies of well-known genera are still needed to provide a framework for MLSA studies. Ideally, such studies should include amino acid-based sequence data Kämpfer and Glaeser (2011). In nucleotide sequence-based studies, the third codon position is problematic because the base compositional bias concentrated at this position can interfere with the results of phylogenetic analyses (as reviewed by Harayama & Kasai, 2006). Consequently, the phylogenetic resolution of MLSA data would be more reliable and stable if only the first and second codon positions were used or alternatively if the corresponding amino acid sequences were considered. However, Chun and Hong (2010) pointed out that nucleotide and respective amino acid sequence-based phylogenetic treeing may sometimes generate contradictory results. Thus, comparison of the sequences of very similar proteins gives less to no resolution at the amino acid level for in such cases sequence variation occurs only at redundant codon positions. More distantly related proteins may contain high variation at the nucleotide sequence level; hence, it is important that aligned amino acid sequences are used. Some of these problems were addressed by Kämpfer and Glaeser (2012) who illustrated them using examples drawn from *Acinetobacter*, *Pseudomonas* and *Streptomyces* systematics.

Refinements are being made to the MLSA approach (see chapter 'Multi-locus Sequence Analysis: Taking Prokaryotic Systematics to the Next Level' by Rong and Huang, in this volume) as exemplified by Mende, Sunagawa, Zeller, and Bork (2013) who proposed a novel MLSA method, called specI, in which 40 universal, single-copy, protein-coding genes were selected to calculate a genome similarity based on an ANI of 40 genes. Within the genus *Vibrio*, an MLSA analysis has been conducted using 1000 genes (Chun et al., 2009; Haley et al., 2010), which represents a ground breaking approach.

2 THE PHENOTYPE

Moore, Mihaylova, Vandamme, Krichevsky, and Dijkshoorn (2010) defined phenotypic traits as "observable characteristics that result from the expression of genes of an organism, which can largely be modulated by environmental or other conditions (e.g. growth conditions, temperature, pH)".

An important prerequisite for the recognition and study of the phenotype of a prokaryote is the fact that the organism (in most cases as a pure culture) has to be cultivated. Overmann (2006) reviewed the basic principles of cultivation and Pham & Kim (2012) recently summarized some novel developments in this area. These included the application of modified and adapted media (Alain & Querellou, 2009; Button, Schut, Quang, Martin, & Robertson, 1993; Connon & Giovannonis, 2002; Hamaki et al., 2005; Janssen, 2008; Janssen, Yates, Grinton, Taylor, & Sait, 2002; Kawanishi et al., 2011; Kim, 2011; Leadbetter, Schmidt, Graber, & Breznak, 1999; Sangwan, Kovac, Davis, Sait, & Janssen, 2005; Stevenson, Eichhorst, Wertz, Schmidt, & Breznak, 2004; Tamaki & Kamagata, 2005; Tamaki et al., 2005; Tyson & Banfield, 2005; Uphoff, Felske, Fehr, & Wagner-Döbler, 2001; Watve et al., 2000), changes in growth conditions (Alain & Querellou, 2009; Davis, Joseph, & Janssen, 2005; Davis, Sangwan, & Janssen, 2011; Janssen, 2008; Janssen et al., 2002; Kataoka, Tokiwa, Tanaka, Takeda, & Suzuki, 1996; Sait, Hugenholtz, & Janssen, 2002; Sangwan et al., 2005; Stevenson et al., 2004; Stott et al., 2008; Watve et al., 2000), culture of communities and coculturing (Crocetti, Banfield, Keller, Bond, & Blackall, 2002; Evstigneeva, Raoult, Karpachevskiy, & La Scola, 2009; Frey-Klett, Garbaye, & Tarkka, 2007; Nadell, Xavier, & Foster, 2009; Nichols et al., 2008; Pagnier, Raoult, & La Scola, 2008; Plugge & Stams, 2002; Tamaki & Kamagata, 2005; Vartoukian, Palmer, & Wade, 2010; West, Diggle, Buckling, Gardner, & Griffin, 2007), the application of optical tweezers and other equipment, such as laser microdissection (Tyson & Banfield, 2005), construction of high-throughput microbioreactors (Amanullah et al., 2010; Connon & Giovannonis, 2002; Leadbetter, 2003; Lewis, Epstein, D'Onofrio, & Ling, 2010) and the use of simulated natural environments using diffusion chambers (Bollmann, Lewis, & Epstein, 2007; Ferrari, Binnerup, & Gillings, 2005; Ferrari, Winsley, Gillings, & Binnerup, 2008; Kaeberlein, Lewis, & Epstein, 2002; Kim, 2011). However, despite such developments, there have been very few pioneering and groundbreaking advances in cultivation procedures. Consequently, phenotypic methods, when compared with the tremendous advances made in genotyping by sequencing methods, have not changed significantly in recent times.

The traditional basis for the characterization of *Archaea* and *Bacteria* remains the phenotype. Phenotypic traits may appear to be easy to recognize, but it is important to realize that they may represent very complex features, such as general growth characteristics (aerobic, anaerobic or CO_2 requirements), colony and cell morphology and physiological features which are not only the result of the expression of genes but also reflect the conditions under which the genes are expressed. In other words, such complex features cannot simply be deduced from the presence or absence of specific genes

(Kämpfer, 2011). At present, only pure culture studies can fulfil the requirements for in-depth studies of microbial physiology with regard to the roles of genes, proteins and metabolic pathways. Again, as pointed out by Tindall et al. (2010), the use of type strains is of central importance in prokaryotic systematics; putatively novel organisms need to be compared to the type strains and the type species of the genus before they can be considered to be a member of that genus.

Phenotypic characters also include structural components of prokaryotic cells, such as cell walls, cell membranes and the cytoplasm. When formed (of course on the basis of expressed underlying genetic information) such properties are stable and predictable. Many taxonomically useful differences may be detected from chemotaxonomic studies of cell peptidoglycans, the presence or absence of teichoic and/or mycolic acids, and from the discontinuous distribution of certain fatty acids, polar lipids, respiratory quinones, pigments and polyamines (Tindall et al., 2010). Combinations of such specific cell constituents cannot usually be deduced from the presence of genes or gene clusters that encode them. Again, the observed phenotype is a result of the conditions under which the underlying genotype is expressed.

As a result of these dependencies, many methods targeting complex phenotypic characteristics of prokaryotes suffer from differing degrees of reproducibility (Moore et al., 2010). Consequently, it is essential that rigorous standardized conditions are used to acquire phenotypic data, notably for the establishment of comprehensive databases (Tindall, De Vos, & Trüper, 2008; Tindall, Sikorski, Smibert, & Krieg, 2008). Recently, more sophisticated phenotyping systems have been introduced to generate high-quality phenotypic data for classification and identification, such as the matrix-assisted-laser desorption/ionization time of flight mass spectrometry (MALDI-TOF MS) (Welker & Moore, 2011), which is considered in chapter 'MALDI-TOF Mass Spectrometry Applied to Classification and Identification of Bacteria' by Schuman and Maier, in this volume and high-field ion cyclotron Fourier transform mass spectroscopy (ICP-FT MS), which can be referred to as metabolomics. However, despite the capacity of such systems to generate large amounts of data, which can be stored in databases, the general restrictions of cultivation-based dependencies still apply.

3 THE ONGOING IMPORTANCE OF THE PHENOTYPE IN AN ORGANISM BASED TAXONOMY

The basic unit in prokaryotic systematics has long been recognized as the "species". However, despite this, there are still no universally accepted definitions of species in bacteriology (Cowan, 1978; Kämpfer, 2011). Cowan (1978) addressed three meanings for the term "species":

- a category (a mental representation),
- a taxonomic group, and
- a concept.

Krichevsky (2011) reminded us that, as the English philosopher John Locke (1632–1704) argued, nature does not make species: "People do, as a mechanism to facilitate communication of a collection of similar ideas under one general term" (cited by Krichevsky, 2011). In a reply to several articles on "Species concepts" published in Microbiology Today in 2007, Sneath (2007) noted: "Historically, the term 'species' was taken by the early botanists and zoologists in the sense of the smallest distinct groups of individual organisms, that is, the members of a group were not only very similar to each other, but the group was also distinct from nearby groups. The earliest usage did not prescribe in what manner groups were distinct, though it was usually based on some form of overall morphological similarity. The groups thus corresponded to primary clusters. These were the smallest clusters that were clearly distinct from others. The same still applies for groups from molecular sequence or other data".

Several practical "circumscriptions" of a "species" can be found in the literature, always reflecting the development of new methods, e.g. "A group of related organisms that is distinguished from similar groups by a constellation of significant genotypic, phenotypic and ecological characteristics" (Colwell, 1970). Notably, Wayne et al. (1987) were the first to use the term "phylogenetic" in a microbial species "definition" when they wrote: "At present, the species is the only taxonomic unit that can be defined in *phylogenetic terms*. In practice, DNA reassociation approaches the sequence standard and represents the best applicable procedure at the present time. The phylogenetic definition of a species generally would include strains with approximately 70% or greater DNA–DNA relatedness and with $5\,°C$ or less ΔT_m. Both values must be considered".

Wayne and his colleagues went on to say: "Phenotypic characteristics should agree with this definition and would be allowed to override the phylogenetic concept of species only in a few exceptional cases. It is recommended that a distinct genospecies that cannot be differentiated from another genospecies on the basis of any known phenotypic property not be named until they can be differentiated by some phenotypic property". Stackebrandt et al. (2002) anticipated the importance of gene and genome sequences for the "definition" of species: "A category that circumscribes a (preferably) genomically coherent group of individual isolates/strains sharing a high degree of similarity in (many) independent features, comparatively tested under highly standardized conditions". They also addressed the importance of comprehensive characterization studies: "More emphasis should be placed on discriminating markers. Description of species should be based on the use of well-documented criteria, laboratory protocols and reagents which are reproducible. Descriptive and diagnostic characters should be described in sufficient detail to permit comparisons between taxa and allow reproduction of observations" (Stackebrandt et al., 2002).

To this, Tindall et al. (2010) added: "The characterization of a strain is a key element in prokaryote systematics. Although various new methodologies have been developed over the past 100 years both the newer methodologies and those considered as being "traditional" remain a key element in determining whether a strain belongs to a known taxon or constitutes a novel one. In the case of a known taxon, a selected set of tests may be used to determine that a strain has been identified as the member of an existing taxon. However, in the case of a strain or set of strains shown to be novel taxa, they should be characterized as comprehensively as possible. The goal of that characterization is to place them within the hierarchical framework laid down by the Bacteriological Code (Lapage et al., 1992), as well as to provide a description of the taxon". This basic principle should not be changed in the era of "omics".

CONCLUSIONS AND CHALLENGES

As pointed out earlier (Kämpfer, 2011), the key question is whether a future taxonomic system for prokaryotes should be "organism"-based or basically "genome-sequence-based". The basic unit of evolution (and hence taxonomy) is the organism with its smallest unit, the cell. In this context, natural selection drives evolution by selecting from existing phenotypes; hence, it is the phenotype that drives this process, both in a cellular and in an environmental context.

Prokaryotic systematics serves many purposes (Moore et al., 2010), and hence, it is essential that classifications are stable and predictable. Many laboratories are still working with cultivation-based approaches and the phenotype. However, comprehensive phenotypic and genotypic characterization (in the framework of a polyphasic approach) is necessary for classification (which is a prerequisite of identification). Hence, both traditional and novel phenotypic approaches are important for the characterization of novel taxonomic categories, such as genera and species. Genomic (and other omic) approaches will of course provide a rich source of additional data, as exemplified by the recent study of Qin et al. (2014), which provides a parameter for delineating bacterial genera. Hence, new approaches to classification should continue to be based upon the polyphasic taxonomic concept, which should encompass new methodological approaches with the genome as the basic underlying information but still considering the phenotype and other essential elements, notably the nomenclatural type concept.

Brenner (2010) pointed out in his review on synthetic biology, "that it is very difficult to predict higher levels of "information" from genome data sets" and went on to say that "Molecules may tell us nothing about cells and their behaviour. In essence, the conversion of data into knowledge (at different levels) constitutes a great challenge for future biological research". The same restraints and challenges apply to prokaryotic systematics!

REFERENCES

Alain, K., & Querellou, J. (2009). Cultivating the uncultured: Limits, advances and future challenges. *Extremophiles*, *13*, 583–594.

Amanullah, A., Otero, J. M., Mikola, M., Hsu, A., Zhang, J., Aunins, J., et al. (2010). Novel micro-bioreactor high throughput technology for cell culture process development: Reproducibility and scalability assessment of fed-bath CHO culture. *Biotechnology and Bioengineering*, *1*, 57–67.

Auch, A. F., Klenk, H. P., & Göker, M. (2010). Standard operating procedure for calculating genome-to-genome distances based on high-scoring segment pairs. *Standards in Genomic Sciences*, *2*, 142–148.

Bapteste, E., & Boucher, Y. (2008). Lateral gene transfer challenges principles of microbial systematics. *Trends in Microbiology*, *16*, 200–207.

Bapteste, E., O'Malley, M. A., Beiko, R. G., Ereshefky, M., Gogarten, J. P., Franklin-Hall, L., et al. (2009). Prokaryotic evolution and the tree of life are two different things. *Biology Direct, 4*, 34. http://dx.doi.org/10.1186/1745-6150-4-34.

Bentley, D. R. (2006). Whole-genome re-sequencing. *Current Opinion in Genetics and Development, 16*, 545–552.

Bollmann, A., Lewis, K., & Epstein, S. S. (2007). Incubation of environmental samples in a diffusion chamber increases the diversity of recovered isolates. *Applied and Environmental Microbiology, 73*, 6386–6390.

Brenner, S. (2010). Sequences and consequences. *Philosophical Transactions of the Royal Society B, 365*, 207–212.

Button, D. K., Schut, F., Quang, P., Martin, R., & Robertson, B. R. (1993). Viability and isolation of marine bacteria by dilution culture: Theory, procedures, and initial results. *Applied and Environmental Microbiology, 59*, 881–891.

Chun, J., Grim, C. J., Hasan, N. A., Lee, J. H., Choi, S. Y., Haley, B. J., et al. (2009). Comparative genomics reveals mechanism for short-term and long-term clonal transitions in pandemic *Vibrio cholerae*. *Proceedings of the National Academy of Sciences of the United States of America, 106*, 15442–15447.

Chun, J., & Hong, S. G. (2010). Methods and programs for calculation of phylogenetic relationships from molecular sequences. In A. Oren & R. T. Papke (Eds.), *Molecular phylogeny of microorganisms* (pp. 23–40). Norfolk: Caister Academic Press.

Chun, J., & Rainey, F. A. (2014). Integrating genomics into the taxonomy and systematics of the *Bacteria* and *Archaea*. *International Journal of Systematic and Evolutionary Microbiology, 64*, 316–324.

Ciccarelli, F. D., Doerks, T., von Mering, C., Creevey, C. J., Snel, B., & Bork, P. (2006). Toward automatic reconstruction of a highly resolved tree of life. *Science, 311*, 1283–1287.

Cole, J. R., Konstanidis, K., Farris, R. J., & Tiedje, J. M. (2010). Microbial diversity phylogeny: Extending from rRNAs to genomes. In W.-T. Liu & J. K. Jackson (Eds.), *Environmental molecular microbiology* (pp. 1–19). Norfolk: Caister Academic Press.

Colwell, R. R. (1970). Polyphasic taxonomy of bacteria. In H. Iizuka & T. Hazegawa (Eds.), *Culture collections of microorganisms* (pp. 421–436). Tokyo: University of Tokyo Press.

Connon, S. A., & Giovannonis, S. J. (2002). High-throughput methods for culturing microorganisms in very-low-nutrient media yield diverse new marine isolates. *Applied and Environmental Microbiology, 68*, 3878–3885.

Cowan, S. T. (1978). *A dictionary of microbial taxonomy*. Cambridge, UK: Cambridge University Press.

Crocetti, G. R., Banfield, J. F., Keller, J., Bond, P. L., & Blackall, L. L. (2002). Glycogen-accumulating organisms in laboratory-scale and full-scale wastewater treatment processes. *Microbiology, 148*, 3353–3364.

Dagan, T., Artzy-Rrandup, Y., & Martin, W. (2008). Modular networks and cumulative impact of lateral transfer in prokaryote genome evolution. *Proceedings of the National Academy of Sciences of the United States of America, 105*, 10039–10044.

Davis, K. E. R., Joseph, S. J., & Janssen, P. H. (2005). Effects of growth medium, inoculum size, and incubation time on culturability and isolation of soil bacteria. *Applied and Environmental Microbiology, 71*, 826–834.

Davis, K. E. R., Sangwan, P., & Janssen, P. H. (2011). *Acidobacteria, Rubrobacteridae* and *Chloroflexi* are abundant among very slow-growing and mini-colony forming soil bacteria. *Environmental Microbiology, 13*, 798–805.

Doroghazi, J. R., & Buckley, D. H. (2010). Widespread homologous recombination within and between *Streptomyces* species. *Multidisciplinary Journal of Microbial Ecology, 4*, 136–1143.

Evstigneeva, A., Raoult, D., Karpachevskiy, L., & La Scola, B. (2009). Amoebae co-culture of soil specimens recovered 33 different bacteria, including 4 new species and *Streptococcus pneumoniae*. *Microbiology*, *155*, 657–664.

Ferrari, B. C., Binnerup, S. J., & Gillings, M. (2005). Microcolony cultivation on a soil substrate membrane system selects for previously uncultured soil bacteria. *Applied and Environmental Microbiology*, *71*, 8714–8720.

Ferrari, B. C., Winsley, T., Gillings, M., & Binnerup, S. (2008). Cultivating previously uncultured soil bacteria using a soil substrates membrane system. *Nature Protocols*, *3*, 1261–1269.

Frey-Klett, P., Garbaye, J., & Tarkka, M. (2007). The mycorrhiza helper bacteria revisited. *New Phytologist*, *176*, 22–36.

Gevers, D., Cohan, F. M., Lawrence, J. G., Spratt, B. G., Coenye, T., Feil, E. J., et al. (2005). Re-evaluating prokaryotic species. *Nature Reviews. Microbiology*, *3*, 733–739.

Goris, J., Konstantinidis, K. T., Klappenbach, J. A., Coenye, T., Vandamme, P., & Tiedje, J. M. (2007). DNA-DNA hybridization values and their relationship to whole-genome sequence similarities. *International Journal of Systematic and Evolutionary Microbiology*, *57*, 81–91.

Haley, B. J., Grim, C. J., Hasan, N. A., Choi, S. Y., Chun, J., Brettin, T. S., et al. (2010). Comparative genomic analysis reveals evidence of two novel *Vibrio* species closely related to *V. cholerae*. *BMC Microbiology*, *10*, 154.

Hamaki, T. M., Suzuki, M., Fudou, R., Jojima, Y., Kajiura, T., Tabuchi, A., et al. (2005). Isolation of novel bacteria and actinomycetes using soil-extract agar medium. *Journal of Bioscience and Bioengineering*, *99*, 485–492.

Harayama, S., & Kasai, H. (2006). Bacterial phylogeny reconstruction from molecular sequences. In E. Stackebrandt (Ed.), *Molecular identification, systematics, and population structure of prokaryotes* (pp. 105–140). Berlin, Heidelberg: Springer.

Janssen, P. H. (2008). New cultivation strategies for terrestrial microorganisms. In K. Zengler (Ed.), *Accessing uncultivated microorganisms* (pp. 173–192). Washington: ASM Press.

Janssen, P. H., Yates, P. S., Grinton, B. E., Taylor, P. M., & Sait, M. (2002). Improved culturability of soil bacteria and isolation in pure culture of novel numbers of the divisions *Acidobacteria*, *Acidobacteria*, *Proteobacteria* and *Verrucomicrobia*. *Applied and Environmental Microbiology*, *68*, 2391–2396.

Johnson, J. L., & Ordal, E. J. (1968). Deoxyribonucleic acid homology in bacterial taxonomy: Effect of incubation temperature on reaction specificity. *Journal of Bacteriology*, *95*, 893–900.

Kaeberlein, T., Lewis, K., & Epstein, S. S. (2002). Isolating 'uncultivable' microorganisms in pure culture in a simulated natural environment. *Science*, *296*, 1127–1129.

Kämpfer, P. (2011). The systematics of prokaryotes: The state of the art. *Antonie Van Leeuwenhoek*, *101*(1), 3–11.

Kämpfer, P., & Glaeser, S. (2011). Prokaryotic taxonomy in the sequencing era and the role of MLSA in classification. *Microbiology Australia*, *32*, 66–70.

Kämpfer, P., & Glaeser, S. P. (2012). Prokaryotic taxonomy in the sequencing era – The polyphasic approach revisited. *Environmental Microbiology*, *14*(2), 291–317.

Kataoka, N., Tokiwa, Y., Tanaka, Y., Takeda, K., & Suzuki, T. (1996). Enrichment culture and isolation of slow-growing bacteria. *Applied Microbiology and Biotechnology*, *45*, 771–777.

Kawanishi, T., Shiraishi, T., Okano, Y., Sugawara, K., Hashimoto, M., Maejima, K., et al. (2011). New detection systems of bacteria using highly selective media designed by SMART: Selective medium-design algorithm restricted by two constraints. *PLoS One*, *6*, 1–10.

Kim, J. (2011). Review and future development of new culture methods for unculturable soil bacteria. *Korean Journal of Microbiology*, *47*, 179–187.

Konstantinidis, K. T., & Tiedje, J. M. (2005). Genomic insights into the species definition for prokaryotes. *Proceedings of the National Academy of Sciences of the United States of America*, *102*, 2567–2572.

Konstantinidis, K. T., & Tiedje, J. M. (2007). Prokaryotic taxonomy and phylogeny in the genomic era: Advancements and challenges ahead. *Current Opinion in Microbiology*, *10*, 504–509.

Koonin, E. V. (2003). Comparative genomics, minimal gene-sets and the last universal common ancestor. *Nature Reviews. Microbiology*, *1*, 27–136.

Krichevsky, M. I. (2011). What is a bacterial species? I will know it when I see it. *The Bulletin of BISMiS*, *2*, 17–23.

Kubatko, L. S., & Degnan, J. H. (2007). Inconsistency of phylogenetic estimates from concatenated data under coalescence. *Systematic Biology*, *56*, 17–24.

Kurtz, S., Phillippy, A., Delcher, A. L., Smoot, M., Shumway, M., Antonescu, C., et al. (2004). Versatile and open software for comparing large genomes. *Genome Biology*, *5*, R12.

Lapage, S. P., Sneath, P. H. A., Lessel, E. F., Skerman, V. B. D., Seeliger, H. P. R., & Clark, W. A. (1992). *International code of nomenclature of bacteria (1990 revision)*. Washington, DC: American Society for Microbiology.

Leadbetter, J. R. (2003). Cultivation of recalcitrant microbes: Cells are alive, well and revealing their secrets in the 21th century laboratory. *Current Opinion in Microbiology*, *6*, 274–281.

Leadbetter, J. R., Schmidt, T. M., Graber, J. R., & Breznak, J. A. (1999). Acetogenesis from H2 plus CO2 by spirochetes from termite guts. *Science*, *283*, 686–689.

Lewis, K., Epstein, S., D'Onofrio, A., & Ling, L. L. (2010). Uncultured microorganisms as a source of secondary metabolites. *Journal of Antibiotics*, *63*, 468–476.

Ludwig, W. (2010). Molecular phylogeny of microorganisms: Is rRNA still a useful marker? In A. Oren & R. T. Papke (Eds.), *Molecular phylogeny of microorganisms* (pp. 65–84). Norfolk: Caister Academic Press.

Ludwig, W., & Klenk, H.-P. (2001). Overview: A phylogenetic backbone and taxonomic framework of prokaryotes. In G. M. Garrity (Ed.), *Bergey's manual of systematic bacteriology* (pp. 49–65) (2nd ed.). New York: Springer.

Maiden, M. C., Bygraves, J. A., Feil, E., Morelli, G., Russel, J. E., Urwin, R., et al. (1998). Multilocus sequence typing: A portable approach to the identification of clones within populations of pathogenic microorganisms. *Proceedings of the National Academy of Sciences of the United States of America*, *95*, 3140–3145.

Margulies, M., Egholm, M., Altman, W. E., Attiya, S., Bader, J. S., Bemben, L. A., et al. (2005). Genome sequencing in microfabricated high-density picolitre reactors. *Nature*, *437*, 376–380.

Meier-Kolthoff, J. P., Auch, A. F., Klenk, H. P., & Göker, M. (2013). Genome sequence-based species delimitation with confidence intervals and improved distance functions. *BMC Bioinformatics*, *14*, 60.

Meier-Kolthoff, J. P., Klenk, H.-P., & Göker, M. (2014). Taxanomic use of DNA G+C content and DNA-DNA hybridization in the genomic age. *International Journal of Systematic and Evolutionary Microbiology*, *64*, 352–356.

Mende, D. R., Sunagawa, S., Zeller, G., & Bork, P. (2013). Accurate and universal delineation of prokaryotic species. *Nature Methods*, *10*, 881–884.

Moore, E. R. B., Mihaylova, S. A., Vandamme, P., Krichevsky, M. I., & Dijkshoorn, L. (2010). Microbial systematics and taxonomy: Relevance for a microbial commons. *Research in Microbiology*, *161*, 430–438.

Nadell, C. D., Xavier, J. B., & Foster, K. R. (2009). The sociobiology of biofilms. *FEMS Microbiology Reviews, 33*, 206–224.

Nichols, D., Lewis, K., Orjala, J., Mo, S., Ortenberg, R., O'Connor, P., et al. (2008). Short peptide induces an 'uncultivable' microorganism to grow *in vitro*. *Applied and Environmental Microbiology, 74*, 4889–4897.

Overmann, J. (2006). Principles of enrichment, isolation, cultivation and preservation of prokaryotes. In M. Dworkin, S. Falkow, E. Rosenberg, K. H. Schleifer, & E. Stackebrandt (Eds.), *The prokaryotes: Vol. 1.* (pp. 80–136) (3rd ed.). New York: Springer.

Pagnier, I., Raoult, D., & La Scola, B. (2008). Isolation and identification of amoeba-resisting bacteria from water in human environment by using an Acanthamoeba polyphaga co-culture procedure. *Environmental Microbiology, 10*, 1135–1144.

Pham, V. H. T., & Kim, J. (2012). Cultivation of unculturable soil bacteria. *Trends in Biotechnology, 30*, 475–484.

Plugge, C. M., & Stams, A. J. M. (2002). Enrichment of thermophilic syntrophic anaerobic glutamate-degrading consortia using a dialysis membrane reactor. *Microbial Ecology, 43*, 379–387.

Qin, Q.-L., Xie, B.-B., Zhang, X. Y., Chen, X.-L., Zhou, B.-C., Zhou, J., et al. (2014). A proposed genus boundary for the prokaryotes based on genomic insights. *Journal of Bacteriology, 196*, 2210–2215.

Richter, M., & Rosselló-Móra, R. (2009). Shifting the genomic gold standard for the prokaryotic species definition. *Proceedings of the National Academy of Sciences of the United States of America, 45*, 19126–19131.

Rosselló-Móra, R. (2006). DNA-DNA reassociation methods applied to microbial taxonomy and their critical evaluation. In E. Stackebrandt (Ed.), *Molecular identification, systematics, and population structure of prokaryotes* (pp. 23–50). Heidelberg, Berlin: Springer.

Sait, M., Hugenholtz, P., & Janssen, P. H. (2002). Cultivation of globally distributed soil bacteria from phylogenetic lineages previously only detected in cultivation-independent surveys. *Environmental Microbiology, 4*, 654–666.

Sangwan, P., Kovac, S., Davis, K. E. R., Sait, M., & Janssen, P. H. (2005). Detection and cultivation of soil Verrucomicrobia. *Applied and Environmental Microbiology, 71*, 8402–8410.

Serrano, W., Amann, R., Rosselló-Móra, R., & Fischer, U. (2010). Evaluation of the use of multilocus sequence analysis (MLSA) to resolve taxonomic conflicts within the genus *Marichromatium*. *Systematic and Applied Microbiology, 33*, 116–121.

Sneath, P. H. A. (2007). The species concept. *Microbiology Today, 34*, 45.

Stackebrandt, E., Frederiksen, W., Garrity, G. M., Grimont, P. A. D., Kämpfer, P., Maiden, M. C. J., et al. (2002). Report of the *ad hoc* committee for the re-evaluation of the species definition in bacteriology. *International Journal of Systematic and Evolutionary Microbiology, 52*, 1043–1047.

Stackebrandt, E., & Goebel, B. M. (1994). Taxonomic note: A place for DNA-DNA re-association and 16S rRNA sequence analysis in the present species definition in bacteriology. *International Journal of Systematic Bacteriology, 44*, 846–849.

Stevenson, B. S., Eichhorst, S. A., Wertz, J. T., Schmidt, T. M., & Breznak, J. A. (2004). New strategies for cultivation and detection of previously uncultured microbes. *Applied and Environmental Microbiology, 70*, 4748–4755.

Stott, M. B., Crowe, M. A., Mountain, B. W., Smirnova, A. V., Hou, S., Alam, M., et al. (2008). Isolation of novel bacteria, including a candidate division, from geothermal soils in New Zealand. *Environmental Microbiology, 10*, 2030–2041.

Sullivan, C. B., Diggle, M. A., & Clarke, S. C. (2005). Multilocus sequence typing: Data analysis in clinical microbiology and public health. *Molecular Biotechnology, 29*, 245–254.

Tamaki, H., & Kamagata, Y. (2005). Cultivation of uncultured fastidious microbes. *Microbes and Environments, 20*, 85–91.

Tamaki, H., Sekiguchi, Y., Hanada, S., Nakamura, K., Nomura, N., Matsumara, M., et al. (2005). Comparative analysis of bacterial diversity in freshwater sediment of a shallow eutrophic lake by molecular and improved cultivation-based techniques. *Applied and Environmental Microbiology, 71*, 2162–2169.

Tindall, B. J., De Vos, P., & Trüper, H. G. (2008). Judicial Commission of the International Committee of Systematics of Prokaryotes. *XIth international (IUMS) congress of bacteriology and applied microbiology. Minutes of the meetings, 23, 24, 27 July 2005, San Francisco, CA, USA. International Journal of Systematic and Evolutionary Microbiology, 58*, 1737–1745.

Tindall, B. J., Rosselló-Móra, R., Busse, H.-J., Ludwig, W., & Kämpfer, P. (2010). Notes on the characterization of prokaryote strains for taxonomic purposes. *International Journal of Systematic and Evolutionary Microbiology, 60*, 249–266.

Tindall, B. J., Sikorski, J., Smibert, R. A., & Krieg, N. L. (2008). Phenotypic characterization and the principles of comparative systematics. In G. M. Garrity (Ed.), *Methods for general and molecular microbiology* (pp. 330–393) (3rd ed.). Washington: ASM Press.

Tyson, G. W., & Banfield, J. F. (2005). Cultivating the uncultivated: A community genomics perspective. *Trends in Microbiology, 13*, 411–415.

Uphoff, H. U., Felske, A., Fehr, W., & Wagner-Döbler, I. (2001). The microbial diversity in picoplankton enrichment culture: A molecular screening of marine isolates. *FEMS Microbiology Ecology, 35*, 249–258.

Vartoukian, S. R., Palmer, R. M., & Wade, W. G. (2010). Strategies for culture of 'unculturable' bacteria. *FEMS Microbiology Letters, 309*, 1–7.

Vinuesa, P. (2010). Multilocus sequence analysis and bacterial species phylogeny estimation. In A. Oren & R. T. Papke (Eds.), *Molecular phylogeny of microorganisms* (pp. 41–64). Norfolk: Caister Academic Press.

Watve, M., Vaishali, S., Charushila, S., Monali, R., Anagha, M., Yogesh, S., et al. (2000). The 'K' selected oligophilic bacteria: A key to uncultured diversity? *Current Science, 78*, 1535–1542.

Wayne, L. G., Brenner, D. J., Colwell, R. R., Grimont, P. A. D., Kandler, O., Krichevsky, M. I., et al. (1987). International Committee on Systematic Bacteriology. Report of the ad hoc committee on reconciliation of approaches to bacterial systematics. *International Journal of Systematic Bacteriology, 37*, 463–464.

Welker, M., & Moore, E. R. (2011). Applications of whole-cell matrix-assisted laser-desorption/ionization time-of-flight mass spectrometry in systematic microbiology. *Systematic and Applied Microbiology, 34*(1), 2–11.

West, S. A., Diggle, S. P., Buckling, A., Gardner, A., & Griffin, A. S. (2007). The social lives of microbes. *Annual Review of Ecology, Evolution, and Systematics, 38*, 53–77.

Williams, D., Andam, C. P., & Gogarten, J. P. (2010). Horizontal gene transfer and the formation of groups of microorgansms. In A. Oren & R. T. Papke (Eds.), *Molecular phylogeny of microorganisms* (pp. 167–184). England: Caister Academic Press.

Young, J. M. (2001). Implications of alternative classifications and horizontal gene transfer for bacterial taxonomy. *International Journal of Systematic and Evolutionary Microbiology, 51*, 945–953.

Index

Note: Page numbers followed by *f* indicate figures and *t* indicate tables.

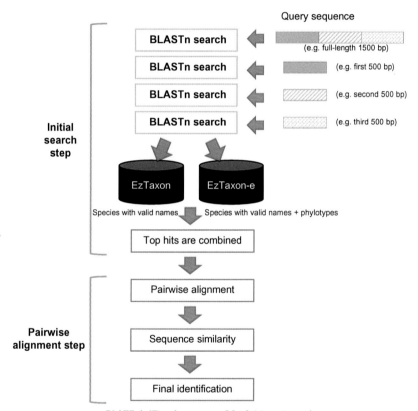

PLATE 1 (Fig. 1 on page 66 of this volume.)

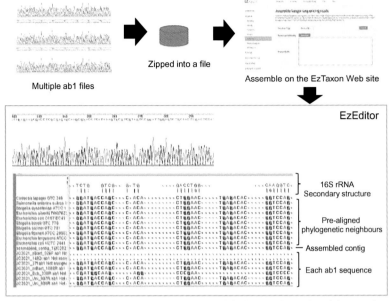

Multiple ab1 files

Zipped into a file

Assemble on the EzTaxon Web site

EzEditor

16S rRNA
Secondary structure

Pre-aligned
phylogenetic neighbours

Assembled contig

Each ab1 sequence

Edit the assembled contig by comparing phylogenetic neighbours, original sequencing reads and the secondary structure

Final 16S rRNA sequence

16S rRNA sequence similarity search with EzTaxon

Phylogenetic analysis using the EzEditor and MEGA programs

PLATE 2 (Fig. 3 on page 70 of this volume.)

Tasks	Rank	Name	Strain	Authors	Accession	Pairwise Similarity (%)	Diff/Total nt	Completeness (%)
	1	Escherichia coli	KCTC 2441(T)	(Migula 1895) Castellani and Chalmers 1919	EU014689	99.66	4/1182	100
	2	Escherichia fergusonii	ATCC 35469(T)	Farmer et al. 1985	CU928158	99.49	6/1183	100
	3	Shigella flexneri	ATCC 29903(T)	Castellani and Chalmers 1919	X96963	99.41	7/1183	100
	4	Shigella sonnei	GTC 781(T)	(Levine 1920) Weldin 1927	AB273732	99.24	9/1183	100
	5	Escherichia coli	O157 EC4115	(Migula 1895) Castellani and Chalmers 1919	CP001164	98.82	14/1183	100
	6	Shigella boydii	GTC 779(T)	Ewing 1949	AB273731	98.82	14/1183	100
	7	Shigella dysenteriae	ATCC 13313(T)	(Shiga 1898) Castellani and Chalmers 1919	X96966	98.56	17/1182	100
	8	Escherichia albertii	TW07627(T)	Huys et al. 2003	ABKX01000030	98.48	18/1183	100
	9	Salmonella enterica subsp. houtenae	DSM 9221(T)	(Le Minor et al. 1985) Le Minor and Popoff 1987	U92195	97.80	26/1182	100
	10	Citrobacter koseri	CDC 3613-63	Frederiksen 1970	AF025372	97.38	31/1183	100

Taxonomic group

Completeness of query

Similarity values

Identification of JC2021_27F.ab1 Edit
Length of sequence :1185bp View query sequence
Completeness :81.79% (81 ~ 1230)

PLATE 3 (Fig. 4 on page 72 of this volume.)

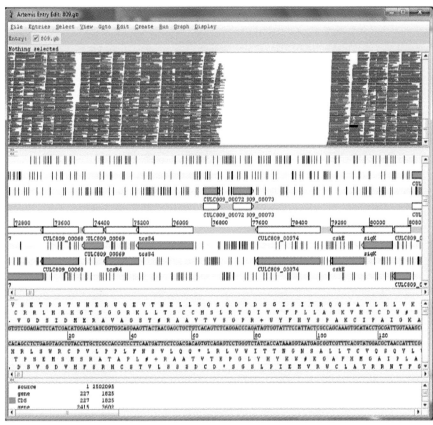

PLATE 4 (Fig. 1 on page 82 of this volume.)

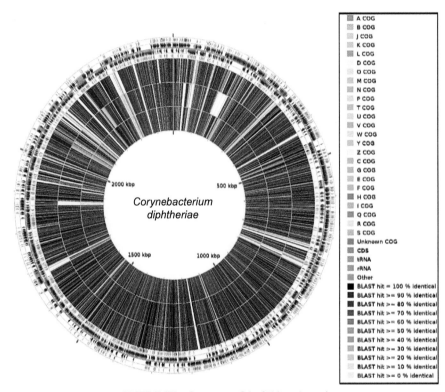

▨	A COG
▨	B COG
▨	J COG
▨	K COG
▨	L COG
	D COG
▨	O COG
▨	M COG
▨	N COG
▨	P COG
▨	T COG
▨	U COG
▨	V COG
▨	W COG
▨	Y COG
	Z COG
▨	C COG
▨	G COG
▨	E COG
▨	F COG
▨	H COG
▨	I COG
▨	Q COG
▨	R COG
▨	S COG
▨	Unknown COG
▨	CDS
▨	tRNA
▨	rRNA
▨	Other
■	BLAST hit = 100 % identical
■	BLAST hit >= 90 % identical
■	BLAST hit >= 80 % identical
■	BLAST hit >= 70 % identical
■	BLAST hit >= 60 % identical
■	BLAST hit >= 50 % identical
▨	BLAST hit >= 40 % identical
▨	BLAST hit >= 30 % identical
▨	BLAST hit >= 20 % identical
▨	BLAST hit >= 10 % identical
▨	BLAST hit >= 0 % identical

PLATE 5 (Fig. 2 on page 84 of this volume.)

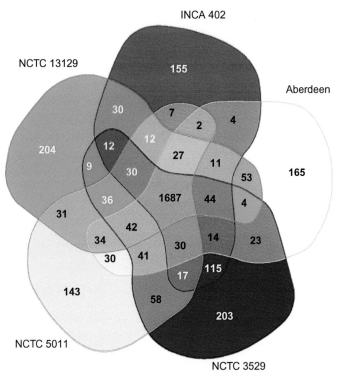

PLATE 6 (Fig. 4 on page 86 of this volume.)

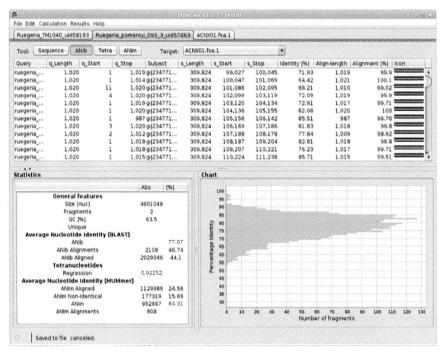

PLATE 7 (Fig. 4 on page 113 of this volume.)

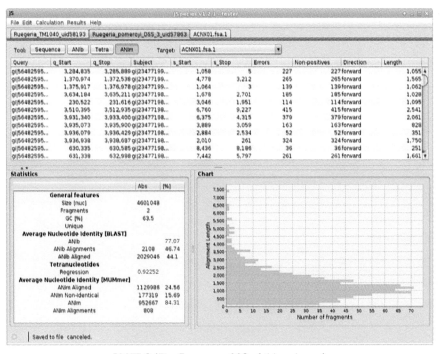

PLATE 8 (Fig. 5 on page 113 of this volume.)

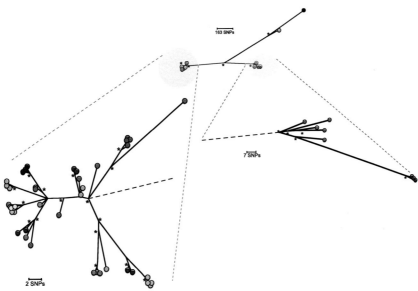

PLATE 9 (Fig. 1 on page 129 of this volume.)

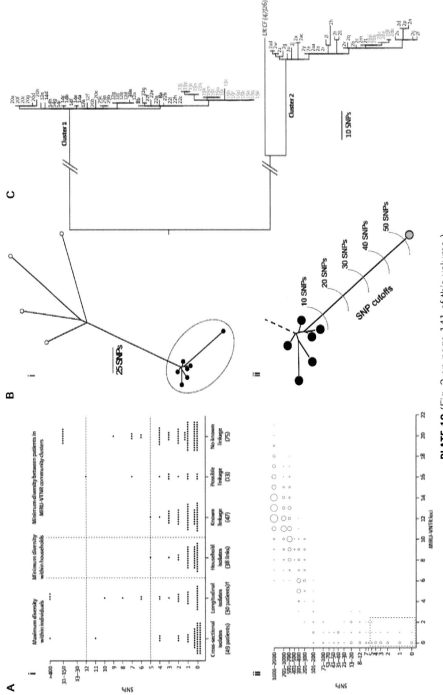

PLATE 10 (Fig. 3 on page 141 of this volume.)

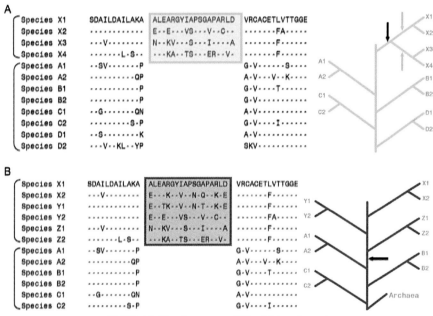

PLATE 11 (Fig. 1 on page 156 of this volume.)

```
                        #1
                        <─────────────────────────────────────────>
AquifexaeolicusVF5    MAKEKFERTKEHVNVGTIGHVDHGKSTLTSAITCVLAAGLVEGGKAKCFKYEEIDKAPEEKERGITINITHVEYETAKRHYAHVDCPGHA
ThermocrinisalbusDSM  MAKEKFVREKEHVNVGTIGHVDHGKSTLTGAITCVLGAGLMPGGKAKCTKYEEIDKAPEERERGITINITHVEYETAKRHYAHVDCPGHA
Hydrogenobactertherm  MAKEKFIREKEHVNVGTIGHVDHGKSTLTGAITCVLAAGVLPGGKAKCTKYEEIDKAPEEKERGITINITHVEYETPKRHYAHVDCPGHA
Hydrogenobaculumsp.Y  MAKEKFVREKEHINVGTIGHVDHGKSTLTGAITCVLGAGVLGGGKAKCYRYEEIDKAPEERERGITINITHVEYETPKRHYAHVDCPGHA
Sulfurihydrogen_Azo   MAKEKFVRGKEHLNVGTIGHVDHGKTTLTAAITYYQGK----KGLAKFVGYADIDKAPEEReRGITINITHVEYETEKRHYAHVDCPGHA
Sulfurihydrogen_YO3   MAKEKFVRGKEHLNVGTIGHVDHGKTTLTAAITYYQGK----KGLAKFVGYAQIDKAPEERERGITINITHVEYETEKRHYAHVDCPGHA
PersephonellamarinaE  MAREKFERKKEHVNVGTIGHVDHGKTTLTAAITYVLGK----KGLAEFIGYGEIDKAPEERDRGITINITHVEYETEKRHYAHVDCPGHA
Thermovibrioammonifi  MAKQKFERTKPHKNVGTIGHVDHGKTTLTAAITHCLAL----QGKAQEVGYDGIDKAPEERERGITIATAHVEYESDKYHYAHVDCPGHA
Desulfurobacteriumth  MAKQKFERTKPHKNVGTIGHVDHGKTTLTAAITHCLAL----QGKAQEVAYDGIDKAPEERERGITIATAHVEYEGDKYHYAHVDCPGHA
ThermotogamaritimaMS  MAKEKFVRTKPHVNVGTIGHIDHGKSTLTAAITKYLSL----KGLAQYIPYDGIDKAPEEKARGITINITHVEYETEKRHYAHIDCPGHA
Fervidobacteriumnodo  MAKEKFVRTKPHINVGTIGHIDHGKTTLTAAITKYCSL----FGWADYTPYEMIDKAPEERARGITINITHVEYQTEKRHYAHIDCPGHA
Bacillussubtilissubs  MAKEKFDRSKBHANIGTIGHVDHGKTTLTAAITTVLHKK---GGKGTAMAYDGIDGAPEERERGITIQTAHVEYETETRHYAHVDCPGHA
Staphylococcusaureus  MAKEKFDRSKEHANIGTIGHVDHGKTTLTAAIATVLAK----NGDGVAQGYDMIDNAPEEKERGITINTGHIEYQTDKRHYAHVDCPGHA
                      **:|** * * *:|*****:|****|****:|         *   *  ** ***|***** |*|*|*|| |****|******

AquifexaeolicusVF5    DYIKNMITGAAQMDGAILVVSAADGPMPQTREHVLLARQVNVPYIVVFMNKCDMVDDEELLELVELEVRELLSKYEYPGDEVPVIRGGAL
ThermocrinisalbusDSM  DYIKNMITGAAQMDGAILVVSAADGPMPQTREHVLLARQVNVPYIVVFMNKCDMVDDAELLDLVELEVRELLSKYEYPGDEVPVIRGGAL
Hydrogenobactertherm  DYIKNMITGAAQMDGAILVVSAADGPMPQTREHVLLARQVNVPYIVVFMNKCDMVDDPELLDLVELEVRELLSKYEFPGDEVPVIRGGAL
Hydrogenobaculumsp.Y  DYIKNMITGAAQMDGAILVVSAADGPMPQTREHVLLARQVNVPYIVVFMNKCDMVDDPELLDLVELEVRDLLNKYEFPGDDVPIIRGGAL
Sulfurihydrogen_Azo   DYIKNMITGAAQMDGAILVVSAADGPMPQTREHVLLARQVNVPYIVVFLNKCDMVDDEELIDLVEMEVRELLSKYDFPGDEVPVIRGGAL
Sulfurihydrogen_YO3   DYIKNMITGAAQMDGAILVVSAADGPMPQTREHVLLARQVNVPYIVVFLNKCDMVDDEELIDLVEMEVRELLSKYDFPGDEVPVIRGGAL
PersephonellamarinaE  DYIKNMITGAAQMDGAILVVSAADGPMPQTREHVLLARQVNVPYIVVFLNKCDMVDDEELLELVELEVRELLNKYEFPGDDVPVIRGGAL
Thermovibrioammonifi  DYIKNMITGAAQMDGAILVVSAADGPMPQTREHVLLARQVNPAIVVFLNKVDMVDDEELLELVELEVREGLSEYGVPGDEVPVIRGGAL
Desulfurobacteriumth  DYIKNMITGAAQMDGAILVVSAADGPMPQTREHVLLARQVNVPYIVVFLNKVDMVDDEELLELVELEVRELLNEYDFPGDEVPVIKGSAL
ThermotogamaritimaMS  DYIKNMITGAAQMDGAILVVAATDGPMPQTREHVLLARQVNVPEVPYIVIVFINKTDMVDDPELIDLVEMEVRDLLGQYGYPGDEVPVIRGGAL
Fervidobacteriumnodo  DYIKNMITGAAQMDGAILVVAATDGPMPQTREHVLLARQVNVPAMIVFINKVDMVDDPELVDLVEMEVRDLLSKYEFPGDELPVIRGGAL
Bacillussubtilissubs  DYVKNMITGAAQMDGAILVVSATDGPMPQTREHILLSKNVGVPYIVVFLNKCDMVDDEELLELVEMEVRDLLSEYDFPGDDVPVIRGGAL
Staphylococcusaureus  DYVKNMITGAAQMDGGILVVSAADGPMPQTREHILLGRNVGVPALVVFLNKVDMVDDEELLELVEMEVRDLLSEYDFPGDDVPVIAGGAL
                      **|***********.|****|*|**********|**|||*  ** ||**|** ***** **|*|***|**|**.|* |***|||*|* ****

                      #2                                          #3
                      <─────────────>                             <────>
AquifexaeolicusVF5    GALQELEQNSPGKWVESIKELLNAMDEYIPTPQREVDKPFLMPIEDVFSISGRGTVVTGRVERGVLRPGDEVEIVGLREEPLKTVATGIE
ThermocrinisalbusDSM  GALQELEGGKPDKWCGSILQLLEAMDEYIPTPVREADKPFLMPIEDVFSISGRGTVVTGAVERGTLRPGEEVEVVGLREEPLKTVATSIE
Hydrogenobactertherm  GALQELEGGKPDRWCGSILQLLEAMDEYVPTPVREADKPFLMPIEDVFSISGRGTVVTGRVERGVLKPGEEVEIVGIREEPLKTVATGIE
Hydrogenobaculumsp.Y  GALEELDKGKPDKWCNAIVDLMKALDDYIPSPQRETDKPFLMPIEDVFTIGGRGTVVTGRVERGVLKPGEEVEIVGLKEEGLKTTATSVE
Sulfurihydrogen_Azo   GALNDDP-----KWFKSVEDLLKAMDEYIPTPPRETDKPFLMAVEDVFTITGRGTVVTGRVERGTLKIGDEVEIVGLSEEKKKTVVTGIE
Sulfurihydrogen_YO3   GALNDDP-----KWFAGVEELLKAMDEYIPTPPRETDKPFLMAVEDVFTITGRGTVVTGRVERGTLKVGDEVEIVGLSEEKKKTVVTGIE
PersephonellamarinaE  GALNDEE-----KWVKSIEELLDAMDNYIPTPERATDKPFLMAIEDVFTIGGRGTVVTGRVERGTLKVGDEVEIVGLGDEIRKTVVTGIE
Thermovibrioammonifi  KALECTDPN---CEWCGPIYELVKALDEYVPEPVREIDKPFLMPIEDVFGISGRGTVVTGRVERGQTLKVGDEVEIVGLRDEPIKTVATGIE
Desulfurobacteriumth  KALECTGPD---CPDCGPIYELVNALDEYVPEPVREVDKPFLMPIEDVFGISGRGTVVTGRVERGKLTVGEEVEIVGLREEPIKTVATGIE
ThermotogamaritimaMS  KAVEAPNDPN--HEAYKPIQELLDAMDNYIPDPQRDVDKPFLMPVEDVFTITGRGTVVTGRIERGRIRPGDEVEIIGLSYEIKKTVVTSVE
Fervidobacteriumnodo  KAVEAPNDPN--HPDLKAIKELLDAMDSYFPDPVREVDKPFLMPVEDVFTITGRGTVVTGRIERGVIKPGVEAEIIGMGYETKKTVITSVE
Bacillussubtilissubs  KALEGDA-----EWEAKIFELMDAVDEYIPTPERDTEKPFMMPVEDVFTITGRGTVVTGRVEGEVEIIGLQDENKKTTVTGVE
Staphylococcusaureus  KALEGDA-----KYEEKILELMEAVDTYIPTPERDSGDKPFMMPVEDVFGITGRGTVATGRVERGQIKVGEEVEIIGLHDTG-KTTVTGVE
                      *|:           |  *|*|.*|*  * * * |***|*|.*|****|*|***** ***|***  * * *.*|:*|    **, *,:|*

                                                                            #4
                                                                            <──────────
AquifexaeolicusVF5    MFRKVLDEALPGDNIGVLLRGVGKDDVERGGVLAQPGGVKAHKRFRAQVYVLGKEEGGRHTPFFVNYRPQFYFRTADVTGTVVKLPEGVE
ThermocrinisalbusDSM  MFRKVLDEALPGDNIGVLLRGVGKDDVERGGVLAKPGGVKAHRKFRAQVYVLGKEEGGRHGPFFAGYRPQFYFRTADVTGVVVKLPEGVE
Hydrogenobactertherm  MFRKILDEALPGDNVGVLLRGVGKDDVERGGVLAKPGTVKPHRFRAQVYVLGKEEGGRHTPFFVNYRPQFYFRTADVTGVVVKLPEGGE
Hydrogenobaculumsp.Y  MFRKILDEALPGDNVGVLLRGVGKDQVERGGVLAKPGGITPHKKFKAQVYVLGKEEGGRHTPFFLNYRPQFYIRTADVTGTVVKLPEGGE
Sulfurihydrogen_Azo   MFRKQLDEAIAGDNVGVLLRGITKDEVERGGVLAKPGTITPHKKFKAQVYVLGKEEGGRHTPFFLGYRPQFYIRTADITGTVVGLPEGGE
Sulfurihydrogen_YO3   MFRKQLDEAIAGDNVGVLLRGITKDEVERGGVLAKPGTITPHKKFKAQVYVLGKEEGGRHTPFFLGYRPQFYIRTADITGTVVELPEGGE
PersephonellamarinaE  MFRKTLDEAVAGDNVGVLLRGIGKDEVERGGVLAKPGGITPHKKFKAQVYVLGKEEGGRHTPFFNGYQPQFYFRTTDVTGVVVELPEGGE
Thermovibrioammonifi  MFRKVLDEALPGDNIGVLLRGVGKDEVERGMVVAKPGGIKPHRKFKAEVYILGKEEGGRHTPFFNGYQPQFYFRTTDVTG-KVKLPEGVE
Desulfurobacteriumth  MFRKVLDEALPGDNVGILLRGVGKDEVERGMVVAKPGGINPHKKFKAEVYILGKEEGGRHTPFFNGYQPQFYFRTTDVTG-KVKLPEGVE
ThermotogamaritimaMS  MFRKELDEGIAGDNVGCLLRGIDKDEVERGQVLAAPGGIKPHTKFKASEVYILGKKEEGGRHTPFTKGYKPQFYIRTADVTGEIVGLPEGVE
Fervidobacteriumnodo  MFRKELDEAMAGDNVGCLLRGVDKDEVERGQVIAKPGGITPHKKFKANIYVLKKEEGGRHTPFTKGYKPQFYIRTADVTGEIVDLPAGVE
Bacillussubtilissubs  MFRKLLDYAEAGDNIGALLRGVGREEIQRGGVLAKPGTITPHGKFKASVYVLGKDEGGRHTPFFGNYRPQFYFRTTDVTG-IIHLPEGVE
Staphylococcusaureus  MFRKLLDYAEAGDNIGALLRGVAREDVQRGGVLAAPGGITPHTEFKAEVYVLGKDEGGRHTPFFGNYRPQFYFRTTDVTG-VVHLPEGTE
                      **** ** .,.***|* ***|!  |!||!**|* |!* **||..* .*!*||!*!*.*|****|***|**|**   *|*|****|**|*|** | ** **

AquifexaeolicusVF5    MVMPGDNVELEVELIAPVALEEGLRFAIREGGRTVGAQVVTKILD
ThermocrinisalbusDSM  MVMPGDNVELEVELIKPVAMEEGLRFAIREGGRTVGAGVVTKILE
Hydrogenobactertherm  MVMPGDNVELEVELIGPVAMEEGLRFAIREGGRTVGAGVVTQILD
Hydrogenobaculumsp.Y  MVMPGDNVEFEVELIHPVAMEEGLRFAIREGGRTVGAGVVTKILD
Sulfurihydrogen_Azo   MVMPGDNVELEVELMVPVAMEEQMRFAIREGGRTVGAGVVTKILD
Sulfurihydrogen_YO3   MVMPGDNVELTVELMVPVAMEEQMRFAIREGGRTVGAGVVTKILD
PersephonellamarinaE  MVMPGDNVELTVELMEPVAIEEQMRFAIREGGRTVGAGVVTQIIE
Thermovibrioammonifi  MVMPGDNVTFEVELLKPVAIEEGLRFAIREGGKTVGAQVVTEILD
Desulfurobacteriumth  MVMPGDNVTFEVELLKPVAIEEGLRFAIREGGKTVGAQVVTEILD
ThermotogamaritimaMS  MVMPGDHVEMBIELIYPVAIEKGQRFAVREGGRTVGAQVVBVIE
Fervidobacteriumnodo  MVMPGDNVEMTIELIYPVAIEKGHRFAVREGGRTVGAGVVGEIIE
Bacillussubtilissubs  MVMPGDNTEMNVELISTIAIEEGTRFSIREGGRTVGSGVVGTITE
Staphylococcusaureus  MVMPGDNVEMTVELIAPIAIEDGTRFSIREGGRTVGSGVVTEIIK
                      ******|. | |**|.|*|*. **||****|***|***|*  |
```

PLATE 12 (Fig. 3 on page 162 of this volume.)

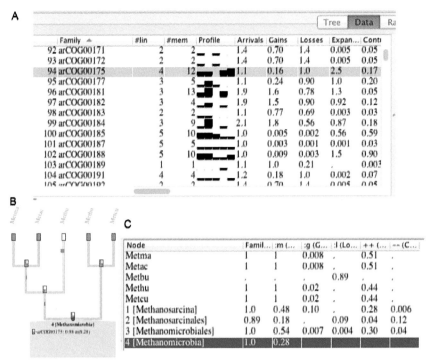

A

Family ▲	#lin	#mem	Profile	Arrivals	Gains	Losses	Expan...	Contr
92 arCOG00171	2	2		1.4	0.70	1.4	0.005	0.05
93 arCOG00172	2	2		1.4	0.70	1.4	0.005	0.05
94 arCOG00175	4	12		1.1	0.16	1.0	2.5	0.17
95 arCOG00177	3	5		1.1	0.24	0.90	1.0	0.20
96 arCOG00181	3	13		1.9	1.6	0.78	1.3	0.05
97 arCOG00182	3	4		1.9	1.5	0.90	0.92	0.12
98 arCOG00183	2	2		1.1	0.77	0.69	0.003	0.03
99 arCOG00184	3	9		2.1	1.8	0.56	0.87	0.18
100 arCOG00185	5	10		1.0	0.005	0.002	0.56	0.59
101 arCOG00187	5	5		1.0	0.003	0.001	0.001	0.03
102 arCOG00188	5	10		1.0	0.009	0.003	1.5	0.90
103 arCOG00189	1	1		1.1	1.0	0.21	.	0.003
104 arCOG00191	4	4		1.2	0.18	1.0	0.002	0.07
105 arCOG00192	2	2		1.4	0.70	1.4	0.005	0.05

B

C

Node	Famil...	:m (...	:g (G...	:l (Lo...	++ (...	−− (C...
Metma	1	1	0.008	.	0.51	.
Metac	1	1	0.008	.	0.51	.
Metbu	.	.	.	0.89	.	.
Methu	1	1	0.02	.	0.44	.
Metcu	1	1	0.02	.	0.44	.
1 [Methanosarcina]	1.0	0.48	0.10	.	0.28	0.006
2 [Methanosarcinales]	0.89	0.18	.	0.09	0.04	0.12
3 [Methanomicrobiales]	1.0	0.54	0.007	0.004	0.30	0.04
4 [Methanomicrobia]	1.0	0.28				

PLATE 13 (Fig. 6 on page 196 of this volume.)

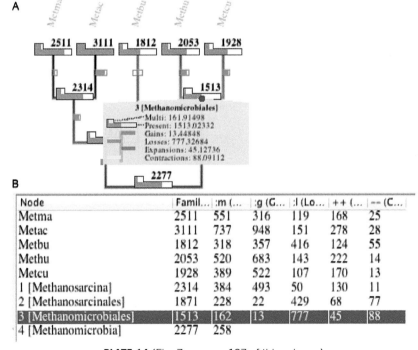

A

B

Node	Famil...	:m (...	:g (G...	:l (Lo...	++ (...	−− (C...
Metma	2511	551	316	119	168	25
Metac	3111	737	948	151	278	28
Metbu	1812	318	357	416	124	55
Methu	2053	520	683	143	222	14
Metcu	1928	389	522	107	170	13
1 [Methanosarcina]	2314	384	493	50	130	11
2 [Methanosarcinales]	1871	228	22	429	68	77
3 [Methanomicrobiales]	1513	162	13	777	45	88
4 [Methanomicrobia]	2277	258				

PLATE 14 (Fig. 7 on page 197 of this volume.)

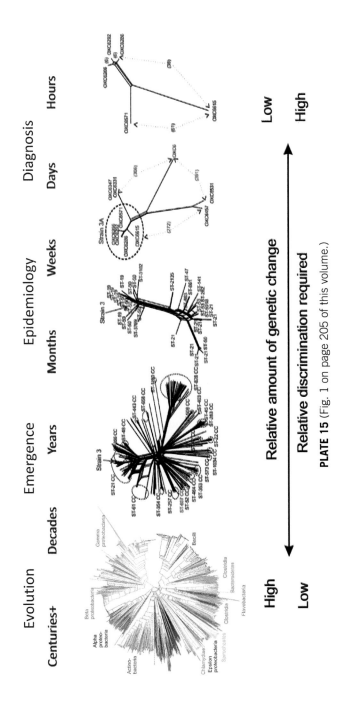

PLATE 15 (Fig. 1 on page 205 of this volume.)

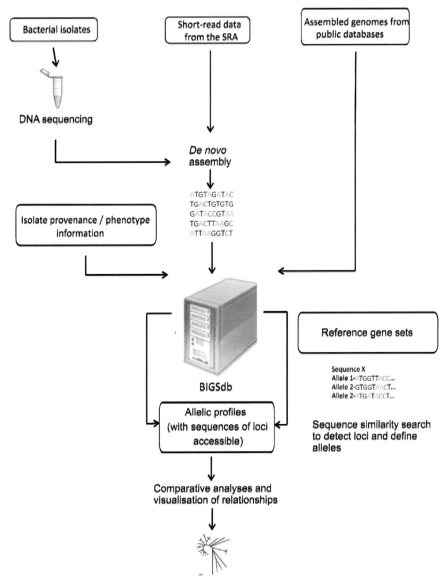

PLATE 16 (Fig. 2 on page 206 of this volume.)

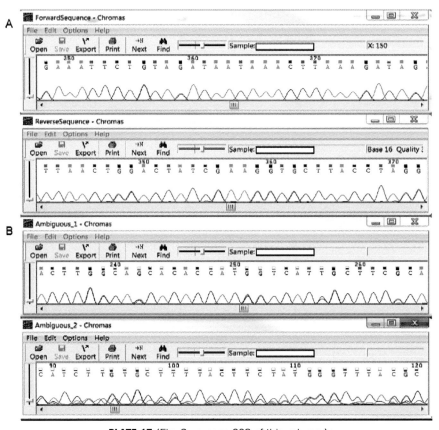

PLATE 17 (Fig. 3 on page 208 of this volume.)

A

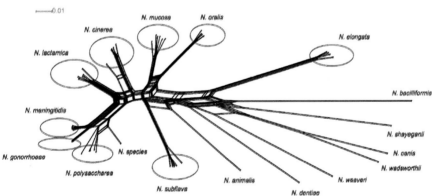

B

PLATE 18 (Fig. 5 on page 212 of this volume.)

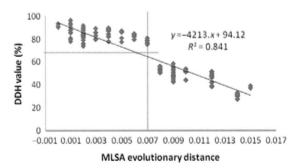

PLATE 19 (Fig. 3 on page 239 of this volume.)

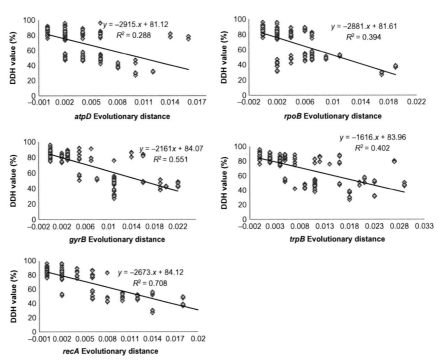

PLATE 20 (Fig. 4 on page 240 of this volume.)

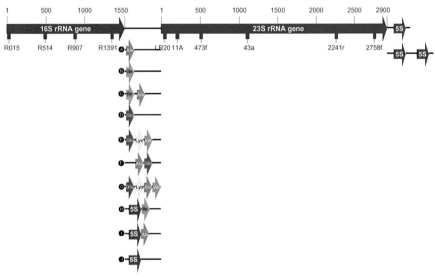

PLATE 21 (Fig. 1 on page 257 of this volume.)

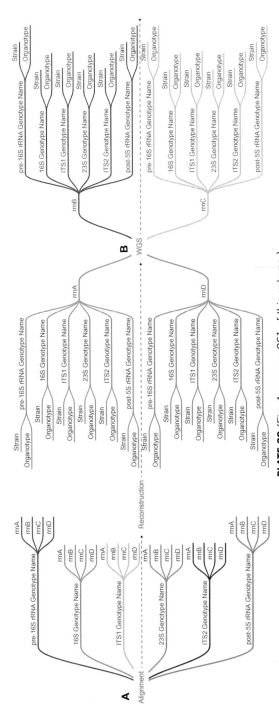

PLATE 22 (Fig. 4 on page 261 of this volume.)

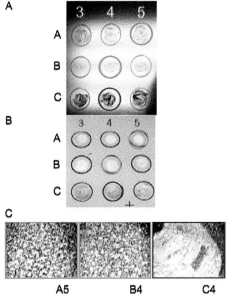

PLATE 23 (Fig. 3 on page 281 of this volume.)

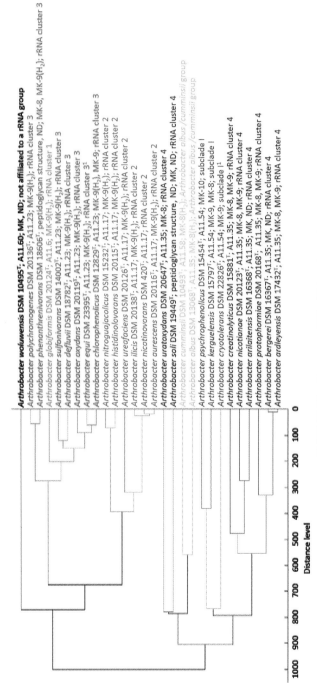

PLATE 24 (Fig. 5 on page 294 of this volume.)

Edwards Brothers Malloy
Thorofare, NJ USA
December 8, 2014